Advances in Molecular Techniques

Advances in Molecular Techniques

Rakesh S. Sengar
Amit Kumar
Reshu Chaudhary
Ashu Singh

CRC Press
Taylor & Francis Group
Boca Raton London New York

CRC Press is an imprint of the
Taylor & Francis Group, an **informa** business

CRC Press
Taylor & Francis Group
6000 Broken Sound Parkway NW, Suite 300
Boca Raton, FL 33487-2742

Printed on acid-free paper

International Standard Book Number-13: 978-0-8153-7075-8 (Hardback)

Visit the Taylor & Francis Web site at
http://www.taylorandfrancis.com

and the CRC Press Web site at
http://www.crcpress.com

There is no teacher equal to mother and there's nothing more contagious than the dignity of a father.

Taking a moment to thank my late beloved father, Dr. S.S. Sengar for inculcating the values for a content life.

Rakesh S. Sengar

Contents

Foreword

Molecular biology is the study of living things at the level of the molecules that control and make up the living things. While traditional biology concentrated on studying whole living organisms and how they interact within populations (a "top-down" approach), molecular biology strives to understand living things by examining the components that make them up (a "bottom-up" approach). The molecules that form the basis of life provide scientists with a more predictable and mechanistic tool for scientists to study. Working with whole organisms (or even just whole cells) can be unpredictable, with the outcome of experiments relying on the interaction of thousands of molecular pathways and external factors. Molecular biology provides scientists with a toolkit with which they may "tinker" with the way life works.

Students of biology are responsible for an ever-growing volume of information, and that amount of detail is reflected in today's textbooks, which are increasingly becoming encyclopedic references as opposed to teaching texts. But students are more likely to succeed with a text that they enjoy reading and that gives them a sense of the wonderment that inspired their own instructors to study biology. For this reason, we have endeavored to strike a balance; thus, I found this manual's presentation with an inviting and accessible writing style and the level of authority and rigor as expected from a majors'-level text, up to the mark.

This laboratory manual is written primarily to benefit undergraduate and postgraduate students who are supposed to have a good understanding of the subject. In this manual, the topics have been chosen to include the basic foundation areas of molecular biology; each chapter starts with an introduction to the basic approach, followed by detailed methods sections with easy-to-follow protocols and comprehensive troubleshooting. The first part of the book introduces DNA Manipulative and Ingratiative Techniques, such as isolation, purification, and quantitative determination of DNA and RNA by different methods. The second part describes Regulative Techniques, including RFLP, AFLP, microsatellite markers, and QTL-based methods for genomic sequencing, mapping, and labeling. The final part encompasses a variety of Recombinant Techniques and deals with gene cloning, expression, determination, and transformation. The manual will be of great use to both the first-timer and the experienced scientist, I hope, and the present work will go a long way to supplement text-book knowledge and help students better understand the subject.

I congratulate Prof. R.S. Sengar, Dr. Amit Kumar, Dr. Reshu Chaudhary, and Er. Ashu Singh for their sincere efforts in contributing their knowledge and expertise to realize this molecular biology manual, a very useful one for undergraduate and postgraduate students.

Dr. Narendra Singh Rathore
Deputy Director General (Agril. Edn.)
Indian Council of Agricultural Research
Krishi Anusandhan Bhavan-II
Pusa, New Delhi

Preface

The term molecular biology chiefly concerns itself with understanding the interactions between the various systems of a cell, including the interactions between the different types of DNA, RNA, and protein biosynthesis, as well as learning how these interactions are regulated. Environmental resources are part of nature that mankind considers to be useful and valuable for sustaining life processes.

Molecular genetics is the discipline that aims to comprehend biological activity at the gene sublevel. These biological activities usually involve the plant or animal cell and the nucleic acids and proteins that are at work within the cell. Molecular biologists use some standard techniques that have been optimized over many years. These basic techniques can also be employed by scientists from many different areas of research and applied science.

The following book would be a guide that introduces these classic molecular biology techniques. The book, entitled *Advances in Molecular Techniques*, is a comprehensive technical book on the title prepared by a consortium of authors. This book has been written to cater to the needs of students dealing with life science. In this book, adequate subject matter has been provided to facilitate an attempt to unveil the obscurity attached to molecular machinery entrails. Through each section, we provide background information, protocols, references to peer-reviewed literature, and help selecting reagents for your application. We have tried our best, with a wide array of fundamental research-based collection of case studies, to consolidate and improve the knowledge and skills on the relevant techniques.

Another unusual feature of this volume is the publication of an account of the current state of nucleic acid research in the international scenario. We have included this to provide information to our readers about a part of the world from which scientific news is very difficult to obtain. The remaining articles follow our usual pattern of attempting to present "essays in circumscribed areas" in which recent developments in particular aspects of the field of nucleic acids and molecular biology are discussed by workers provided with an opportunity for more personal expression than is normally met in review articles. To this end, it is our policy to encourage discussion, argument, speculation, and the expression of points of view that are individualistic and perhaps even controversial, yet important to take into account for technical reasons. We seek to provide a forum for discussion and debate, and we welcome further suggestions from readers as to how this end may best be served. Indeed, we should like, again, to remind readers that we wish them to write to us with their comments.

The present book will be very helpful to anyone in the scientific community desiring a compact handbook encompassing the recent molecular and genetic research done so far with standard protocols.

We have received generous help from many professors and scientists. The preparation of this book was under the guidance of Prof. H.B. Singh, Department of Plant Pathology, Banaras Hindu University (BHU), Varanasi; Prof. Rita Goyal, Department of Microbiology, G.B. Pant University of Agriculture and Technology,

Pantnagar, U.S. Nagar; Prof. V.P. Singh, Vice Chancellor, M.J.P. Rohilkhand University, Bareilly; Dr. Anil Kumar Gupta, Professor and Head of the Department of Molecular Biology and Genetic Engineering, College of Basic Science, G.B. Pant University of Agriculture and Technology, Pantnagar; Prof. N.K. Singh, National Professor (DRBP Pal) Biotech National Research Centre on Plant Biotechnology, New Delhi; Prof. R.L. Singh, Professor and Head, Ram Manohar Lohia Anadha University, Faizabad; and Dr. Rakesh Singh and Dr. Sandeep Kumar Sharma, National Bureau of Plant Genetic Resources (NBPGR), New Delhi, and we thank them for their extraordinary help in shaping this book. It would be rather impossible to list all those who have provided encouragement and help in the preparation of this laboratory manual. We are extremely thankful to all of them who made this a reality.

We are thankful to CRC Press, Taylor and Francis Group, New Delhi (India), for giving us this opportunity to make such a book that will help the students, as well as our scientific community, to get 100% success in decoding the enigma attached to genetics. We would like to thank Renu Upadhyay and Shikha Garg, for giving us full support during this project. Any suggestion for the enhancement of the quality for this book is most welcome.

Prof. Rakesh S. Sengar
Dr. Amit Kumar
Dr. Reshu Chaudhary
Er. Ashu Singh

Authors

Professor Rakesh S. Sengar, the author of the book, is a professor and head of the Department of Ag. Biotechnology, College of Agriculture in Sardar Vallabhbhai Patel University of Agriculture and Technology, Meerut. Dr. Sengar has 24 years of teaching, research, and extension experience to his credit. He remained associated with the teaching for a period of about 10 years at G.B. Pant University of Agriculture and Technology, Pantnagar, U.S. Nagar (Uttarakhand). Prof. Sengar possesses a career of leadership in improving agrarian livelihood, food, nutrition, and ecological security in cutting-edge research and technology development, in higher education and human resources development, and in policy and program formulation, execution, and appraisal, each pursued in national and global capacities. He has supervised 11 doctoral students and published more than 118 research papers in India and foreign journals of repute. Dr. Sengar contributed more than 200 abstracts or papers to different conferences, symposiums, and workshops organized at national and international forums. He also participated in several international programs, seminars, and conferences as chairperson of technical sessions, group leader, invited speaker, and delegate. He has published 897 popular articles in Hindi and English languages in important journals and magazines. He has also published 16 books for the scientific community and farmers. Dr. Sengar has delivered more than 217 talks at All India Radio and Television at Delhi, Lucknow, Rampur, and Bareilly centers. Dr. Sengar is a lifetime member of several professional societies. He is a member of the editorial boards and review committees of a few journals. He is also chief editor of *International Journal of Biological Sciences Biotech Today* and founder of the Society of Green World for Sustainable Environment. He has been the principal and co-investigator of 17 projects financed by various government departments and has had wide exposure to international programs and also in handling international projects. He has traveled widely and has visited the United Kingdom, Nepal, and Dubai under various programs. Dr. Sengar is the recipient of the Best Writer award from Vishwa Agro Marketing and Communication, Kota, Rajasthan; the Kunwar Saxena Bahadur SRDA award from the Society for Recent Development in Agriculture, Meerut; the Man of the Year and Research Board of Advisors awards from the American Biographical Institute, Inc., U.S.A.; and the Aryabhat award from Vigyan Bharti, New Delhi. Dr. Sengar also received the Gold Medal 2011 from the Hi-Tech Horticulture Society, Meerut; the Dr. J.C. Edward Medal 2012 from the Bioved Research Society, Allahabad; the Scientist of the Year 2013 award from the Academy of Environmental Biology, Lucknow; the Life Time Achievement Award 2014 from the Dr. Babasaheb Ambedkar National Institute of Social Sciences, Dr. Ambedkar Nagar (Mhow) M.P.; the Dr. R.B. Lal Vishisht Sikshak Puruskar 2015 from Uttar Pradesh Academy of Agricultural Sciences, Lucknow; the Farm and Food Award 2016 from Farm and Food, New Delhi; and the

Young Achiever Award 2016 from the Society of Plant Research, Bareilly. He is a fellow of the Society of Plant Research, Bareilly; the Society for Recent Development in Agriculture, Meerut; and Gramin Vikas Avam Siksha Prasar Samiti, Agra and Technical Society for Basic and Allied Sciences, New Delhi. Dr. Sengar has made significant contributions in the fields of plant physiology, biotechnology, and crop production and management systems besides providing leadership in shaping up national level research institutions and agricultural universities. Dr. Sengar has assisted several developing countries in enhancing their human resources, research, and technology development capacities, in augmenting productivity and sustainability to enhance farmers' income and to alleviate hunger and poverty crises.

Dr. Amit Kumar, PhD in agriculture biotechnology from Sardar Vallabhbhai University of Agriculture and Technology, Meerut and MPhil in Biotechnology from Vinayaka Missions University, Tamilnadu. He is currently working as an assistant professor in the Department of Biotechnology, Faculty of Science, Swami Vivekanand Subharti University, Meerut. Dr. Kumar has 4 years teaching experience as an assistant professor. Dr. Kumar has contributed more than 30 abstracts to different national and international conferences. He has published more than 22 research papers/review papers. He has published 10 Hindi popular articles in national magazines. He has published 3 international and 2 national book chapters. Dr. Kumar has co-authored *Objective Biotechnology* with Dr. Rakesh S. Sengar. He received the National Fellowship during his doctoral work. Dr. Kumar has qualified for the ICAR's All India Examination, ICAR-SRF (PGS), and the NET (National Eligibility Test) examination. His areas of interest are plant tissue culture and molecular biology.

Dr. Reshu Chaudhary, the co-author of this book, received her MSc in biochemistry and her PhD from Chaudhary Charan Singh University, Meerut, and is currently working as a women scientist in the Department of Agricultural Biotechnology, College of Agriculture, Sardar Vallabhbhai Patel University of Agriculture and Technology, Meerut. She has nearly 6 years of research experience. She is a good researcher and has published 3 books and more than 43 scientific articles and research papers in different national and international journals/seminars of repute. Apart from this she is also an editor of *International Journal of Biological Sciences Biotech Today*, published under the Society of Green World for Sustainable Environment, along with being an active reviewer at many reputed journals. Dr. Chaudhary is recipient of the Young Scientist Award 2012 from Bioved Research Society, Allahabad, and the Fellow Award 2016 from The Society for Plant Research,

Bareilly. She is working for the sustainable development of different crops for the last 2 years.

Er. Ashu Singh was born in Allahabad, India. She received a BTech degree in genetic engineering from the Allahabad Agricultural Deemed University, Allahabad, India, in 2008, and the MTech degree in biotechnology engineering from the Sardar Vallabhbhai Patel University of Agri. and Tech (State Agriculture University) Modipuram Meerut, in 2011. In 2013, she joined the Department of Ag. Biotechnology, Sardar Vallabhbhai Patel University of Agri. and Tech (State Agriculture University) Modipuram Meerut, as a women scientist (WOS-A). Since April 2013, she has been with the Department of Ag. Biotechnology, College of Agriculture, where she is working on varietal improvement of aromatic rice. Her current research interests include molecular characterization and gene mapping of salinity and drought resistance of aroma rice (*Oryza sativa* L.). Miss Singh is a National Woman Scientist Fellow of the Department of Science and Technology, New Delhi, India. She has extensive experience at 8 trainings of more than 3 months at different reputed institutes and research laboratories with technical expertise in molecular breeding and also has made significant contributions in the vast arenas of stress physiology, biotechnology, and basmati rice production while exploring the applicability of research in enhancing productivity. She also holds a diploma from the Industry Program in Bioinformatics (BII, Noida) and submitted a research study on "Importance of Protein folding for protein structure." Apart from this, she is also an associate editor at *International Journal of Biological Sciences Biotech Today*, published under the Society of Green World for Sustainable Environment, along with being an active reviewer at many reputed journals. Miss Singh contributed 12 research papers in reputed indexed journals worldwide and attended different conferences/symposiums/workshops organized in national and international forums.

1 DNA Manipulative and Ingratiative Techniques

ISOLATION OF TOTAL GENOMIC DNA

Isolation of DNA from different tissues, either plant or plasmid, is at the heart of molecular biology because all the genetic information is carried in the linear sequence of nucleotides in DNA. Several protocol have been developed for the isolation of genomic DNA, but selective protocol are described below for plant, plasmid, yeast, and bacteria.

ISOLATION OF PLANT DNA USING THE CTAB METHOD

Introduction

Isolation of DNA from plant tissues is at the heart of plant molecular biology Because plant cells are surrounded by rigid cell walls and because plant tissues often contain a variety of secondary metabolites that can damage DNA, isolation from plants can present some particular difficulties. Among the many protocols developed for DNA isolation from plants, the method presented here is one of the simplest and most effective for a variety of plant species.

DNA isolated by methods such as the one presented here represents total cellular DNA. In plants, this means the isolation of 3 distinct genomes:

1. The nuclear genome
2. The chloroplast genome
3. The mitochondrial genome

Generally, when we talk about plant DNA, we mean nuclear DNA, but it is important to remember that chloroplasts and mitochondria each have distinct genomes. These genomes, like the nuclear genome, are targets of research in molecular biology and genetic engineering. Total plant DNA isolated here will be compared with DNA isolated from chloroplasts.

Experimental Outline

Day 1: DNA isolation (2 h).

Grind tissue and suspend in extraction buffer.
(EB) incubate at 65°C for 1 h.
Extract with chloroform.
Precipitate DNA with isopropyl alcohol.
Resuspend the DNA in buffer and store.

Day 2: Compare this DNA preparation with DNA isolated from chloroplasts by gel electrophoresis.

Materials Required

Chemicals

- *For EB (extraction buffer):* 50 mM Tris (pH 8.0), 1% CTAB, 50 mM EDTA, 1 Mm 1-10-o-phenanthroline, 0.7 M NaCl, 1% beta-mercaptoethanol
- Chloroform
- Isopropyl alcohol
- 80% ethanol
- 15 mL ammonium acetate
- *For TE buffer:* 10 mL Tris (pH 8.0), 1 mL EDTA (pH7.5)

Glassware and Other

- Measuring bottle
- Conical flask
- Measuring cylinder
- Mortar and pestle
- Centrifuge tubes (50 mL capacity, capped)

Equipment

- Water Bath
- Cooling Centrifuge
- Gel doc system
- Electrophoresis unit
- Power pack
- Micropipette

Pre-Lab Preparation

Pea tissue used in this protocol should be air-dried or freeze-dried. Grow the plants either in sand or perillite. Tissue can be air-dried in the laboratory by spreading the cut shoots on a sheet of absorbent paper and turning daily. Tissue treated in this way will typically dry in 5–7 days. Dried tissue can be saved for use at a later date by sealing in a tightly closed jar and freezing.

Method (Figure 1.1)

Day 1: DNA isolation.

1. Grind 1 g dried (air-dried or freeze-dried) pea shoots to a fine powder in a mortar with pestle.
2. Mix the powder with 25 mL EB (extraction buffer) in a 50 mL capped centrifuge tube.
3. Place the tube into a 65°C water bath and incubate for 1 h; during the h, mix the tube's contents by several inversions.
4. Remove the tube from the water bath and allow it to cool for several minutes on the bench.

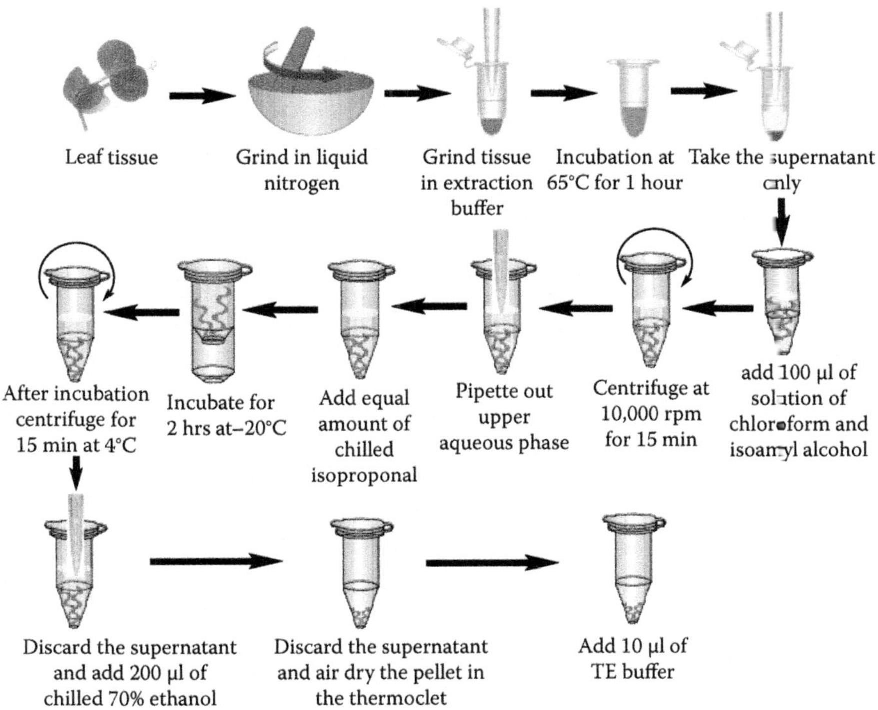

FIGURE 1.1 Isolation of genomic DNA from leaf tissue.

Note: Do not skip this step; contents will boil out of the tube if added at 65°C.

5. Add 20 mL chloroform to the tube, screw the cap, and mix by inversion until the contents are thoroughly mixed. When mixed, the extraction buffer and the chloroform will form a thick emulsion.
6. Centrifuge the tube at >3500 × g for 10 min to break the emulsion and separate the tube contents into 2 phases.
7. Upon removal from the centrifuge, the contents of the tube form 3 distinct layers. At the bottom is a green layer of chloroform. Often, bits of plant material are pelleted under this layer. In the middle is an interphase consisting largely of denatured protein and bits of leaf tissue. This layer is whitish or yellowish in color. At the top of the tube is the straw-colored aqueous layer. This top layer contains the majority of the DNA in the preparation. Using a pipette, remove this top layer into a small flask or beaker. Avoid transferring any of the interphase material from the centrifuge tube.
8. What remains in the centrifuge tube, the interphase and organic (chloroform) layers, is hazardous waste. Follow your instructor's direction for proper disposal.
9. To the aqueous phase in the flask or beaker, add 2/3 volume of isopropyl alcohol (e.g., if you have transferred 24 mL to the flask, add 2/3 × 24 or 16 mL of isopropyl alcohol). Mix by swirling the contents. DNA will precipitate to form a cottony mass.

10. Using a glass rod or a Pasteur pipette, transfer the DNA to a clean flask. Add 10 mL of 80% ethanol and 15 mM ammonium acetate and swirl to wash.
11. After about 20 minutes, transfer the precipitated DNA to a microcentrifuge tube and centrifuge briefly to drive the DNA to the bottom of the tube. Using a Pasteur or capillary pipette, remove the residual ethanol. Allow the DNA to dry in the uncapped tube for about 10 minutes on the bench top.
12. Add 0.75 mL of TE buffer to the DNA to dissolve the precipitate.

Note: Large quantities of DNA may require some time to dissolve completely. Leave the capped tube in the refrigerator until the next lab class.

Day 2: Cut 2 μL of the DNA with a restriction endonuclease as directed by your instructor.

On an agarose gel, compare total DNA (both uncut and cut) with DNA isolated from chloroplasts (both uncut and cut).

This protocol is generally trouble-free. The only difficulties we have experienced with it come from (i) using incompletely dried material and (ii) using heat-dried material. Rarely, DNA prepared by this method forms a flocculent rather than cotton precipitate. Such a precipitate will not spool on a glass rod and must be collected by centrifugation. If desired, RNA can be removed from the DNA preparation by adding 20 μg RNase A (heat-treated to destroy DNase), along with the TE buffer in step 12 of the protocol.

Note on the use of chloroform in this exercise: If you want to avoid the use of chloroform in this exercise, make the following modifications to the lab protocol.

1. Grind tissue as before, mix with EB, and incubate in the water bath.
2. Centrifuge to pellet the undigested plant tissue.
3. Transfer the supernatant to a fresh tube, flask, or beaker and precipitate DNA with isopropanol as before.

Note: The supernatant will be colored dark green in this preparation making the precipitation of DNA somewhat more difficult to see. If this is a problem, the supernatant can be diluted with additional EB or TE buffer before the precipitation step. DNA prepared in this way should not be expected to cut well with restriction end nucleases, but it may be satisfactory.

Precautions

Reagents used in this protocol are potentially dangerous.

1. CTAB (cetyltrimethylammonium bromide, in the extraction buffer) is a strong detergent and can cause burns to the skin.
2. Chloroform is toxic by inhalation or on contact with skin. Follow your instructor's direction for proper handling and disposal of these materials.
3. All glassware/plasticware and solutions (except organic solvents) should be sterilized/autoclaved.
4. Gloves should be worn to avoid contamination of the experimental material and apparatus with nucleases, which occur in fair abundance in skin exudates.

5. Phenol causes severe burns and phenol-containing solutions should, therefore, be handled with care. Thoroughly rinse burns with large volumes of water. Do not use ethanol.
6. For work with RNA, rinse all glassware/plasticware with 1% diethyl carbonate solution to inactivate RNase and autoclave them.

Isolation of Plant DNA Using the SDS Method

Introduction

This method can be used to isolate DNA from a wide variety of plants, including pine, tomato, juniper, cypress, sorghum, maize, soybean, cycad, fern, prickly pear cactus, cabbage, and cotton. It is specially designed for plants in which polyphenols are a problem, although it has provided good results for every plant species on which it has been tested.

The solution contains SDS and other disruption agents required for complete cell lysis. In addition to aiding in cell lysis, SDS is an anionic detergent that breaks down proteins, fatty acids, and lipids associated with the cell membranes. If it gets cold, it will form a white precipitate in the bottle. Heating to 60°C will dissolve the SDS and will not harm the SDS or the other disruption agents. Solution PB1 can be used while it is still warm. The solution is 99% isopropanol and will precipitate DNA along with some organic contaminants. Most of the co-extracted impurities will be removed at this step. Isopropanol will precipitate and pellet the DNA. The pellet at this point contains relatively pure DNA along with some organic contaminants (mostly polysaccharides and phenolics depending on the plant tissues processed), leaving a majority of the contaminants in solution.

Materials Required

Chemicals

- Extraction buffer: 50 mM Tris (pH 8.0), 1% CTAB, 50 mM EDTA, 1 Mm 1-10-o-phenanthroline, 0.7 M NaCl, 1% beta-mercaptoethanol
- 10% SDS
- Ammonium acetate
- Isopropanol
- 70% ethanol
- TE Buffer

Glassware and Other

- Measuring bottle
- Conical flask
- Measuring cylinder
- Mortar and pestle
- Centrifuge tubes

Equipment

- Centrifuge
- Incubator
- Micropipette
- Electrophoresis unit

Method (Figure 1.2)

1. Take 1 g of plant sample and wash with distilled water.
2. Remove excess water with tissue paper.
3. Grind the sample with 4 mL of extraction buffer.
4. Incubate on ice for 5 min.
5. Centrifuge at 6,000 rpm for 5 min.
6. Resuspend the pellet in 3 mL of extraction buffer and 0.2 mL of 10% SDS.
7. Incubate at 60°C–65°C for 30 min.
8. Add 1.3 mL of ammonium acetate (7.5 M).
9. Keep ice cold for 10 min.
10. Centrifuge at 10,000 rpm for 10 min.
11. Take the supernatant and add an equal volume of ice-cold isopropanol.
12. Incubate at 4°C overnight.
13. Centrifuge at 12,000 rpm for 10 min.
14. Discard the supernatant and add 1 mL of 70% ethanol.
15. Centrifuge at 12,000 rpm for 5 min.
16. Discard the supernatant, air dry the pellet, and dissolve the pellet in 50 μL of TE buffer.

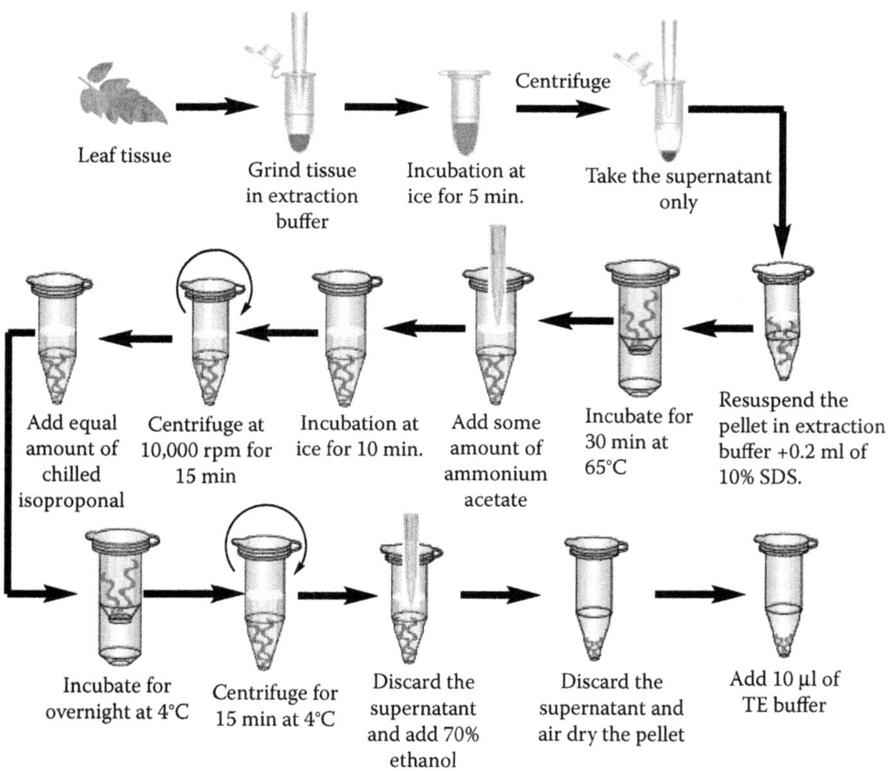

FIGURE 1.2 Isolation of DNA from SDS method.

Precautions

1. All glassware/plasticware and solutions (except organic solvents) should be sterilized/autoclaved.
2. Gloves should be worn to avoid contamination of the experimental material and apparatus with nucleases, which occur in fair abundance in skin exudates.
3. Phenol causes severe burns and phenol-containing solutions should, therefore, be handled with care. Thoroughly rinse burns with large volumes of water. Do not use ethanol.
4. For work with RNA, rinse all glassware/plasticware with 1% diethyl carbonate solution to inactivate RNase and autoclave them.

To Isolate Plasmid DNA from Bacteria

Introduction

Plasmids are double-stranded, circular, self-replicating extrachromosomal DNA, and their size varies from 1–100 kb. They confer antibiotic resistance to bacteria enabling them to grow in selective antibiotic medium. Plasmids are used as cloning vectors and in various molecular biology techniques, such as restriction enzyme digestion, ligation, and transformation. Many methods have been used to isolate plasmid DNA, essentially involving the following steps:

1. Growth of bacteria
2. Harvest and lysis of bacterial cells
3. Extraction and isolation of plasmid DNA

Plasmids are generally isolated by the alkaline lysis method. Cells are lysed by treatment with alkali (NaOH) and sodium dodecyl sulfate (SDS). SDS denatures bacterial proteins and NaOH denatures the plasmid DNA renatured rapidly because of its small size and covalent conformation, while chromosomal DNA and proteins are trapped in a complex formed between sodium and SDS.

Most of the chromosomal DNA and bacterial proteins precipitate along with the SDS complex and are removed by centrifugation. The reannealed plasmid DNA, present along with RNA, in the supernatant is then concentrated with isopropanol and finally plasmid DNA is precipitated and separated by treatment with ethanol.

Materials Required

Chemicals

1. For plasmid extraction and isolation
   a. *Solution I:* TEG (resuspension) buffer
      i. 15% glucose; prevents immediate lysis of cells
      ii. 25 mM Tris (pH 8.0); maintains pH
      iii. 10 mM EDTA; chelates metal ions and weakens cell wall
   b. *Solution II:* lysis buffer
      i. 0.2 N NaOH; denatures chromosomal and plasmid DNA
      ii. 1%SDS; denatures bacterial membrane proteins and disrupts cell membrane

   c. *Solution III:* 3 M sodium acetate; renatures DNA
   d. *Solution IV:* isopropanol; precipitates plasmid DNA
   e. *Solution IV:* TE buffer (pH 8.0); 25 mM Tris and 1 mM EDTA
2. For growth of bacterial culture
   a. *Strain:* pUC 18 in DHα (*E. coli* strain). pUC has a size of 2686 bp with an ampicillin resistance gene as a marker. It is commonly used as a cloning vector.
   b. Luria broth (LB) and agar medium
   c. Ampicillin (100 μg/ml): Dissolve commercially available ampicillin injection in required amount of D/W.
   d. Bacterial cells

Glassware and Other

- Petri plates
- Culture tubes
- Beakers
- Microcentrifuge tubes
- Conical flask

Equipment

- Vortex
- Micropipettes
- Electrophoresis assembly

Method (Figure 1.3)

Revival of cells

Day 1:

1. Activate the pUC 18 culture by growing on LB medium containing ampicillin 100 μg/mL.
2. After activation, streak loop full of activated culture on LB agar plate containing ampicillin 100 μg/mL. Incubate the plate overnight at 37°C.

OR

1. Break and open the lyopholized vial.
2. Add 0.1 mL of LB broth containing ampicillin with a concentration of 100 μg/mL. Resuspend until the lyopholized sample goes into solution.
3. Streak a loop full of this suspension on an LB agar plate containing ampicillin 100 μg/mL (fresh plates can be stored for 7 days). Incubate the plate overnight at 37°C.

Day 2: Pick a single colony from the plate and inoculate into 10 mL LB broth containing ampicillin 100 μg/mL and incubate in shaker at 37°C overnight. Use this culture or extraction and isolation of plasmid from bacterial culture.

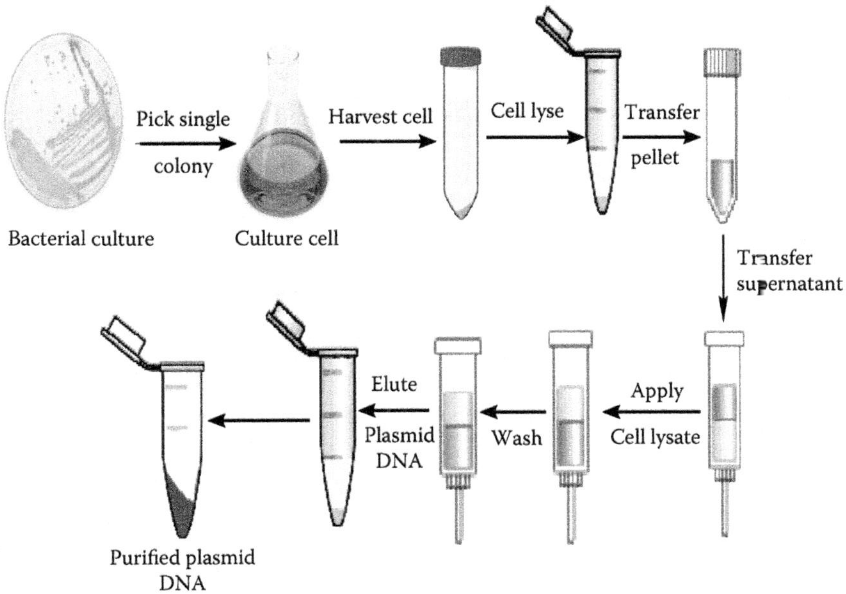

FIGURE 1.3 Isolation of plasmid DNA.

Day 3:

1. Withdraw 1.5 mL culture and place in a 1.5 mL tube.
2. Centrifuge for 10 min at 6,000 rpm. Discard the supernatant.
3. Resuspend the pellet in 100 μL of solution I. Vortex gently. Place on ice for 5 min.
4. Add 200 μL of solution II at room temperature. Mix gently by inverting the tube.
5. Add 150 μL of solution III. Mix gently by inverting the tube. Place on ice for 5 min.
6. Centrifuge for 10 min at 8,000 rpm.
7. Transfer the supernatant to a fresh tube and add 450 μL of isopropanol. Mix by inverting the tube to precipitate the DNA. Keep for 15 minutes at room temperature.
8. Centrifuge for 20 min at 10,000 rpm. Decant the supernatant. Invert the vial on blotting paper to drain leftover supernatant. White precipitate will be visible, sticking to the side of the vial wall. Dry 10–15 min at 37°C.
9. Resuspend the pellet in 20 μL of 1 × TE (add along the slides); mix by tapping tube with finger so that DNA goes into the solution. Be sure mixing is complete with intermittent tapping for 20 min.
10. Add 5 μL of RNase A (to remove RNA) and incubate for 30 min at 37°C.
11. Add 3 μL of gel loading buffer to the prepared DNA.
12. Perform electrophoresis in 1% agarose gel as in the first experiment.

Precautions

1. Handle all the solutions carefully.
2. Maintain cooling conditions properly wherever mentioned.
3. All glassware/plasticware and solutions (except organic solvents) should be sterilized/autoclaved.
4. Gloves should be worn to avoid contamination of the experimental material and apparatus with nucleases, which occur in fair abundance in skin exudates.

Isolation of Chloroplasts

Introduction

Chloroplasts are the subcellular sites of photosynthesis, the process by which green plants using energy from light to produce carbohydrate and oxygen from carbon dioxide and water. Under the microscope, chloroplasts are recognized as bean-shaped, membrane-bound, green (chlorophyll-containing) organelles. In this experiment, chloroplasts are isolated from a cell homogenate by density gradient centrifugation. A measurement of the amount of chlorophyll in the preparation is used to asses yield of intact chloroplasts. Finally, the chloroplasts are saved for the isolation of DNA.

Density gradient centrifugation will be used to isolate chloroplasts from pea seedlings. The objectives are

1. Isolate chloroplasts from pea.
2. Assay chlorophyll content of isolated chloroplasts.

Experimental Outline

1. Prepare sucrose gradients (30 min).
2. Harvest pea tissue and process sample (30 min).
3. Apply sample to gradient and centrifuge (30 min).
4. Retrieve chloroplasts from gradient and assay yield (30 min).

Materials Required

Chemicals

- Pea seedlings
- Homogenization buffer
- 10 mM KCl
- 1 mM $MgCl_2$
- 1% (w/v) dextran T_{40}
- 1% (w/v) Ficoll
- 0.1% (w/v) bovine serum albumin
- Make to volume with 30% (w/v) sucrose in 0.1 M tricine buffer (pH 7.5)
- Sucrose (gradient) solutions (w/w) prepared in 0.1 M tricine buffer (pH 7.5): 60%, 50%, 40%, 30%.

Glassware and Others

- Beakers
- Conical flask
- Measuring cylinder
- Test tubes
- Ice bucket
- Cheesecloth and miracloth (Calbiochem)
- Centrifuge tubes
- Pasteur pipettes

Equipment

- Blender
- Centrifuges (clinical and super speed)
- Spectrophotometer

Pre-Lab Preparation

Time table of events: The exercise can be divided into 4 parts: (i) preparation of sucrose gradient, (ii) sample preparation, (iii) centrifugation, and (iv) analysis of results. Each of these activities will require approximately 30 minutes.

Equipment requirements: The protocol described here is written to be used with a refrigerated, super-speed centrifuge (e.g., Sorvall RC series) equipped with the swinging bucket rotor (e.g., Sorvall HB4). Other comparable centrifuges will perform equally well, but specified speeds and times of centrifugation might have to be changed slightly. The operation manual for your particular centrifuge and rotor will explain these changes.

The assay of chlorophyll content in gradient fractions requires the use or spectrophotometer or colorimeter.

A blender is required for tissue homogenization. A standard household blender will work well, but the blades should be as sharp as possible (some laboratories have been known to modify their blenders by replacing the standard blades with razor blades).

Plant Material

The protocol in this experiment calls for the use of pea seedlings as starting material. The seed is relatively inexpensive and easy to grow. For best results, use fresh seed. Seedlings should be about 7 days old. To plant, soak seeds overnight in a large container. Sow the seeds on a layer of about 1.5 in of wet horticultural grade vermiculite in a standard nursery flat (21 × 10 × 2 in). The seed can be sown thickly, nearly touching one another. (We usually use a 500 mL beaker full of dry seed to plant 1 flat. At harvest, you can expect about 300 g of shoot tissue from such a planting). Cover the seed with 0.5–1 in of vermiculite and water well. Cover the flat with plastic wrap to hold in moisture until the seedlings begin to emerge. Once the seedlings have emerged and the plastic has been removed, keep well-watered. The seedlings can be grown in the lab on a window sill, in

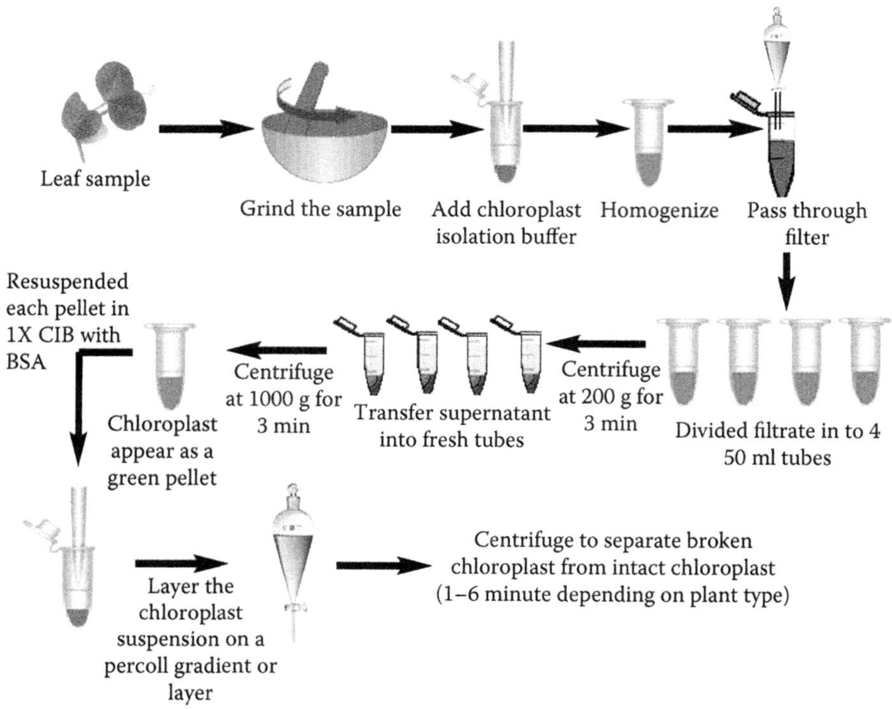

FIGURE 1.4 Procedure of chloroplast isolation.

a growth chamber, or in the greenhouse, for best results. However, do not grow the seedlings under intense light. Under very bright light, chloroplasts tend to accumulate large granules of starch, and these can do damage during blending and centrifugation.

Note: Left over shoots can be air-dried or freeze-dried and used for isolation of chloroplast genomic DNA in the next experiment.

Method (Figure 1.4)

1. Prepare 2 sucrose gradients in 50 mL centrifuge tubes.
   a. Pipette out 5 mL of 60% sucrose solution into bottom of each tube.
   b. Layer 5 mL 50% sucrose, then 10 mL 40% sucrose into each tube. Layers should be distinct from one another if you are careful.

Hint: Tip the tube as you add each layer of sucrose. Let the tip of the pipette just touch the surface of the liquid in tube.

   c. With the tip of a Pasteur pipette or stirring rod, gently mix at the interface of the 50% and 40% layers to disturb slightly.
   d. Layer 5 mL 30% sucrose on top of the gradient.
   e. Keep the gradients on ice while you prepare the tissue sample.

2. Harvest 5 g of 7-day-old pea seedlings at the soil line with a razor blade.
3. Chop the tissue into small pieces with a razor blade or scissors and transfer them to a chilled blender containing 20 mL ice-cold homogenization buffer.

Note: More than one 5 g batch of seedlings can be blended at a time. If several lab groups are sharing the blender, they should use 20 mL of buffer for each 5 g of seedlings homogenized and divide the homogenate.

4. Homogenize with 5 2–3 s bursts of the blender at high speed.
5. Filter the homogenate into a beaker (on ice) through 4 layers of cheesecloth, squeezing the cloth gently to remove most of the liquid (wear gloves).
6. Refilter through 1 layer of Miracloth, moistened in homogenization buffer, by gravity. Do not squeeze, you may want to prepare a wet-mount slide of the residue left in the cheesecloth or Miracloth for the microscopic examination. What have you removed from the homogenate by filtration?
7. Layer 10 mL of the filtrate onto the top of each of your gradients (prepared in step 1). Check to see that the 2 gradients are balanced against one another. If necessary, add homogenization buffer to make the tubes balance. Centrifuge at 4°C in a HB4 rotor, 4,000 rpm for 5 min, then increase speed to 10,000 rpm for 10 min. Allow the centrifuge to come to a stop. Carefully remove your gradients from the rotor.
8. You should see 2 green bands in the gradients. The green band toward the bottom of the tube is the fraction containing intact chloroplasts. Remove the top of the gradient carefully with a Pasteur pipette. Save the 2 chlorophyll-containing fractions in clean tubes on ice.
9. Prepare wet-mount slides of the chlorophyll-containing fractions and examine under the microscope. What differences do you notice between them?
10. Assay chlorophyll content of the 2 green bands. For each sample,
    a. Into a clean centrifuge tube, pipette 50 μL of the gradient fraction to be assayed and 0.95 mL distilled water.
    b. Add 4 mL acetone.
    c. Centrifuge in a clinical centrifuge, 5 min.
    d. Measure the absorbance of the solution in a spectrophotometer at 652 nm. (The appropriate blank for this measurement is 80% acetone in water.)
    e. Calculate chlorophyll content:
    $A_{652} \times 29 = \mu g$ chlorophyll/10 μL chloroplast fraction.
11. Freeze the intact chloroplast fraction to save for DNA isolation.

Results

Make a diagram of your gradient. Have materials other than chloroplasts banded in the tube? Where are they? What do they look like? Why was the cell homogenate filtered before being loaded on the gradient? What was removed by filtration? What fraction of the chloroplasts in the homogenate have you isolated intact?

After centrifugation, the gradients should show 2 bands of chlorophyll. The upper of the 2 contains broken chloroplasts and membrane fragments. The lower of the bands (about 3/4 of the distance to the bottom of tube) contains intact chloroplasts. Probably the most frequent cause of a poor yield of intact chloroplasts is overblending. Using any type of homogenizer to disrupt tissue is a trade-off between efficiency in breaking cells open and generating so much shear in the solution that organelles are also disrupted.

If you plan to use the chloroplasts from this prep for DNA isolation, avoid contaminating the intact chloroplast fraction with broken chloroplasts. Also, try to retrieve the intact chloroplast fraction from the gradient in as small a volume as possible.

Precautions

1. Handle all the solutions carefully.
2. Maintain cooling conditions properly wherever mentioned.
3. All glassware/plasticware and solutions (except organic solvents) should be sterilized/autoclaved.
4. Gloves should be worn to avoid contamination of the experimental material and apparatus with nucleases, which occur in fair abundance in skin exudates.

Isolation of Chloroplast DNA

Introduction

The DNA of plant cells is found in 3 distinct genomes. First, there is nuclear DNA, familiar as the DNA that makes up the chromosomes. But mitochondria and chloroplasts each have DNA of their own. These genomes are closed, circular DNA molecules encoding many of the enzymes necessary for the function of the organelles. Because of the importance of mitochondria and chloroplasts to the cell, their DNA is of interest to molecular biologists and biotechnologists. The chloroplast DNAs of several species of plants have been cloned and sequenced in their entirety. In at least one organism (the green alga chlorella), chloroplasts as well as nuclear genomes, have been genetically transformed. In this laboratory exercise, DNA will be isolated from chloroplasts and compared with total DNA. The objectives of this exercise are

1. Isolate DNA from chloroplasts.
2. Compare chloroplast DNA to genomic DNA.

Experimental Outline

Day 1: DNA isolation (2 h)

Add EB buffer to chloroplast preparation.
Incubate 1 h at 65°C.
Extract with chloroform.
Precipitate DNA.
Resuspend in buffer.

Day 2: Compare DNA from chloroplast to genomic DNA.

Restriction digest of DNA (optional).
Pour gel; load cut and uncut DNAs.

Materials Required

Chemicals

- *EB (extraction buffer):* 50 mM Tris (pH 8.0), 1% CTAB, 50 mM EDTA, 1 mM phenanthroline, 0.7 M NaCl, 0.1% beta-mercaptoethanol
- Chloroform
- Isopropyl alcohol
- *TE buffer:* 10 mM Tris (pH 8.0), 1 mM EDTA

Glassware and Others

- Conical flask
- Measuring Cylinder
- Centrifuge tubes

Equipment

- Water bath
- Cooling Centrifuge
- Electrophoresis Unit
- Micropipette

Pre-Lab Preparation

Chloroplasts are extracted using extraction buffer and reagents used elsewhere in this unit. No special preparation beyond the requirements for those exercises is required for this experiment.

Method (Figure 1.5)

Start with the frozen chloroplast preparation from the previous experiment. Typically, this sample will have a volume of several milliliters. For each mL of chloroplasts, add 4 mL EB. If necessary, transfer the mixture to a capped centrifuge tube of at least twice the volume of the chloroplasts and EB.

1. Incubate the mixture at 65°C for 1 h.
2. Remove the tube from the water bath and allow to cool on the bench top for several min before proceeding.
3. Add an approximately equal volume of chloroform to the tube, recap, and mix by inversion.
4. Centrifuge the tube at $> 3500 \times$ g for 10 min.
5. Upon its removal from the centrifuge, the tube contents will have separated into 2 distinct layers. Using a pipette, transfer the upper (aqueous) layer into a fresh centrifuge tube. (This tube should be the same size that was used in the first step). The lower (organic) layer is hazardous waste. Follow your instructor's directions for proper disposal.

6. Add 0.6 mL of isopropanol for each ml of DNA-containing extract in the centrifuge tube. Mix by inversion.
7. Centrifuge at > 10,000 × g for min.
8. After centrifugation, decant the liquid in the tube away from the DNA-containing pellet. Stand the tube upside down on a paper towel or Kimwipe for several minutes to allow the liquid to drain. The tube's inside can be wiped carefully to remove liquid, but take care not to dislodge the DNA pellet.

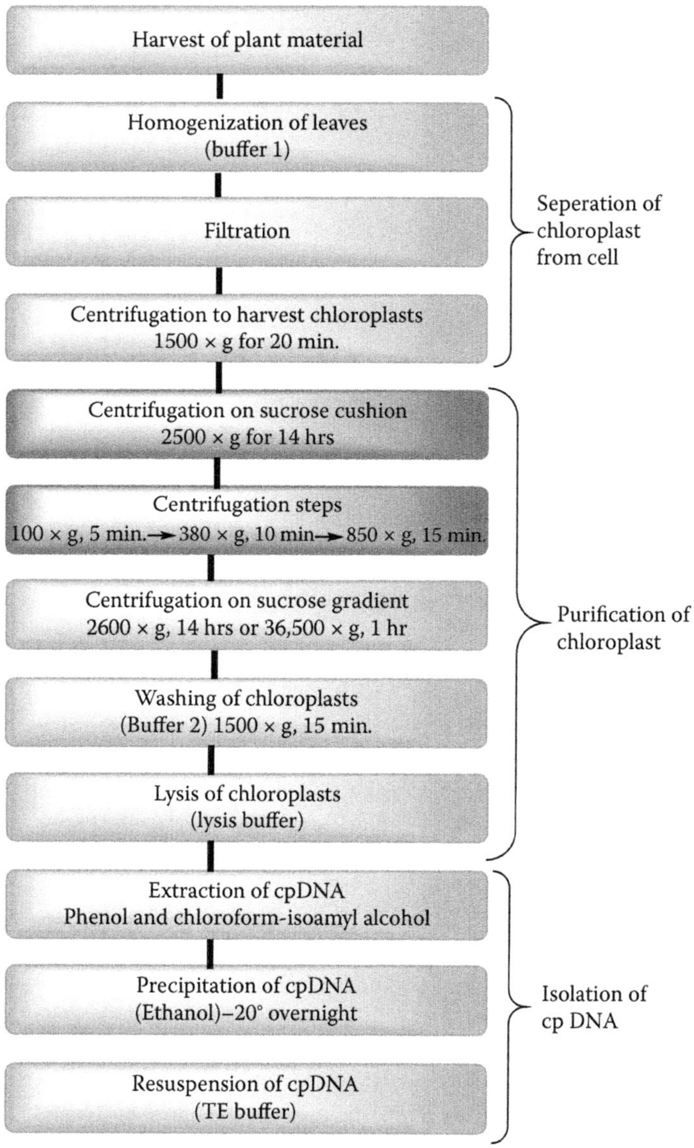

FIGURE 1.5 Isolation of chloroplast DNA.

Results

The yield of chloroplast DNA is expected to be low and will depend in some measure on the quality of the chloroplast preparation produced in the previous experiment. Also note that chloroplast DNA may not cut well with restriction end nucleases. Still, it is likely that at least 1 lab group or individual will get that which can be shared with the rest of the class. Some of the students will accept this activity as a challenge. Since it is teamed with an examination of total DNA, none of the students will be completely without results. The exercise is worth a try if for no other reason than it will give the students some experience in working with very small quantities of DNA.

Precautions

1. Some reagents used in this protocol are potentially hazards. CTAB (cetyltrimethylammonium bromide) is a strong detergent and can cause burns to the skin. Follow the instructor's direction for proper handing/disposal of these materials.
2. Handle all solutions carefully.
3. Maintain cooling conditions properly wherever mentioned.
4. All glassware/plasticware and solutions (except organic solvents) should be sterilized/autoclaved.
5. Gloves should be worn to avoid contamination of the experimental material and apparatus with nucleases, which occur in fair abundance in skin exudates.

Isolation of Yeast Genomic DNA

Introduction

The yeast *Saccharomyces cerevisiae* is a simple eukaryotic organism that is well-suited as a tool for analysis of mammalian gene regulation. Development of yeast artificial chromosomes (YACs) greatly facilitated analysis of complex mammalian genetics loci by allowing cloning, maintenance, and manipulation of larger stretches of exogenous DNA in yeast. More recently, YACs have been used to generate transgenic mice. Introduction of a mutation in YACs by homogenous recombination, coupled with the ability to produce transgenic mice, provided a system to analyze the effect of the mutation in the context of the whole locus with an animal model.

The first step concentrates and pellets the microbial cells. In some cases it may take longer to completely pellet the cells. It is important to pellet the cells completely and remove all the culture media in this step. A lysis buffer stabilizes and homogeneously disperses the microbial cells prior to lysis. This step can be beneficial for hard-to-lyse organisms. Heat can help denature proteins and aid in cell lysis. The cell debris is sent to the bottom of the tube while DNA remains in the supernatant by centrifuge. 70% ethanol wash removes all contamination in the DNA sample. It is important to avoid any traces of the ethanol-based wash solution.

Materials Required

Chemicals

- Lysis buffer
- 9.1 mL 1 M K_2HPO_4
- 920 μL 1 M KH_2PO_4
- 2 mL 10% Triton X-100
- 88 mL DD water
- 100× Mg^{2+} solution
- 0.1 M $MgCl_2$
- 4.5 M β-mercaptoethanol (Just before use, add an appropriate amount of β-mercaptoethanol from a 14.7 M stock.)
- Chloroform
- Isopropanol
- 70% ethanol
- TE buffer: add 0.4 mL or 0.25 M EDTA solution to 1.0 mL of 1 M Tris-HCI buffer (pH 8.0). Make volume to 100 mL with distilled water.

Glassware and Other

- Conical flask
- Measuring Cylinder
- Centrifuge tubes

Equipment

- Centrifuge
- Water bath
- Micropipette
- Laminar flow

Method (Figure 1.6)

1. Place 5 mL of the overnight yeast culture in a centrifuge tube.
2. Centrifuge at 6,000 rpm for 5 min.
3. Discard the supernatant and add 200 μL of lysis buffer to the pellet.
4. Incubate in ice-cold conditions for 5 min.
5. Transfer to 95°C water bath for 1 min and immediately place in ice-cold conditions for 5 min.
6. Vortex; add 200 μL of chloroform and mix it.
7. Centrifuge at 10,000 rpm for 10 min.
8. Transfer the upper aqueous phase into a fresh Eppendorf tube contain 500 μL of ice-cold isopropanol.
9. Incubate at 4°C overnight.
10. Centrifuge at 12,000 rpm for 10 min.
11. Discard the supernatant and add 1 mL of 70% ethanol.
12. Centrifuge at 12,000 rpm for 5 min.
13. Discard the supernatant, air dry the pellet, and dissolve the pellet in 50 μL of TE buffer.

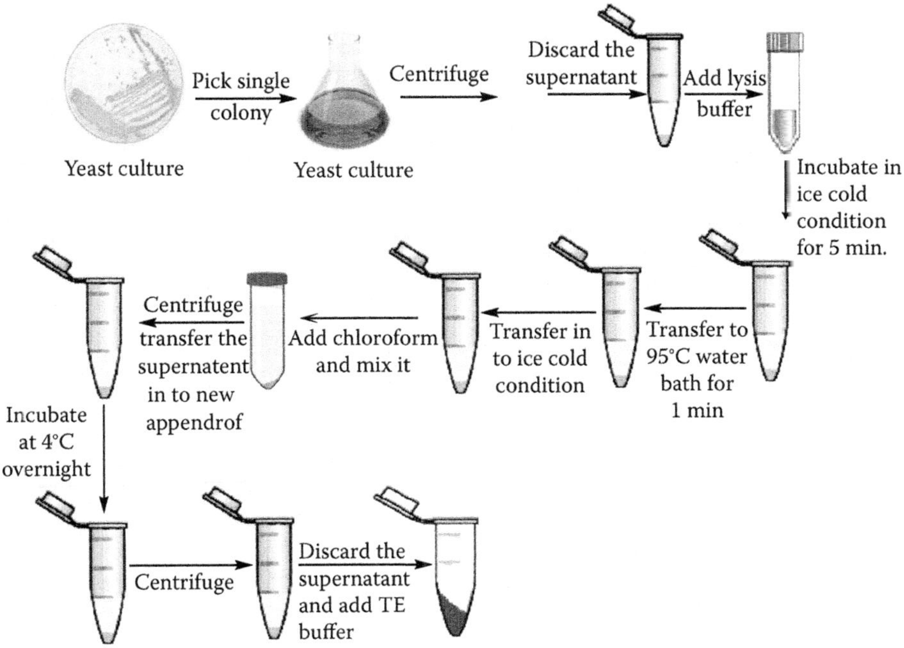

FIGURE 1.6 Isolation of yeast DNA.

Isolation of Bacterial Genomic DNA

Introduction

The isolation of DNA from bacteria is a relatively simple process. The organism to be used should be grown in a favorable medium at an optimal temperature and should be harvested in the late-log to early-stationary phase for maximum yield. The cells are then lysed and the DNA isolated. Following lysis, the cellular constituents are selectively removed. One this is accomplished, DNA is precipitated from solution with alcohol and dissolved in an appropriate buffer.

The lysis of the bacteria is initiated by resuspending the bacterial pellet in a buffer containing lysozymes and EDTA. The EDTA disrupts the other membrane of the gram-negative envelop by removing the Mg^{+2} from the lipopolysaccharide layer and additionally inhibits DNases. This allows the lysozymes access to the peptidoglycan. After partial disruption of the peptidoglycan by the enzyme, a detergent such as SDS is added to lyse without lysozyme. Once the cells are lysed, the solution should be treated gently to prevent breakage of the DNA strands. Subsequent steps involve the separation of the DNA from other macromolecules in the lysate. Phenol (that has been equilibrated with Tris buffer) is commonly used to dissociated proteins from nucleic acids. This reagent also removes lipids and some polysaccharides. Proteolytic enzymes such as pronase or proteinase K are often added to further remove proteins.

Materials Required

Chemicals

- Lysozyme
- Proteinase K
- Ammonium acetate
- Isopropanol
- 70% Ethanol
- TE buffer: add 0.4 mL or 0.25 M EDTA solution to 1.0 mL of 1 M Tris-HCI buffer (pH 8.0). Make volume to 100 mL with distilled water.

Glassware and Other

- Conical flask
- Measuring Cylinder
- Centrifuge tubes

Equipment

- Laminar Flow
- Centrifuge
- Micropipette
- Incubator

Method

1. Use 1 mL of overnight bacterial culture (*E. coli*).
2. Centrifuge at 12,000 rpm for 5 min.
3. Resuspend the pellet in 100 μL of TE buffer.
4. Add 10 μL of lysozyme (10 mg/mL).
5. Incubate at room temperature for 30 min.
6. Add 20 μL of 10% SDS and 15 μL proteinase K (20 mg/mL).
7. Incubate at 55°C for 1 h.
8. Add 400 μL of 6 M ammonium acetate; vortex thoroughly.
9. Take the supernatant and add equal volume of ice-cold isopropanol.
10. Incubate at –20°C for 30 min.
11. Centrifuge at 10,000 rpm for 10 min.
12. Discard the supernatant and add 1 mL of 70% ethanol.
13. Centrifuge at 12,000 rpm for 5 min.
14. Discard the supernatant, air dry the pellet, and dissolve the pellet in the 50 μL of TE buffer.

Large-Scale Isolation of Cosmid DNA

The method used for the isolation of large-scale, double-stranded DNA is a modification of an alkaline lysis procedure followed by equilibrium ultracentrifugation in cesium chloride-ethidium bromide gradients. Briefly, cells containing the desired plasmid or cosmid are harvested by centrifugation, incubated in a lysozyme buffer, and treated with alkaline detergent. Detergent solubilized proteins and membranes are precipitated with sodium acetate, and the lysate is cleared first by filtration of precipitate through cheesecloth and then by centrifugation. The DNA-containing supernatant is transferred to a new tube, and the cosmid DNA is precipitated by the addition of polyethylene glycol and collected by centrifugation. The DNA pellet is resuspended in a buffer containing cesium chloride and ethidium bromide, which is loaded into polyallomer tubes and subjected to ultracentrifugation overnight. The ethidium bromide-stained cosmid DNA bands, equilibrates within UV light, and the lower band is removed with a 5 mL syringe.

Isolation

1. Pick a colony of bacteria harboring the cosmid DNA of interest into a 12 × 75 mm Falco tube containing 2 mL of LB media supplemented with the appropriate antibiotic (typically ampicillin at 100 µg/mL) and incubate at 37°C for 8–10 h with shaking at 250 rpm.
2. Transfer the culture to an Erlenmeyer flask containing 50 mL of similar media, and incubate further for 8–10 h; transfer 12.5 mL of the culture to each of 4 L of similar media and incubate for an additional 8–10 h.
3. Harvest the cells by centrifugation at 7,000 rpm for 20 min in 500 mL bottles.
4. Resuspend the cell pellets in old media and transfer to 2 bottles, centrifuge as before, and decant the media.
5. The cell pellets can be frozen at –70°C at this point.
6. Resuspend the cell pellets in a total of 7 mL GET/lysozyme solution (3.5 mL for each bottle) by gently teasing the pellet with a spatula and incubate for 10 min at room temperature.

Note: Do not vortex the lysate at any time because this may shear the chromosomal DNA.

7. Add a total of 140 mL of alkaline lysis solution (70 mL for each bottle), gently mix, and incubate for 5 min in an ice-water bath.
8. Add 105 mL of 3 M sodium acetate, pH 4.8 (52.5 mL for each bottle), cap tightly, gently mix by inverting the bottle a few times, and incubate in an ice-water bath for 30–60 min.
9. Clear the lysate of precipitated sodium dodecyl sulphate (SDS), proteins, membranes, and chromosomal DNA by pouring through a double layer of cheesecloth. Transfer the lysate into a 250 mL centrifuge bottle, centrifuge at 10,000 rpm for 30 min at 4°C.

Purification

For Cesium Chloride-Gradient Purification

1. Pour the cleared supernatants into a clean beaker, add one-fourth volume of 50% PEG and 0.5 M NaCl and swirl to mix.
2. Incubate in an ice-water bath for 1–2 h.
3. Collect the PEG-precipitated DNA by centrifugation in 250-mL bottles at 7,000 rpm for 20 min at 4°C.

4. Dissolve the pellets in a combined total of 32 mL of 100:10 TE buffer, 5 mL of ethidium bromide, and 37 g cesium chloride (final concentration of cesium chloride should be 1 g).
5. Transfer the sample into 35-mL, polymer centrifuge tubes, remove air bubbles, seal with rubber stoppers, and crimp properly.
6. Centrifuge at 60,000 rpm for 16–20 h at 15°C–20°C in an ultracentrifuge.
7. Visualize the ethidium bromide-stained DNA under long-wave UV light.
8. Remove the lower DNA band using a 5 mL syringe with a 25-gauge needle. It may be helpful first to remove and discard the upper band. To remove the ethidium bromide, load the DNA sample onto an equilibrated 1.5-mL Dowex column, and collect 0.5-mL fractions.
9. Equilibrate the Dowex AG resin by successive centrifugation, resuspension, and decanting with 1 M NaOH, water, and add 1 M Tris-HCl (pH 7.6) until the Dowex solution has a pH of 7.6.
10. Pour those fractions with an A260 of 1 or greater into 35 Corex glass tubes; add 1 volume of double distilled water and ethanol precipitate by adding 2.5 volumes of cold 95% ethanol.
11. Incubate at least 2 h at –20°C and centrifuge at 10,000 rpm for 45 min.
12. Gently decant the supernatant. Add 80% ethanol and centrifuge as before.
13. Decant and dry the DNA pellet in a vacuum oven.
14. Resuspend the DNA in 10:0.1 TE buffer.

For Diatomaceous Earth-Based Purification

1. Pour the supernatants obtained after isolation into 500-mL bottles and add DNase-free RNase A and RNase Ti such that the final concentration of RNase A is 40 mg/mL and RNase T1 is 40 mg/mL incubate in a 37°C water bath for 30 min.
2. Add an equal volume of isopropanol and precipitate at room temperature for 5 min.
3. Centrifuge at 9,000 rpm for 30 min. Decant the supernatant and drain the DNA pellet.
4. Resuspend each DNA pellet in 20 mL 10:1 TE buffer and add 40 mL of defined diatomaceous earth in guanidine-HCl (100 mg/mL) to each bottle. Allow the DNA to bind at room temperature for 5 min with occasional mixing. Centrifuge at 9,000 rpm for 10 min.
5. Decant the supernatant, resuspend each pellet in 40 mL of diatomaceous earth-wash buffer, and centrifuge as above.
6. Decant the supernatant, resuspend each pellet in 40 mL of acetone, and centrifuge as above.
7. Decant the supernatant and dry the pellet in a vacuum oven. Resuspend the pellet in 20 mL of 10:1 TE buffer and elute the bound DNA by incubation at 65°C for 10 min with intermittent mixing.
8. Remove the diatomaceous earth by centrifugation at 9,000 rpm for 10 min. Repeat if necessary.
9. Combine the DNA-containing supernatants and precipitate the DNA in 35 mL. Corex glass tubes adding 2.5 volumes of cold 95% ethanol.

10. Resuspend the dried DNA pellet in 2 mL of 10:1 TE buffer and assay for concentration by absorbance readings at 260 nm or by agarose gel electrophoresis.

Miniprep Double-Stranded DNA Isolation

Principle

A miniprep double-stranded DNA isolation has been developed to generate a sufficient amount of template DNA for several Sequenase-catalyzed fluorescent terminator reactions.

Isolation

1. Pick a colony of bacteria harboring the plasmid DNA of interest into a 17 × 100 mm Falcon tube containing 6 mL of TB media supplemented with the appropriate antibiotic (typically ampicillin at 100 μg/mL) and incubate at 37°C for 16–18 h with shaking at 250 rpm.
2. Harvest the cells by centrifugation at 3000 rpm for 5 min in a centrifuge and decant the supernatant. The cell pellets can be frozen at –70°C at this point.
3. Resuspend the cell pellets in 0.2 mL of TE-RNase solution (50:10 TE buffer containing 40 1 g/rni RNase A; some also add RNase T1 to a final concentration of 10 U/1.11) by gentle vortexing, add 0.2 mL of alkaline lysis solution, gently mix, and incubate for 15 min at room temperature.
4. Add 0.2 mL of 3 M sodium acetate (pH 4.8), gently mix by swirling, transfer to 1.5-mL microcentrifuge tubes, and incubate in an ice-water bath for 15 min.
5. Clear the lysate of precipitated SDS, proteins, membranes, and chromosomal DNA by centrifugation at 12,000 rpm for 15 min in a microcentrifuge at 4°C.
6. Transfer the supernatant to a fresh 1.5-mL microcentrifuge tube, incubate in an ice-water bath for 15 min, and centrifuge as above for an additional 15 min.
7. Transfer the supernatant to a clean 1.5-mL tube.

Standard Alkaline Lysis Purification

Precipitate the DNA by adding 1 mL of 95% ethanol, and resuspend the dried DNA pellet in 100–200 41'10:0.1 TE buffer. Electrophorese an aliquot of the DNA sample on a 0.7% agarose gel to determine the concentration and purity.

Diatomaceous Earth-Base Purification

1. Add 1 mL of defined diatomaceous earth in guanidine-HCl (20 mg/mL) to each tube. Allow the DNA to bind at room temperature for 5 min with occasional mixing. Meanwhile, soak the Prep-A-Gene nitrocellulose membrane in isopropanol for at least 3 min.
2. Turn on the vacuum pump of the Prep-A-Gene manifold and adjust the vacuum level to 8-in Hg, let the membrane dry for 1 minute, and then release the vacuum. Pour and mix well the samples into the wells of the Prep-A-Gene manifold and filter at 8-in Hg until all the liquid is filtered through.
3. Wash the samples 4 times with 250 mL of diatomaceous earth-wash buffer, using a repeat pipette, allowing all of the liquid to filter through between washes.

4. Reduce the vacuum to 5-in Hg before turning the vacuum off. Without unscrewing the black clamps, release the white clamps and place the collection tack on the table top\screw-capped tubes into the manifold. Clamp the manifold with the white clamps, and apply 300 μL of 10:1 TE buffer heated to 65°C and pull the eluted in the vacuum tube through Hg column. After the liquid has filtered through, raise the vacuum to 10- to12-in Hg and let the membrane dry for 1 min.
5. Turn off the vacuum at the stopcock and remove the collection rack containing the tubes. Use ethanol to precipitate the DNA and resuspend the dried DNA pellet in 3.0 μL of 10:0.1 μL TE buffer.

Plasmid Miniprep by Birnboim Method

Materials Required

Chemicals

- Stock solutions
- 3 M potassium acetate (pH 4.8): add 60 mL of 5 M potassium acetate and 11.5 mL glacial acetic acid. Bring the volume up to 100 mL.

Method

1. Inoculate 3 mL LC with antibiotic with 1 colony using a toothpick. Grow overnight at 37°C. Remove toothpick and spin down in a tabletop centrifuge for 10 min at maximum rpm.
2. Suck off supernatant and resuspend the pellet in 200 μL TEG.
3. Transfer to microvial. Add 400 μL freshly prepared 0.2 M NaOH, 1% sodium dodecyl sulphate (SDS); mix well.
4. Add 300 μL 3 M potassium acetate (0–14.8), mix well and spin for 5–10 min.
5. Transfer 800 μL supernatant to a clean microvial and precipitate with 600 μL isopropanol.
6. Spin for 10–20 min; remove supernatant carefully.
7. Dissolve the pellets in 200 μL Tris-EDTA containing 20 μg/mL RNase.
8. Incubate for 10 min at 37°C.
9. Extract once with phenol/sevag and once with sevag, add 10 μL 7.5 M acetate, and precipitate with 750 μL absolute ethanol.
10. Spin for 20–30 min; carefully remove the supernatant.
11. Dry the pellets briefly under vacuum.
12. Resuspend the pellets in 50 μL water or Tris-EDTA. Store at –20°C.
13. Use 5–10 μL for restriction analysis.

Plasmid Miniprep by Boiling Method

Materials Required

Chemicals

- Stock solutions
- STET buffer

- 10 mM Tris (pH 8.0)
- 50 mM EDTA
- 0.5% Triton X100
- 8% sucrose

Method

1. Inoculate 3 mL LC in which an antibiotic is present with 1 colony using a toothpick. Grow overnight at 37°C.
2. Remove toothpick, spin down in tabletop centrifuge and resuspend cells in 700 µL STET buffer. Transfer to microvials.
3. Add 125 µL freshly prepared lysozyme (10 mg/mL), place in boiling water for 40 s and spin down for 10 min.
4. Remove the pellets from the microvials using a toothpick.
5. Extract the aqueous layer with phenol/sevag and sevag.
6. Add 350 µL 7.5 M acetate and precipitate with 600 µL isopropanol during 30 min at room temperature.
7. Spin for 20–30 min; wash the pellets with 70% ethanol, dry briefly under vacuum, and resuspend in 50 µL Tris-EDTA containing 50 µ/mL RNase.
8. Use 5–10 µL for gel analysis.

Isolation of Lambda (λ) DNA

Materials Required

- Stock solutions
- Lambda-dil
- 10 inM Tris (pH 7.5), 10 mM $MgSO_4$

Method

1. Grow LE 392 (or the appropriate host) in LC with 0.2% maltose and 10 mM $MgC1_2$.
2. Prepare confluent lysis plates of lambda on LC with $MgC1_2$. Use LC-agarose with $MgC1_2$ and maltose as top.
3. Cool the plates for a few h at 4°C. Layer 5 mL PSB on the plate and incubate at 4°C overnight. Spin lysate for 15 min at 10,000 rpm. Transfer the supernatant to a clean tube, add 1 µL/m1DNase I (1 mg/mL) and incubate for 1 h at 37°C.
4. Transfer to centrifuge tubes and sediment phages for 3 h at 25,000 rpm. Pour off the supernatant, dry the tubes with Kleenex, and resuspend the pellet in 0.5–1 mL lambda-dil. Perform this step carefully.
5. Transfer to microvial and spin for 15 s. Transfer supernatant to store at 4°C.
6. Extract the phage DNA, after the addition of 1/10 volume 0.5 M EDTA and 1150 volume 10% SDS, with phenol/sevag and sevag.
7. Precipitate the DNA with ethanol and resuspend in water or TE.

Quick Method for Isolation of Lambda (λ) DNA

Materials Required

- Stock solutions
- PSB buffer
- 10 mM Tris (pH 7.6), 10 in $MgCl_2$, 100 mM NaCl, 0.05% gelatin

Procedure

1. Grow LE 392 (or the appropriate host) in LC with 0.2% maltose and 10 mM $MgCl_2$.
2. Prepare confluent lysis plates of lambda on LC with $MgCI_2$. Use LC-agarose $MgCl_2$ and maltose as top. Cool the plates for a few h at 4°C. Layer 5 mL PSB on the plate and incubate at 4°C overnight.
3. Transfer the lysate to a tube, add a drop of chloroform, and store in refrigerator.
4. Incubate 400 μL lysate with 10 μL DNase I (10 mg/mL) and 10 μL RNase 7 (10 mg/mL) for 30 min at 37°C.
5. Add 1 μL DEPC (diethylpyrocarbonate), shake for 1 min, and add 10 μL of 10% SDS. Mix carefully.
6. Add 50 μL 2 M Tris (pH 8.5), 0.2 M EDTA. Heat for 5 min at 37°C.
7. Add 50 μL 5 M potassium acetate and incubate for at least 30 min on ice.
8. Spin for 15 min and precipitate the supernatant with 1 mL ethanol at room temperature.
9. Spin for 15 min, remove the supernatant, dry the pellet briefly, and resuspend in 50 μL Tris-EDTA or water. Use 5 μL for gel analysis.

Single-Stranded M13 DNA Isolation Using Phenol

This isolation procedure is the method of choice for preparation of M13-based templates to be used in Sequenase-catalyzed dye-terminator reactions. A pre-incubated early log phase JM101 culture is prepared by transferring a thawed glycerol stock in 50 mL of liquid media and incubating for 1 h at 37°C with no shaking. M13 plaques are picked with a sterile toothpick and placed into 1.5-mL aliquots of the early log phase JM 101 culture, which are incubated in a 37°C shaker for 4–6 h. After incubation, the bacterial cells are pelleted by centrifugation and the viral-containing supernatant is transferred to a clean tube. The phage particles are precipitated with PEG, collected by centrifugation, and the pellet is resuspended in buffer. The phage protein coat is denatured and removed by 1 phenol and 2 ether extractions. After ethanol precipitation, the dried DNA pellet is resuspended in buffer, and the concentration and purity are assessed by agarose gel electrophoresis against known standards.

Method

1. Prepare an early log phase culture of JM101, as above, and pick M13-based plaques with sterile toothpicks into 12 × 75 mm Falcon tubes containing 1.3-mL aliquots of cells. Incubate for 4–6 h at 37°C with shaking at 250 rpm.

2. Transfer the culture to 1.5-mL microcentrifuge tubes and centrifuge or 5 min at 1200 rpm at 4°C.
3. Pipette the top 1 mL of supernatant to a fresh 1.5-mL microcentrifuge tube containing 0.2 mL 20% PEG/2.5 M NaCl to precipitate the phage particles. Mix by inverting several times and incubate for 15–30 min at room temperature.
4. Centrifuge for 15 min at 12,000 rpm at 4°C to collect the precipitate phage. Decant the supernatant and remove residual PEG supernatant by suctioning twice.
5. Resuspend the pellet in 100 μL of 10 mM Tris-HCl (pH 7.6) by vortexing and add 50 μL TE-saturated phenol.
6. Extract the DNA with phenol and twice with ether, as discussed above, and then by precipitating by adding ethanol.
7. Resuspend the dried DNA in 6 μL 10:0.1 TE for use in single-stranded Sequenase-catalyzed dye-terminator sequencing reactions.

Biomek-Automated Modified-Eperon Isolation Procedure for Single-Stranded M13 DNA

This semi-automated method is a modification of a previously reported procedure and allows the simultaneous isolation of 43 single-stranded DNAs per Biomek robotic Workstation within 3 h. Basically, MO plaques are picked with sterile toothpicks into aliquots of early log phase IM101, prepared as discussed above. The phage-infected cultures are incubated in a 37°C shaker for 4–6 h, transferred into microcentrifuge tubes, centrifuged to separate bacterial cells from the viral supernatant, and then carefully placed on the Biomek table. For each sample, 2 250-μL aliquots are robotically distributed into 2 wells of a 96-well microtitre plate, and this process is repeated for each of the 48 samples until the all 96 wells are filled. A solution of polyethylene glycol (PEG) is then added robotically to each well and mixed. The microtitre plate is covered with acetate to seal, incubated at room temperature to precipitate the phage particles, and then centrifuged. The supernatant then is removed by inverting the plate and gently draining on a paper towel, without dislodging the pellet. After placing the microtitre plate back on the Bionik a more dilute PEG solution is robotically added to each well. The plate is then covered with another sealer and centrifuged again. This rinse step aids in the removal of contaminating proteins and RNA. After removing the supernatant, as before, and placing the microtitre plate back on the Biornek, a Triton X-400 detergent solution is robotically added to each plate. The plate is agitated gently and the sample from each pair of wells is robotically transferred to microcentrifuge tubes, which then are capped and placed in an 80°C water bath for 10 min to aid in the detergent solubilization of phage coat proteins. After a brief centrifugation to collect the condensate, the single-stranded DNA is precipitated, dried, and resuspended. An aliquot from each DNA sample is subjected to agarose gel electrophoresis to assay the concentration and purity. The yield of single-stranded template is approximately 2–3 per sample.

Method

The entire procedure requires 9 rows of P250 tips (counting from the center of the Biomek table toward the left) for the isolation of 48 templates (48ISOL). The reagent module should contain PEG-2000, Triton-X100, Tris-EDTA, and ethanol-acetate, respectively.

1. Prepare an early log phase JM101 culture in 50 mL of 2X Tris, as above.
2. Using sterile toothpicks, transfer individual M13 plaques into 12 × 75 mm Falcon tubes containing 1 mL early log phase cell cultures.
3. Incubate for 4–6 h at 37°C with shaking at 250 rpm. (Growth for longer than 6 h results in cell lysis and contamination of the phage DNA by cellular proteins and nucleic acids.)
4. Separate the bacterial cells from the viral-containing supernatant by centrifugation at 12,000 rpm for 15 min at 4°C.
5. Carefully open the tubes and place on the Biomek table.
6. The Biomek will distribute 2 250-μL aliquots of viral supernatant per sample into the wells of a 96-well flat-bottomed microtitre plate (Dynatech).
7. The Biomek then will add 50 μL of 20% PEG/2.5 M NaCl solution to each well, and mix by pipetting up and down.
8. Cover the plate with an acetate plate sealer and incubate at room temperature for 15 min.
9. Pellet the precipitated phage by centrifuging the plate at 2400 rpm for 20 min. Remove the plate sealer and drain the PEG from the plate by gently draining upside down on a Kimwipe.
10. Return the plate to the table, and the Biomek will robotically add 200 μL of PEG:TE rinse solution to each well.
11. Cover the plate with a plate sealer, centrifuge, and drain, as above.
12. Return the plate to the table, and the Biomek will add 70 μL of TTE solution to each well. Remove and gently agitate to resuspend.
13. The Biomek will then robotically pull the contents from each pair of wells into 1.5 mL microcentrifuge tubes.
14. Incubate the tubes at 80°C for 10 min to denature the viral protein coat and then centrifuge briefly to reclaim condensation. Precipitate the DNA by adding 5U of ice cold ethanol/acetate to each tube, as described above. Resuspend the DNA templates in 20 Ill of 10:0.1 TE buffer.

Isolation of *Aspergillus* DNA

Materials Required

Chemicals

- Stock solutions
- Triisopropyl naphthalene sulphonic acid (TNS) solution: 20 mg/mL TNS in water, freshly prepared.
- 4-aminosalicylic acid (PAS) solution: 120 mg/mL PAS in water, freshly prepared.

- 5× RNB
- 60.55 g Tris, 36.52 g NaCl, 47.55 g ethyleneglycol-bis-(β-aminoethylether)-N,N,N′,N tetra acetic acid (EGTA) in 500 mL water (pH 8.5)

Method

1. Grind 0.5–1.0 g mycelium under liquid nitrogen using the membrane disrupter.
2. Place polypropylene tubes with 1.5 mL water-saturated phenol, 1 mL TNS, 1 mL PAS and, 0.5 mL 5× RNB in a water bath at 55°C; add the frozen mycelium to the tubes and vortex every 20 s for 2–4 min.
3. Add 1 mL sevag and vortex with intervals for another 1–2 min.
4. Centrifuge for 10 min at 4°C at maximum velocity.
5. Extract the aqueous-phase once with phenol-sevag and twice with sevag.
6. Precipitate the DNA with 2 volumes ethanol. Centrifuge for 10 min.
7. Drain the tube, dry it with Kleenex, and resuspend the pellet in 500 μL Tris-EDTA. Transfer to a microvial.
8. Extract with phenol-sevag until interface stays clean. Then extract once with sevag.
9. Precipitate with 2 volumes ice-cold ethanol, spin down, and resuspend the pellet in 100–200 μL TE with 50 μg/mL RNase.

Isolation of *Aspergillus* DNA by the Qiagen Method

Materials Required

Chemicals

- Stock solutions
- Buffer B: 750 mM NaCl, 50 mM MOPS, 15% ethanol (pH 7.0)
- Buffer C: 1000 mM NaCl, 50 mM MOPS, 15% ethanol (pH 7.0)
- Buffer F: 1500 mM NaCl, 50 mM MOPS, 30% ethanol (pH 7.0)
- Buffer G: 1500 mM NaCl, 50 mM MOPS, 15% ethanol (pH 7.0)

Method

1. Grind 0.5–1.0 g mycelium under liquid nitrogen using the membrane disrupter.
2. Place the polypropylene tubes with 1.5 mL water-saturated phenol, 1 mL TNS1 mL PAS and 0.5 mL 5× RNB in a water bath at 55°C, add the frozen mycelium to the tubes, and vortex every 20 s for 2–4 min.
3. Add 1 mL sevag and vortex with intervals for 2 min. Spin for 15 minutes in a tabletop centrifuge at 4°C with maximum velocity. Extract the water-phase. Rinse once with pheranol-sevag and twice with sevag.
4. Precipitate the DNA with 2 volumes ethanol. Spin directly for 10 min in the tabletop centrifuge. Drain the tube, dry with Kleenex, resuspend the pellet in 750 μL 50 mM MOPS (pH 7.0), and add 750 μL buffer G.
5. Equilibrate a QIAGEN-tip with 1 mL buffer B, add the sample and allow to flow through. Wash the resin 3–5 times with 1.5 mL buffer C.

6. Elute the DNA with 500 μL buffer F and precipitate the DNA with 0.8 volume isopropanol; spin for 15 min, dry pellet briefly under vacuum.
7. Resuspend the DNA gently in a small volume of TE or water (preferably overnight).

PURIFICATION OF GENOMIC DNA

Introduction

Purification of nucleic acids is the first step in most molecular biology studies and in all recombinant DNA techniques. DNA is extracted from samples such as bacteria or plant and animal cells by dissolving the sample in a buffer, which is then added to an absorbent material. The buffer is removed from the absorbent material and the DNA can be obtained by elution. Several commercial kits are available for the extraction and isolation of DNA from a wide range of samples. Proteins are removed from nucleic acid solutions by treating with proteolytic enzymes, such as pronase or proteinase K. Methods for purifying nucleic acids from cell extracts are usually a combination of 2 or more of the following techniques: extraction/ precipitation, chromatography, centrifugation, affinity separation.

Principle

Solvent extraction is often used to eliminate contaminants from nucleic acids. For example, a combination of phenol and chloroform is frequently used to remove proteins. Precipitation with isopropanol or ethanol is generally used to concentrate nucleic acids. If the amount of target nucleic acid is low, an inert carrier (such as glycogen) can be added to the mixture to increase precipitation efficiency. Other precipitation methods of nucleic acids include selective precipitation using a high concentration of salt ("salting out") or precipitation of proteins using changes in pH.

Materials Required

Chemicals

- Chloroform
- Isoamylalcohol
- Phenol
- 5M NaCl
- Isopropanol

Glassware and Other

- Beakers
- Conical Flask
- Eppendorf tube

Equipment

- Centrifuge

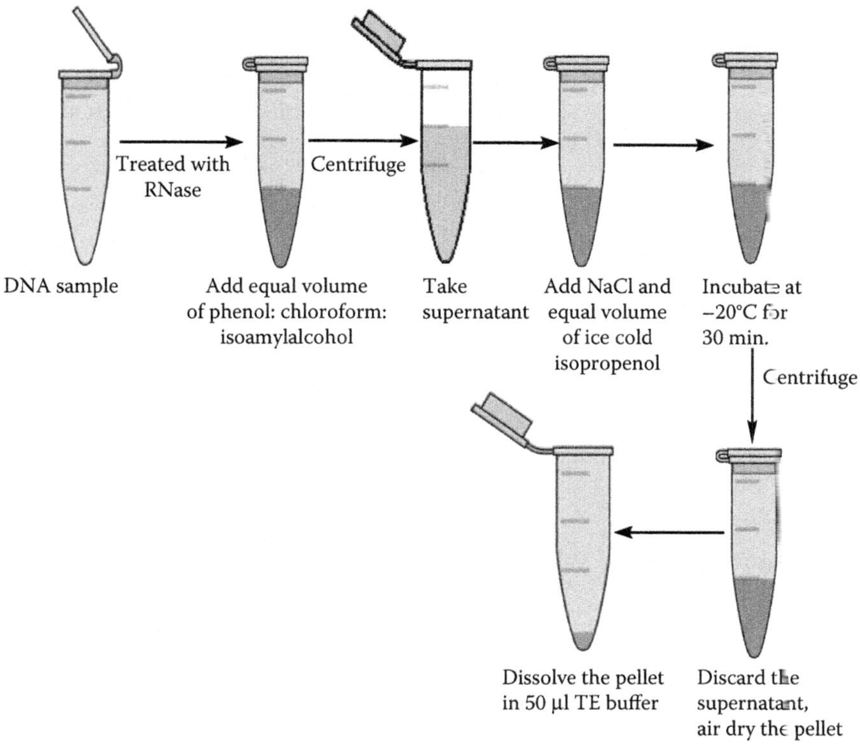

FIGURE 1.7 Purification of DNA.

Method (Figure 1.7)

1. Take 100 µL of the DNA sample and add an equal volume of phenol:chloro form:isoamylalcohol (25:24:1).
2. Centrifuge at 10,000 rpm for 10 min.
3. Transfer the upper aqueous layer to a fresh Eppendorf.
4. Add an equal volume of chloroform:isoamylalcohol (24:1).
5. Centrifuge at 10,000 rpm for 1 min.
6. Take the upper aqueous phase and add 2 µL of 5 M NaCl and an equal volume of ice-cold isopropanol.
7. Incubate at −20°C for 30 min.
8. Centrifuge at 10,000 rpm for 10 min.
9. Discard the supernatant, and air dry the pellet.
10. Dissolve the pellet in 50 µL TE buffer.

Purification of Plasmid DNA

Principle

Equilibrium centrifugation in cesium chloride-ethidium bromide gradients has been the method of choice to prepare large amounts of plasmid DNA. For some purposes, it is essential to have DNA preparations that are free from contaminating RNA. Although the weight of such contaminants in bromide gradients is small, the number of RNA molecules can be relatively large and can contribute significantly to the total number of 5′ termini in a restriction digest. RNA can be removed from plasmid preparations by the following 2 methods: (1) purification by cesium chloride centrifugation and (2) chromatography through Bio-Gel A-150 m or Sepharose CL-4B.

Materials Required

Chemicals

- Cesium chloride
- Ethidium bromide
- Isopropyl alcohol
- 0.1 M saline sodium chloride
- Plasmid DNA
- SDS
- TE buffer

Glassware and Other

- Measuring bottle
- Conical Flask
- Measuring cylinder
- Centrifuge tubes (50 mL capacity, capped)

Equipment

- Cooling Centrifuge
- Spectrophotometer
- Electrophoresis unit
- Power pack
- Micropipette

Purification by Cesium Chloride Centrifugation

Procedure

1. Dissolve 3.9 g cesium chloride in the preparation completely and add 0.4 mL ethidium bromide.
2. Centrifuge at 4000 rpm for 40 h at –20°C in a swing-out rotor.
3. The DNA-ethidium bromide complex fluoresces under long-wave ultraviolet light. The top band consists of linear plasmids and fragments of nuclear DNA while the more intense lower hand is formed by circular plasmids.
4. Draw out the plasmid band using a sterile syringe (fitted with a long needle) and the top DNA band using a Pasteur pipette.
5. Extract plasmic fraction 3 times, using 2 volumes each of isopropyl alcohol to remove ethidium bromide.
6. Cesium chloride and ethidium bromide are removed by exhaustive dialysis for 16 h against several changes of 0.1 M saline sodium chloride solution.
7. Transfer the plasmid solution to sterile tubes and measure the absorbance spectrophotometrically at 260 and 280 nm. The A260 should be double of A280 for a good preparation.

8. Calculate the concentration of plasmid DNA using the relationship that if A260 is 1.0, then equivalent concentration of DNA is 50 µg/mL. The preparation can be stored for several weeks at –20°C. Perform chromatography through Bio-Gel A-150 m or Sepharose CL-413.

Chromatography through Bio-Gel A-150 m or Sepharose CL-4B

Procedure

1. Add 0.1 volume of 3 M sodium acetate (pH 5.2) and precipitate the nucleic acid with 2 volumes of ethanol for 30 min at 4°C.
2. Recover the nucleic acids by centrifugation at higher than 10,000 rpm for 15 min at 4°C. Drain off as much of the supernatant as possible and then store the open tube on the bench to allow the last traces of ethanol to evaporate. Dissolve the pellet in TE buffer (pH 8.0) at a concentration of at least 100 µg/mL. Add DNase-free pancreatic RNase to a final concentration of 10 µL. Incubate the mixture for 1 r at room temperature.
3. Extract the solution once with an equal volume of phenol equilibrated in TE buffer (pH 8.0).
4. Layer up to 1 mL of the aqueous phase on a column of Bio-Gel A-150 m or Sepharose CL-4B equilibrated TE buffer (pH. 8.0) and 0.1% sodium dodecylsulphate (SDS).
5. Wash the DNA into the column; apply a reservoir or TE buffer (pH 8.0) containing 0.1% SDS and immediately begin to collect 0.5-mL fractions.
6. When 15 fractions have been collected, clamp off the bottom of the column, analyze 10 µL of each fraction by electrophoresis through a 0.7% agarose gel or by ethidium bromide fluorescence in order to locate the plasmid DNA.
7. Pool the fractions that contain plasmid DNA, recover the DNA by precipitating with volumes of ethanol for 10 min at 4°C and centrifugation at higher than 10,000 g for min at 4°C.

Precautions

Avoid any contamination while handling the live bacterial cells. For successful lysis of the bacterial cells, the bacterial suspension should become viscous after treatment with detergents. DNA precipitate, even if it is invisible in the centrifuge tubes, should be dissolved very gently to avoid shearing of DNA molecules.

ISOLATION OF RNA FROM PLANT TISSUES

Introduction

Most plant material contains relatively high levels of RNase activity, which is normally located in the vacuoles. During the RNA extraction procedure, RNA should be protected against this endogenous RNase. In this experiment, we describe 2 procedures for the isolation of RNA. In both procedures, a high pH of the extraction buffer and the presence of a chelating agent (EDTA and EGTA, respectively) are used to prevent RNA degradation. In addition, during the isolation of total RNA, a detergent (SDS) is used, and the pulverized material is directly thawed in a mixture

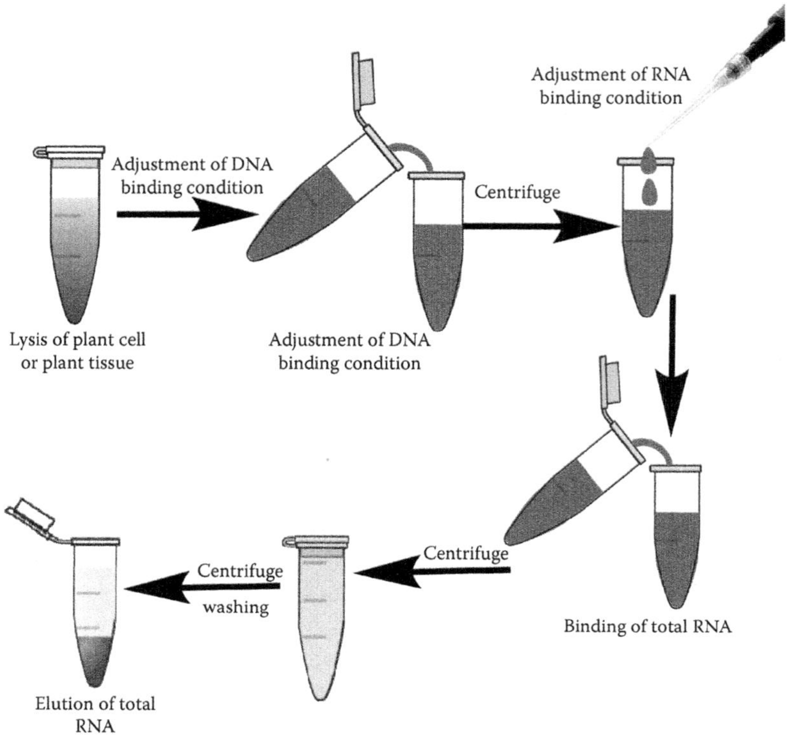

FIGURE 1.8 RNA isolation of plant tissue.

of phenol and extraction buffer (denaturing the RNase). For both RNA extraction procedures, we found that the addition of RNase inhibitions was unnecessary, thereby omitting complicated and expensive buffers. (Figure 1.8)

The extraction of total RNA is the most simple and convenient method of the 2 and yields RNA that is directly suitable for *in vitro* translation and northern blot analysis. After isolation of the poly (A)-containing RNAs, it can be used for cDNA synthesis. This RNA isolation procedure was originally developed for the extraction of RNA form fungi that are notorious for their high RNase activities.

Polysomal RNA will only be used in specific experiments that require the exclusive presence of actively translated mRNAs. The procedure is outlined to simplify existing procedures for obtaining undegraded polysomes. We briefly describe 3 methods to test whether the isolated RNA is undergraded.

The objective of this experiment is to isolate total RNA and polysomal RNA.

Extraction of Total RNA

The following procedure has been applied to a wide variety of plant tissues (including adult leaves and roots, light- and dark-grown seedlings, and callus or suspension cells), invariably yielding large amounts of undegraded to total RNA suitable for *in vitro* translation and cDNA synthesis.

Materials Required

Chemicals

- RNA extraction buffer
- 100 mM LiCl
- 1% SDS
- 100 mL Tris-NaOH (pH 9.0)
- 10 mL of EDTA (storage at room temperature for a maximum of 6 mo)
- Distilled phenol with 0.1% hydroxyquinoline (storage at –20°C for a maximum of 6 mo)
- Chloroform (pro analyse)
- 8 M LiCl the solution should be filtered through sterile Whatman 1 mm filter paper after autoclaving and standing overnight at room temperature. The Whatman filter paper can be autoclaved wrapped in aluminium foil (storage at –20°C for a maximum of 6 mo)
- 2 M LiCl can be prepared from the filtrated 8 M LiCl (storage at 4°C for a maximum of 1 mo)
- 80% ethanol
- Liquid nitrogen
- Double-distilled sterile water

Glassware and Other

- Measuring bottle
- Conical flask
- Measuring cylinder
- Mortar and pestle
- Centrifuge tubes

Equipment

- High-speed centrifuge
- Polycarbonate tubes
- Spectrophotometer
- Gyratory shaker

Method

1. Harvest material in plastic tubes filled with liquid N_2 placed in a container filled with liquid N_2. Determine fresh weight.
2. Prepare a 1:1 mixture of RNA extraction buffer and phenol with hydroxyquinolone, and heat the mixture to 90°C in a water bath in fume hood.
3. Grind material in liquid N_2 in a precooled mortar and pestle until a fine homogeneous power is obtained.
4. Transfer the frozen powder to an Erlenmeyer flask of suitable size kept in ice-salt mixture, precooled at –80°C with the aid of a metal or plastic spatula cooled in liquid N_2.
5. Add 2 mL of the well-mixed phenol-extraction buffer per g fresh weight of plant material. Swirl the flask vigorously. Occasionally heat in the water bath (90°C) until a milky suspension is obtained that is devoid of clumps of

frozen material. The final temperature of his mixture should be 25°C–30°C. Place the Erlenmeyer flask on gyratory shaker at 300 rpm for 5 min at room temperature.

6. Add 1 mL chloroform per g fresh weight of plant material and continue shaking for 15–30°C.
7. Transfer the milky suspension to centrifuge tubes, and centrifuge at 20,000 g for 30 min at 25°C.
8. Remove the aqueous upper phase with a pipette and transfer to an Erlenmeyer flask. Add 1 mL chloroform per g fresh weight of plant material, and place the mixture on a gyratory shaker for 15 min at 300 rpm.
9. Transfer the mixture to chloroform-resistant glass or plastic tubes and centrifuge for 15 min at 12,000 g at 25°C.
10. Remove the aqueous phase as under step 8. Transfer to measuring cylinder and subsequently transfer to centrifuge tubes. Add 1/3 volume of 8 M LiCl, mix well, and precipitate the RNA for 16–48 h at 4°C.
11. Centrifuge at 12,000 *g* for 30 min at 4°C. The resulting RNA pellet is washed once with 2 M LiCl at 4°C and twice with 80% ethanol, and finally vacuum-dried. The pellet is dissolved in double-distilled water stored at −20°C. Generally up to 20 mg of RNA per mL can be dissolved.

Precautions

1. All equipment used during the RNA isolation procedure should be made RNase free. Glassware can be baked at 180°C for 6 h, minimally. Plastic tubes and pipettes should be autoclaved for 20 min at 120°C. In addition, disposable plastic gloves should be worn during all manipulations because active RNases are present on the skin.
2. The material can be stored at −80°C.
3. When 2 g (or less) of plant material is used, the extraction procedure can be done in phenol- and chloroform-resistant plastic tubes (e.g., 14-mL polypropylene tubes).
4. The idea is to thaw the material in the presence of hot phenol, so care must be taken to prevent the phenol from crystallizing due to the cold, powdered plant material.
5. An extraction with phenol only results in a significant loss of poly (A)-containing RNA. This problem is obliterated by the combined chloroform/phenol extraction.
6. The faster and longer the centrifugation, the more solid the interphase will be, reducing the amount of impurities carried over.
7. Ensure that a milky, foaming suspension is obtained.
8. Carefully determine the final volume. When the lithium chloride is higher than 2 M, DNA will also precipitate. It is important that the solution of 8 M LiCl is not contaminated with DNA as it will also be precipitated witNA preparation.
9. In most cases the RNA preparations obtained are suitable for *in vitro* translation and cDNA synthesis. However, sometimes the RNA preparation is contaminated with polysaccharides. The most convenient way to remove this contamination is by dissolving the pellet in a small volume of water,

heating at 60°C for 5 min and centrifugation at 40,000 g for 30 min. The resulting gel-like pellet contains most of the impurities, and almost no loss of RNA is encountered. Poly RNA can be isolated from this total RNA preparation according to standard procedures.

10. All solutions must be autoclaved for 20 min at 120°C. Solutions that are not able to be autoclaved are made RNase-free by adding diethyl pyrocarbonate (DEP) of 0.1% (v/v) (Sigma). The solutions are stirred for 10–16 h. DEP and ethanol and CO_2 formed DEP are remove by heating at 60°C–80°C for 10 min. 96% ethanol and chloroform (proanalyse) and phenol with hydroxyquinoline can be used without any pretreatment.
11. Equipment that cannot be sterilized by heating should be made RNase-free by keeping it overnight in 0.1% DEP. DEP can be removed by rising with sterile distilled water.

EXTRACTION OF POLYSOMAL RNA OF NUCLEAR SEQUENCES

This procedure has been developed partly from published procedures; it has been found to be better than or equal to all of the different procedures described, both in yield and quality of the polysomal RNA obtained. (Figure 1.9)

1. Harvest material in liquid N_2 and determine fresh weight.
2. Grind plant material in liquid N_2 in a mortar and pestle until a fine powder results.
3. Transfer the frozen powder to a second mortar (precooled at 0°C–4°C) and add 4 mL per g of plant material in ice-cold polysome buffer (PB). Gently suspend the powder with a pestle until the slurry does not contain frozen material.
4. Centrifuge for 10 min at 0°C and 27,000 g (e.g., in an SS34 or GSA rotor). Filter the supernatant solution through a 3G-1 glass filter into a measuring cylinder in a cold room.
5. Transfer filtrate to Type 42.1 or Type 65 (Beckmann) polycarbonate tubes and fill to approximately 80%. Under layer the filtrate with 2 mL (Type 65) or 5 mL (Type 42.1) of 60% w/v sucrose in gradient buffer (GB) with a 2-mm inner diameter Hamilton needle fitted to a 10-mL syringe and fill up tubes with filtrate. Centrifuge for 3 h (Type 42.1) at 0°C and 40,000 rpm or 105 min (Type 65) at 49,000 rpm.
6. After centrifugation, decant the tubes carefully and place the tubes inverted on sterile tissues to drain off remaining liquid from the opalescent polysome pellets. The pellets can now either be resuspended directly or be quickly frozen in liquid N_2 and stored at −80°C.
7. Prepare linear 10%–40% w/v sucrose gradients in GB in allo-polymer SW 28 or SW 40 tubes (Beckman) and keep at 4°C.
8. Carefully resuspend the polysome pelltets by pipetting or low-speed vortexing in GB to a final concentration of approximately 2.5 mg/mL of polysomal RNA (60 A_{260} until), which is about 5 mg/mL of polysomes.
9. Carefully load 16–20 mg or up to 5 mg of polysomes on SW 28 or SW 40 gradients, respectively. Centrifugation for 30 min at 0°C and 40,000 rpm

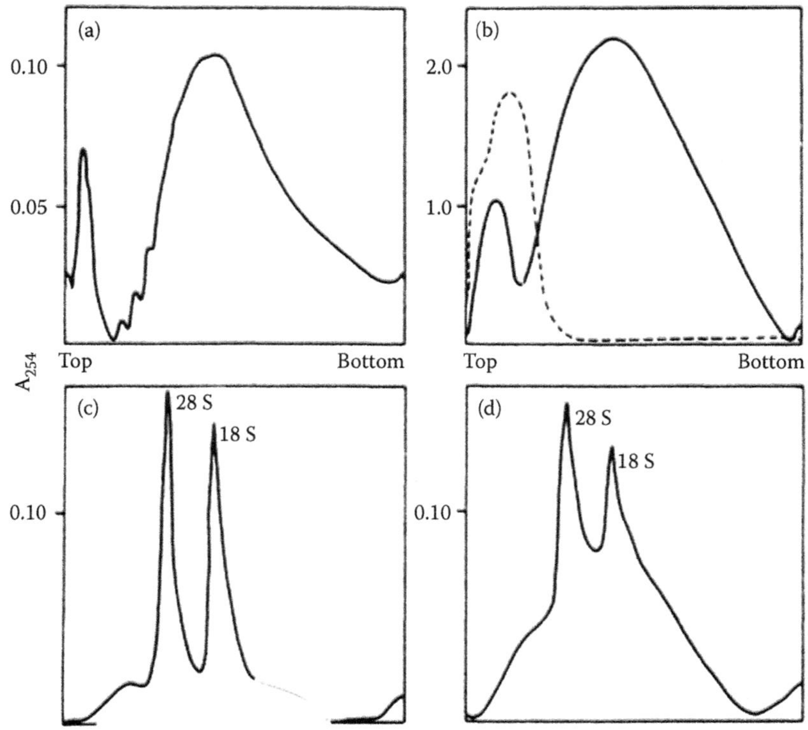

FIGURE 1.9 Polysome preparation and RNA profiles from etiolated shoots of pea: (a) Analytical sucrose gradient centrifugation of polysomes. (b) Preparative sucrose-gradient centrifugation of polysomes. Polysomes sedimenting ≥ 100 S were collected and the RNA extracted. Alternatively, the polysomes were treated with EDTA, sedimented in sucrose gradients and the RNA extracted from ribonucleo-protein sedimenting ≤ 80 S (broken line). (c) Polysomal RNA profile after electrophoresis in 2.4% polyacrylamide gels. S values were calculated from the migration of 28 S and 18 S chicken liver rRNAs and *E. coli* 4 S tRNA. (d) Same as (c) after enrichment for poly(A) RNA.

(SW 40 Ti) without a brake, or 70 min at 25,000 rpm (SW 28) is sufficient to yield clear separation between residual monosomes and most of the contaminating mRNA and polysomes > 100 S (see Figure 1.9). Usually it is sufficient to monitor completely only 1 out of 6 gradients by A_{280} or A_{260} extinction in the flow cell and to fractionate the remainder after identification of the monosome peak and polysomes > 100 S.

10. Polysome-containing fractions are now collected and precipitated overnight at –20°C by addition of 1/10 volume of 3 M sodium acetate (pH 7.0) and 2 volumes of 96% ethanol. The precipitate is collected by centrifugation at 20,000 g for 30 min at 4°C, washed once with 70% ethanol and dried under vacuum. The precipitated polysomes can now be extracted to yield polysomal RNA or can be EDTA-released to remove residual traces of mRNA.
11. Extraction of RNA from ethanol-precipitated polysomes or monosomes is performed by the phenol/chloroform/LiCl procedure (steps 5–11) as

described for the extraction of total RNA. Before RNA extraction, the polysomes or monosomes are first dissolved to approximately 1 mg/mL.

Materials Required

Chemicals

- Polysome buffer (PB)
- 1% Nonidet P40 (shell)
- 50 mM $MgCl_2$
- 25 mM EGTA
- 50 mM Tris-NaOH (pH 9.0)
- 250 mM NaCl (storage at + 4°C for a maximum of 1 month)
- *For Gradient buffer (GB):* 10 mL $MgCl_{2,}$ 5 mL EDTA, 10 mL Tris-NaOH (pH 8.5), 50 mL NaCl (storage at 4°C for a maximum of 1 mo)
- GB + 0.2 M EDTA

Glassware and Other

- Measuring bottle
- Conical flask
- Measuring cylinder
- Mortar and pestle
- Centrifuge tubes

Equipment

- Centrifuge
- Electrophoresis unit
- Hot plate with magnetic stir

Method

1. 75% w/v sucrose in GB. Dissolve 590 g of sucrose (pro analyse) in 410 mL GB to obtain a solution with a density of 1.280 and an n_D of 1.4396 at 20°C the resulting volume is approximately 750 mL after dissolution of all the sucrose check the refractive index and add 0.1% v/v of DEP, stir slowly overnight and heat at 60°C for 1 h (storage at + 4°C for a maximum of 2 mo).
2. RNA-extraction buffer, 8 M LiCl, phenol, chloroform, ethanol 96% and 70% (see solutions for extraction of total RNA). Solutions should be made RNase-free indicated for extraction of total RNA.

Analysis of RNA Quality

Several methods are available to check RNA preparations for degradation. The procedures will be briefly reviewed and discussed.

a. In vitro *translation in cell-free-reticulocyle or wheat-germ-derived extracts* By comparing incorporation levels to known RNA standards of good quality, the integrity of mRNA can be checked. The *in vitro* system is very sensitive to mRNA degradation, resulting in an immediate drop in incorporation level. However, although this method is very fast (3–4 h), it is also expensive, necessitating radioactively labelled amino acids. In addition, if impurities of the RNA (polysaccharides, salt) are present, this will result in reduced incorporation and may lead to inadvertently discarded RNA preparations.

b. *Electrophoresis in polyacrylamide gels*
RNA samples of 2–10 μg can be subjected to electrophoresis through either cylindrical or slab 2.5% polyacrylamide gels in 40 mM Tris-HCl, 40 mM NaH_2PO_4, 2 mM EDTA (pH 7.5), 0.2% SDS, 10% glycerol and subsequently scanner. Although laborious, this is the most reliable method for checking RNA integrity because it will show very limited degradation, resulting in a shift in relative abundance between the 25 and 16 S ribosomal RNA peaks.

c. *Electrophoresis in Agarose Gels*
RNA samples (approximately 1 μg) can be subjected to electrophoresis through 1.0%–1.5% agarose gels (in 40 mM Tris, 20 mM hydrogen acetate, 2 mM EDTA [pH 8.1]) and conveniently stained with ethidium bromide (1 μg/mL). This method is very fast and allows a gross indication of the presence of both ribosomal RNA peaks.

Precautions

1. The extent of cell disruption determines the final polysome yield.
2. Schott Duran 3G-1 sintered-glass filter.
3. Run times should be adjusted to recover all the polysome >100 S while leaving the monosomes on top of the sucrose cushion. The run times needed should be checked before large-scale preparation. Do not use a brake in this centrifugation step.

   For calculating run times in other angle rotors, it is sufficient to calculate the k factor of that rotor at a particular speed, keeping in mind that rotor g values over 450,000 may severely damage the polysomes.

   To calculate k for particular rotor, use the following relationship

$$K = \frac{\text{In}(r_{max}/r_{min}) \times 10^{13}}{\text{w2} \qquad 3600}$$

   where w = 0.10472 × rpm
   ^{r}max = maximum radial distance from centrifugal axis
   ^{r}min = minimum radial distance from centrifugal axis

   From the relationship $t_1 = k_1 t_2 / k_2$ and the fact that k at 40,000 rpm of the type 42.1 rotor is 134, the time required for a particular rotor is now given by

$$t_1 = k_1 3(\text{h})$$
$$134$$

   Since storage involves dipping of the entire centrifuge tube in liquid N_2, this may result in cracking of the tubes after prolonged use. It is therefore advisable to use new tubes only or rigorously check for damage before use.
4. It will take about 1 h to obtain an opalescent solution.

5. EDTA release is applied only when even very low levels of contamination with mRNA must be prevented as is the case in single-copy DNA/mRNA saturation hybridizations.

   For EDTA release, dissolve the dried pellets in GB with 0.2 M EDTA to a concentration of approximately 5 mg/mL of polysomes, incubate at 25°C for 15 min and recentrifuge on a linear 10%–40% sucrose gradient. Protein RNA complexes now sediment at the position of the monosomes. This fraction now contains the mRNAs, which can be purified by oligo-dT chromatography.

   Typical yields obtained are 50%–90% of the amount of cellular total RNA in a particular tissue. The RNA obtained was shown to be suitable for liquid hybridization, *in vitro* translation and cDNA synthesis. The 2D electrophoretic pattern of *in vitro* translation products of polysomal RNA obtained this way is fully comparable to that of total RNA from the same tissue, indicating that no loss of specific sequences occurs.

ISOLATION OF TOTAL RNA FORM BACTERIAL CELLS

INTRODUCTION

Total RNA isolation is based on using the guanidine thiocyanate phenol:chloroform procedure of Chomoczynski and Sacchi. Samples are lysed in highly denaturing conditions, inactivating RNases, followed by phenol:chloroform extraction in which DNA and proteins are removed and then precipitated with isopropanol. An additional step of suspending and reprecipitating ensures further purity of the RNA.

Materials Required

Chemicals

- Bacterial cells
- Water-saturated phenol
- Chloroform
- Isoamyl alcohol
- DEPC water
- TE buffer
- Denaturing solution
- Isopropanol
- 75% ethanol (prepared with DEPC-treated water)

Glassware

- Measuring bottle
- Conical flask
- Measuring cylinder
- Mortar and pestle
- Centrifuge tubes

Equipment

- Centrifuge
- Incubator
- Ice bucket

Method (Figure 1.10)

1. Homogenize the sample in 1 mL of denaturing solution.
2. Add 1 mL of water-saturated phenol followed by 200 μL of chloroform:isoamyl alcohol mix (freshly prepared 49:1).
3. Mix thoroughly and incubate on ice for 15 min.
4. Centrifuge at 10,000 rpm for 20 min at 4°C.
5. Transfer upper aqueous phase carefully to another tube.
6. Add 1 mL of isopropanol to precipitate RNA. Incubate at –20°C for 30 min.
7. Centrifuge at 10,000 rpm for 20 min at 4°C; discard the supernatant.
8. Resuspend the pellet in 0.3 mL of denaturing solution, precipitate RNA by adding 0.3 mL of 100% isopropanol, incubate at –20°C for 30 min, centrifuge at 10,000 rpm for 20 min at 4°C, and discard the supernatant.
9. Resuspend the RNA pellet in 75% ethanol. Incubate at room temperature for 10–15 min to dissolve residual amounts of guanidine.

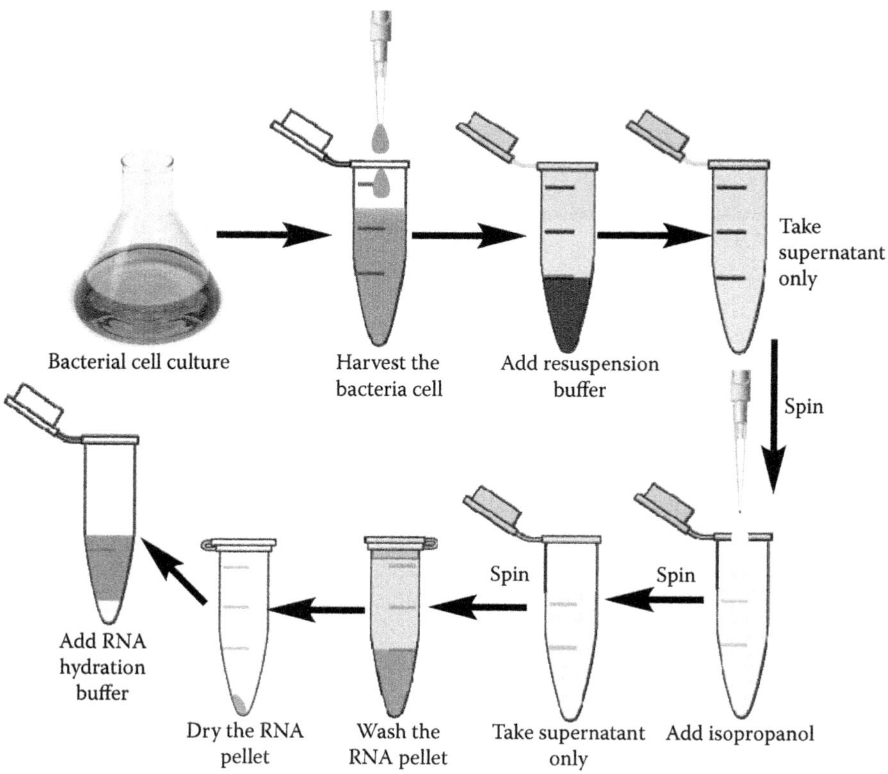

FIGURE 1.10 Isolation of total RNA form bacterial cell.

10. Centrifuge at 10,000 rpm for 20 min at 4°C; discard the supernatant.
11. Do not overdry the pellet. Dissolve in 100–200 μL of DEPC water. Incubate for 10–15 min at 55°C to ensure complete solubilization of RNA.
12. Store the total RNA at –70°C.

Precautions

1. The RNase should be removed completely. It is very important to inactivate RNase as quickly as possible and minimize contamination. They are stable and do not require any cofactors to function. Therefore, a small amount of RNase in an RNA preparation will create a real problem.
2. Water or any salt solution used should be treated with diethyl pyrocarbonate (DEPC) and then autoclaved. This chemical inactivates ribonuclease by covalent modification. Solutions containing Tris cannot be effectively treated with DEPC because Tris reacts with DEPC to inactivate it.
3. Glassware and plasticware should be treated to get rid of any RNase, but autoclaving will not fully inactivate RNase. Glassware can be baked at 300°C for 4 h. Plasticware can be rinsed with chloroform to inactivate RNase.
4. Wear gloves while handling as hands are major source of contamination.
5. Clean the work area with 70% ethanol before use.
6. Water-saturated phenol is supplied with an aqueous layer. When pipetting, be sure not to take the upper aqueous layer.

QUANTITATIVE DETERMINATION OF DNA AND RNA BY A SPECTROPHOTOMETRIC METHOD

Introduction

Both DNA and RNA exhibit strong absorption of ultraviolet light because of the presence of conjugated double bonds of the constituent purine and pyrimidine bases. These have a characteristic absorbance maximum of 260 nm, which is linearly related to the concentration of DNA or RNA in the solution, up to an O.D. value of 2.0. This spectrophotometric method can also be employed for judging purity of DNA and RNA preparations. Proteins are usually the major contaminants in nucleic acid extracts, and these have an absorption maximum of 280 nm. The ratio of absorbance at 260 and 280 nm, hence provides a rough idea about the extent of contaminants in the preparations. A ratio between 1.8 and 2.0 is indicative of fairly pure DNA and RNA preparations, but values less than 1.8 signify presences of proteins as impurities.

Materials Required

Chemicals

- DNA and RNA samples
- Saline sodium citrate (SSC) solution (prepare 0.015 M solution of sodium citrate (pH = 7.0) and dissolve NaCl so that the final concentration in solution is 0.15 M)
- NaCl

Glassware and Other

- Measuring cylinder
- Test tube
- Beaker
- Conical flask

Equipment

- UV spectrophotometer
- Micropipette

Procedure

1. Switch "ON" the spectrophotometer and allow it to warm up for about 10 min. Adjust the wavelength to 260 nm and turn "ON" the UV lamp.
2. Set the instrument at zero absorbance with SSC solution.
3. Read absorbance of solution of the provided sample. If O.D. is too high, appropriately dilute the sample solution with SSC and take the reading again.
4. Calculate concentration of DNA and RNA in the sample from the following formula:
   a. *For standard DNA*
      Concentration of DNA in sample solution (μg/mL) = 50 × A260 × dilution factor
   b. *For RNA*
      Concentration of RNA in sample solution (μg/mL) = 40 × A260 × dilution factor

Precautions

1. Wear gloves while handling as hands are a major source of contamination.
2. Clean the work area with 70% ethanol before use.
3. Water-saturated phenol is supplied with an aqueous layer. When pipetting, be sure not to take the upper aqueous layer.

AGAROSE GEL ELECTROPHORESIS

INTRODUCTION

Agarose gel electrophoresis is a routinely used laboratory technique in which DNA fragments are separated based on their molecular weight.

The different steps involved are

1. Preparation of agarose gel
2. Electrophoresis of DNA fragments
3. Visualization of DNA fragments

Preparation of Agarose Gel

Agarose is a linear polymer extracted from seaweed. Purified agarose is a powder that is insoluble in water or buffer at room temperature but which dissolves on heating.

The solution is then poured into a mold and allowed to solidify. As it cools, agarose undergoes polymerization (sugar polymers cross-link with each other), and the pore size of the gel is determined by the concentration of agarose.

Electrophoresis of DNA Fragments

Electrophoresis is a technique used to separate charged molecules. DNA is negatively charged at neutral pH and, when an electric field is applied across the gel, DNA migrates toward the anode. Migration on the gel depends on

a. Molecular size of DNA
b. Agarose concentration
c. Conformation of DNA
d. Applied current

Agarose gel is a matrix with pores through which DNA fragments move when an electric current is applied. In high concentration gels, the pore size is small. So, small size is monitored by observing the movement of a tracking dye (e.g., xylene cynol and bromophenol blue). The tracking dyes are negatively charged, low molecular weight compounds that are added in the loading dye, which is added with each sample at the start of run. When the tracking dye reaches toward the anode, the run is stopped.

Visualization of DNA Fragments

Since DNA does not have color, it is not visible on the gel. Hence, dye is required to visualize it. The gel is stained with a dye when the electrophoresis is over or the dye is added to the gel while casting it.

Generally an intercalating dye like ethidium bromide is added to agarose gel and the band is visualized by examining the gel under UV light, wherein DNA fluoresces.

Materials Required

Chemicals

- Agarose
- 50 × TAE buffer
- Ethidium bromide (0.5 μg/mL)
- Distilled water

Glassware and Other

- Measuring cylinder
- Test tube
- Beaker
- Conical flask
- Reagent bottle

Equipment

- Micropipettes
- Gel casting unit

- Comb
- Microwave oven
- Electrophoretic tank
- Power supply

Method (Figure 1.11)

1. Prepare 1x TAE by diluting an appropriate amount of 50 × TAE buffer with D/W.
2. Take 50 mL of 1 × TAE in a 250 mL of conical flask and add 0.5 g of agarose. Heat dissolve agarose till a clear solution is obtained.
3. Cool to a warm liquid (60°C).
4. Place the comb of the electrophoresis set such that the comb is about 2 cm away from the cathode.
5. Ethidium bromide can be used for visualizing DNA fragments. Add ethidium bromide to molten agarose to a final concentration of 0.5 μg/mL (stock = 10 mg/mL in water), when temperature is around room temperature, mix and cast the gel, taking care that no air bubbles get trapped in the gel and it is 0.5–0.9 cm thick. Keep the set undisturbed till the agarose solidifies.
6. Gently lift the comb.
7. Connect the power cords, taking care that the negative terminal of the battery is attached to the well side of the gel.
8. Load the samples into the wells. Record the samples loaded in the wells as Lane 1,2,3 .... After loading, start the power connection. Set the voltage at 50 V or 5 v/cm.

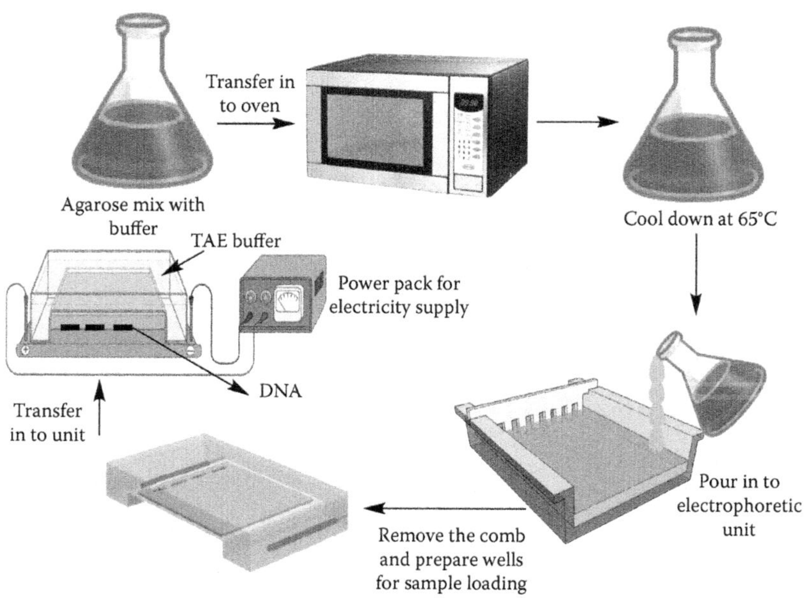

FIGURE 1.11 Preparation of agarose gel.

FIGURE 1.12 DNA pattern in agarose gel.

9. Run till the sample has reached ¾ the length of the gel.
10. Switch off the power supply, and disconnect the electric cords from the electrodes.
11. After electrophoresis, view the gel in UV transilluminator.

Visualizing the DNA (Figure 1.12)

12. Wear latex gloves. Lift the gel casting tray. Slide the gel on your palm without breaking. Pour the buffer into a conical flask; it is reusable.

Precautions

1. Wear latex gloves while handling ethidium bromide and tank buffer after the run.
2. Clean the work area with 70% ethanol before use.
3. Water-saturated phenol is supplied with an aqueous layer. When pipetting, be sure not to take the upper aqueous layer.

ELECTROPHORESIS OF DNA: LINEAR, CIRCULAR, AND SUPER COILED

Introduction

Several techniques have been developed for the analysis of nucleic acids (both RNA and DNA), but agarose gel electrophoresis is an ideal technique for analysis of nucleic acids. The mobility of nucleic acids in agarose gels is influenced by the agarose concentration and the molecular size and molecular conformation of the nucleic acid. Agarose concentrations of 0.3%–2.0% are most effective for nucleic acid separation. The lower limit of 0.3% agarose in the gel allows analysis of linear double-stranded DNA within the range of 5–60 kbp (up to 150 × 106 in molecular weight). Gels with an agarose concentration of 0.8% can separate DNA in the range of 0.5–10 kbp, and 2% agarose gels are used to separate smaller DNA fragments (0.1–3 kbp). For most routine analysis of restriction enzyme fragments, agarose concentrations in the range 0.7%–1% are most appropriate. Nucleic acids migrate in agarose medium at a rate that is inversely proportional to their size (kbp or molecular weight). A linear relationship exists between mobility and the logarithm of kbp (or molecular weight) of a DNA fragment. The DNA conformations most frequently encountered are super helical circular (form-I), nicked circular (form-II), and linear (form-III). The small, compact, super coiled form-I molecules usually have the greatest mobility, followed by the rod-like, linear form-III; form-II migrate more slowly. The relative electrophoretic mobility of the 3 forms of DNA also depends on the ionic strength. Agarose gel electrophoresis is a valuable tool for separation and characterization of nucleic acids, plasmid mapping, and DNA recombination experiments. This method has several advantages, including that it is simple, rapid, inexpensive, and easy to perform; it uses sensitive staining procedures; it has high resolution; and a wide range of molecular weights (0.6–100 × 106) can be analyzed.

Experimental Outline

DNA molecules are negatively charged at neutral or alkaline pH and migrate toward the anode when an electric field is applied. Charge/mass ratio in nucleic acids is unity; thus migration occurs largely on the basis of molecular size of the DNA molecules.

Materials Required

Chemicals

- Tris-acetate buffer (TAE) stock solution (5×). A 5-fold concentrated TAE buffer stock solution is prepared as below:

Tris-base	24.2 g
Glacial acetic acid	5.71 mL
0.5 M EDTA	10.0 mL

Note: Adjust pH of the above to pH 8.0 and add water to make 1 L. Dilute 5 times before use to obtain the working buffer (1 × buffer).

- 1% agarose in 1 × TAE buffer: dissolve 0.75 g agarose in 75 mL of 1 × TAE buffer (working TAE buffer) by boiling and maintain it at 50°C until it is used
- Gel loading solution: 10% glycerol and 0.025% bromophenol blue in water
- Ethidium bromide: dissolve 10 mg of ethidium bromide per mL of the 1 × TAE buffer
- Plasmid DNA preparation: to 20 μL of plasmid DNA preparation, add 10 μL of gel loading solution and mix properly
- Standard DNA marker: take 20 μL of the Hind III DNA digest, add 10 μL of gel loading solution and mix well

Glassware and Others

- Reagent bottles
- Conical flask
- Beakers
- Measuring cylinder

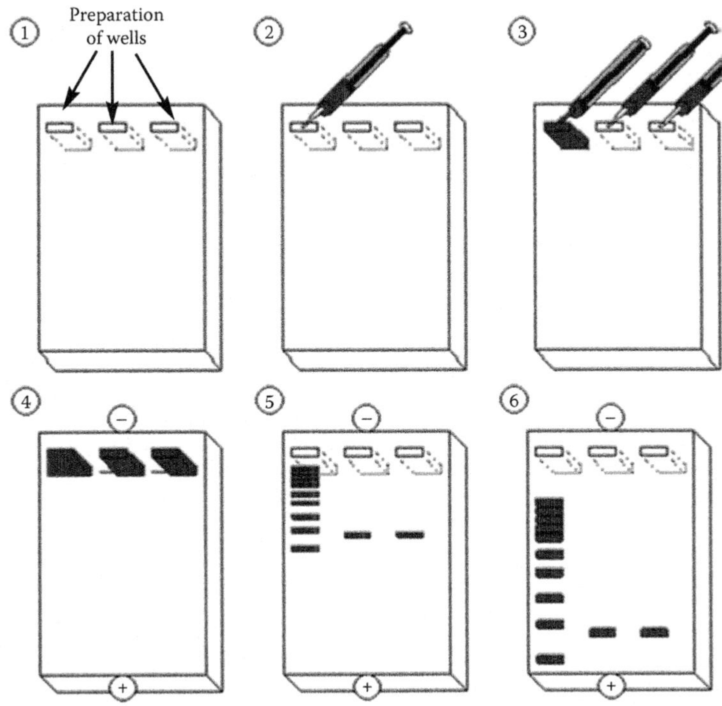

FIGURE 1.13 Running of sample in agarose gel through following steps: (1) Preparation of wells in agarose, (2) Pour the marker in first well, (3) Pour the DNA sample, (4) Electrophoresis unit with power pack, (5) Running the sample, (6) Autoradiography after running the sample.

Equipment

- Microwave oven
- Pipette
- Electrophoretic unit
- Gel casting plate
- Comb
- Power supply
- Gel documentation system

Method (Figure 1.13)

1. Take a clean, dry gel casting plate and make a gel mold using adhesive tape along the sides of the plate to prevent the material to be poured on the plate from running off.
2. Pour 50 mL of 1% agarose solution kept at 50°C on to the casting plate. Immediately place the supplied comb about 1 cm from one end of the plate ensuring that teeth of the comb do not touch the glass plate. Wait till a firm layer of gel is formed.
3. Remove the comb and the tape surrounding the plate carefully and transfer the gel plate to the electrophoretic tank such that wells are toward the cathode.

4. Pour 1 × TAE buffer into the tank until the gel is completely submerged. Connect the electrodes to the power supply.
5. Load the plasmid DNA preparation and the standard DNA markers into separate wells with a micropipette or a syringe.
6. Turn "ON" the power supply and run at 100 V (10–15 mA). Monitor the progress of the fast-running tracking dye (bromophenol blue) during electrophoresis.
7. Turn "OFF" the powder supply when the tracking dye has reached the opposite edge of the gel.
8. Transfer the gel from the casting plate onto a UV-transparent thick plastic sheet and place it in a staining tray containing ethidium bromide solution. Stain for 20–30 min.
9. After staining the gel, place it in water for 20 min to let it cool down to room temperature.
10. Now place the gel, along with the UV-transparent sheet, on a UV transilluminator and view the gel in UV light for the presence of orange-colored bands. For a permanent record, the gel may be photographed

Caution: UV light is extremely injurious for eyes. Wear UV-protectant glasses while viewing gel.

11. Measure the distance moved by each band from the edge of the loading well. Draw a graph between $\log_{10}$ of molecular weight of standard DNA marks vs the distance travelled by each of them.
12. From the distance travelled by the supplied plasmid DNA preparation, determine its molecular weight using the calibration curve prepared in step 11.

Precautions

1. High purity chemicals and water should be used.
2. Ethidium bromide is carcinogenic. Gloves must be worn while handling gels in ethidium bromide solution.
3. Wear safety glasses while viewing the gel on the UV transilluminator.
4. At the time of gel casting, ensure that no air bubbles are entrapped in the gel.
5. Experimentation with bacterial cells presents a potential biohazard.
6. Do not pipette any solutions by mouth. Wash your hands with hot water and soap before eating or drinking.
7. Use gloves.

ELUTION OF PLANT DNA FRAGMENT FROM AGAROSE GEL

Introduction

Purification of DNA is the first step in most molecular biology studies and in all recombinant DNA techniques. DNA is extracted from samples like bacteria or plant and animal cells by dissolving the sample in a buffer, which is then added to an absorbent material. The buffer is removed from the absorbent material, and the DNA can be obtained by elution, which is the technique described below.

Freezing and thawing break agarose clumps, releasing DNA, which, upon centrifugation, appears in the aqueous phase and gets precipitated with 2 volumes of cold ethanol.

Materials Required

Chemicals

- Tris-saturated phenol
- Redistilled phenol
- 0.1 M Tris-HCl (pH 8.0)
- Chloroform
- Isoamyl alcohol
- Tris-saturated butanol
- 3 M sodium acetate buffer (pH 5.5)
- Absolute ethanol
- 70% ethanol (v/v)
- TE

Glassware and Other

- Reagent bottles
- Conical flask
- Beakers
- Measuring cylinder
- Knife or a razor blade

Equipment

- Microcentrifuge
- Micropipette

Method (Figure 1.14)

1. Cut the agarose gel with a razor blade across the region of your choice.
2. Transfer the gel piece to a centrifuge tube and add Tris-saturated phenol.
3. Freeze at –80°C for a minimum of 2 h.
4. Thaw out the frozen gel pieces gradually to 4°C–10°C by keeping the tube in the refrigerator.
5. Centrifuge at 4°C at 10,000 rpm for 10 min.
6. Remove the aqueous layer and add an equal volume of chloroform:isoamyl alcohol to it. Centrifuge as in step 5.
7. Collect the upper aqueous phase and add an equal volume of Tris-saturated butanol to remove ethidium bromide. Centrifuge as above and collect the lower layer.
8. Add 3 M sodium acetate (pH 5.5) so that its concentration in the solution is 0.25 M.
9. Precipitate DNA by adding 2 volumes of chilled absolute ethanol. Keep overnight at –20°C.

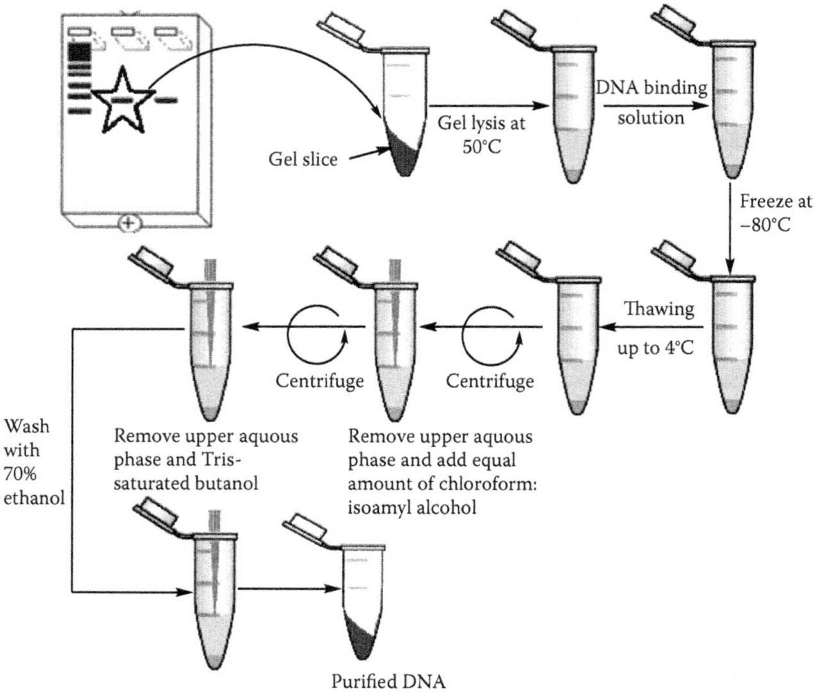

FIGURE 1.14 Elution of plant DNA fragment from agarose gel.

10. Centrifuge at 4°C at 10,000 rpm for 10 min. Wash the precipitates once with 70% ethanol, air-dry, and dissolve in a suitable volume (10 μL/1L) of TE buffer. Perform 1% agarose gel electrophoresis.

Precautions

1. High purity chemicals and water should be used.
2. Ethidium bromide is carcinogenic. Gloves must be worn while handling gels in ethidium bromide solution.
3. Wear safety glasses while viewing the gel on the UV transilluminator.
4. Do not pipette any solutions by mouth. Wash your hands with hot water and soap before eating or drinking.
5. Use gloves.

HYBRIDIZATION AND AUTORADIOGRAPHY OF DNA

Introduction

Complementary strands of DNA hybridize with each other under suitable conditions of temperature and ionic strength. After blotting, the membrane is treated with a radioactively labelled probe, which hybridizes with the specific plant DNA fragment bearing sequence homology with the probe, and the radioactive spot on the membrane is detected by autoradiography. Prior to hybridization, the membrane is prehybridized with salmon-sperm DNA to block the nonspecific binding sites where the probe may otherwise get bound spuriously.

Experimental Outline

1. Preparation of DNA.
2. Hybridization of specific DNA fragments.

Materials Required

Chemicals

- *HSB (5×):* this mixture is composed of 3M NaCl, 100 mM PIPES, and 25 mM Na_2 EDTA. Prepare by dissolving 175.3 g of NaCl, 30.3 g of PIPES, and 7.5 g of Na_2 EDTA in distilled water and adjusting the pH to 6.8 with 4 M NaOH. Make the final volume to 1 L with distilled water.
- *Carrier DNA:* to 0.5 g salmon-sperm DNA, add 100 mL distilled water and autoclave. Store in refrigerator.
- *Denardt's III:* this contains 2 g gelatin, 2 g ficoll, 2 g polyvinyl pyrrolidine, 10.9 g, and 5 g tetrasodium pyrophosphate. Dissolve the components in water and make the volume to 100 mL.
- *TES solution:* prepare 10 mM Tris-HCl-containing ImM EDTA and 0.1% SDS. 10 × SSC. It is composed of 1.5 M NaCl and 0.15 M trisodium citrate. Make 0.5 ×, 1 ×, and 2 × SSC solutions by diluting the above solution accordingly.
- *Prehybridization solution:*

5 × HSB	8.0 mL
Denhardt's III	4.0 mL
Salmon sperm DNA	4.0 mL (freshly denatured by boiling for 5 min)
Sterile distilled water	4.0 mL

Note: Mix well and incubate at 65°C for 5 min.

- *Hybridization buffer:* Prepare by mixing the following:

5 × HSB	4.0 mL
Denhardt's III	2.0 mL
25% dextran sulphate	8.0 mL
Salmon-sperm DNA	0.5 mL (freshly denatured by boiling for 5 min)
Sterile distilled water	5.5 mL

Glassware and Other

- Reagent bottles
- Conical flask
- Beakers
- Measuring cylinder
- Nylon membrane containing DNA fragments
- Labelled probe

- X-ray cassette
- X-ray film
- Plastic bags and boxes

Equipment

- Boiling water bath
- Shaking water bath at 65°C
- Ultra freezer (–70°C)
- Portable Geiger Muller counter

Method

The experimental procedure consists of 2 distinct phases:

1. Hybridization, during which blotted DNA on nylon membrane is incubated with labelled probe.
2. Autoradiography for location of any radioactive zones on the membrane.

Various steps involved in the above mentioned 2 phases are described as follows:

Hybridization (Figure 1.15)

1. Place the nylon membrane containing the immobilized DNA fragments in a plastic bag, add 20 mL prehybridization solution, and seal the bag without leaving any entrapped air bubbles.

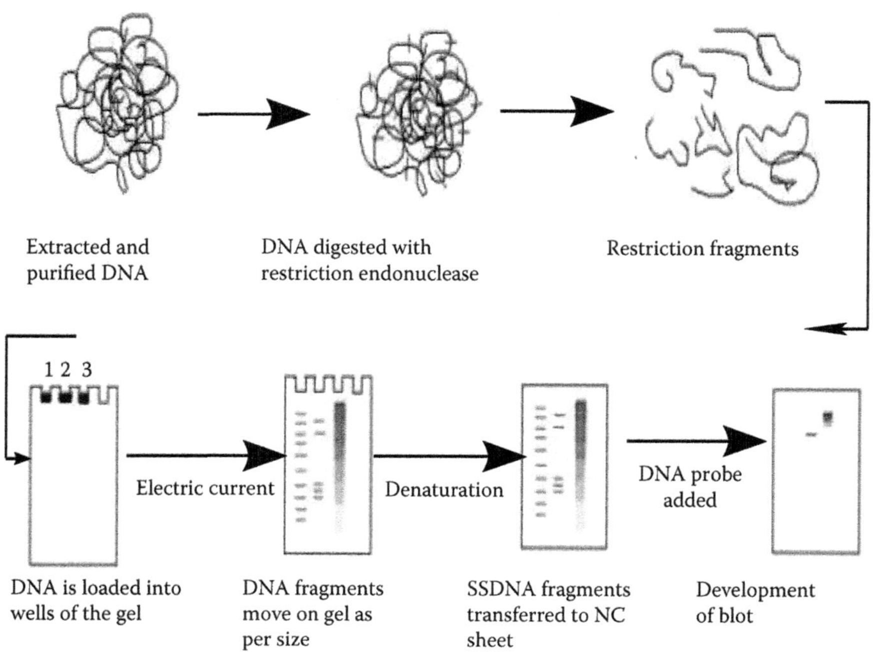

FIGURE 1.15 Hybridization of DNA.

2. Incubate the bag at 65°C in a water bath with gentle shaking for at least 2 h.
3. Denature the probe by incubating in a boiling water bath for 5 min and then quickly chill it on ice.
4. Cut open the plastic bag and drain prehybridization solution. Add labelled probe (freshly denatured by boiling for 5 min) to 20 mL hybridization solution and pour the mixture into the plastic bag. Seal the bag, taking utmost care that no air bubbles are entrapped.
5. Incubate at 65°C overnight with gentle shaking for hybridization to take place.
6. The next day, cut open the bag, remove the membrane with forceps and place it in a box with a lid. Dispose of the plastic bag and the solution in radioactive waste reservoirs.
7. Wash the membrane successively with 200 mL of the following buffers in a vigorously shaking water bath.

First wash: 2 × SSC, 0.1% SDS 20 min at room temperature
Second wash: 1 × SSC, 0.1% SDS, 20 min at 65°C
Third wash: 0.5 × SSC, 0.1% SDS, 20 min at 65°C

Autoradiography (Figure 1.16)

8. For autoradiography, dry the washed membrane between folds of filter paper and then wrap in Saran Wrap. Measure the strength of the radioactivity signal with a hand Geiger Muller counter.
9. In a dark room (with red safety light), place the membrane in a film cassette with the DNA-containing surface facing upward, and place a sheet of X-ray film on top of it. Carefully close the cassette and keep at –70°C for 2–3 d.
10. Develop the film in the dark room (with red safety light) for 3 min in X-ray developer, rinse briefly with deionized water, and fix for 3 min in X-ray fixer. Rinse with deionized water for 5 min and let the film air-dry.

Precautions

1. Radioactive substances are hazardous to health. Hence, necessary precautions such as use of gloves and protective clothing must be observed while handling these materials. Never handle radioactive material with bare hands.

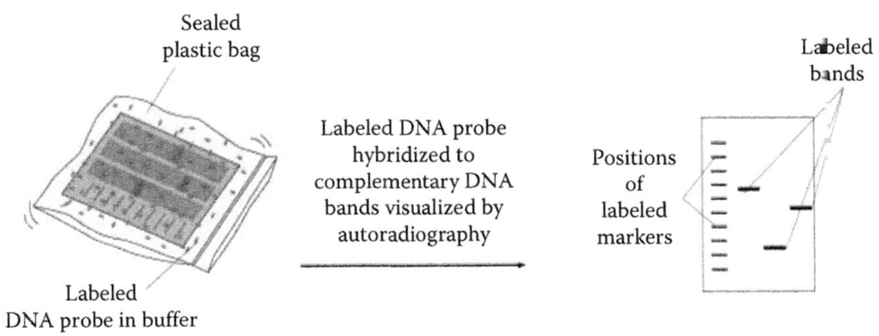

FIGURE 1.16 Autoradiography of DNA.

2. Carry out radioactive work only behind a radioactive shield or screen.
3. Always wash your hands and monitor yourself before leaving a radioactive area.
4. Dispose of radioactive waste in approved containers provided for the purpose.
5. To keep probe DNA in denatured form, after boiling, keep the Eppendorf tube immediately on ice without any undue delay.

PCR IN OUTLINE (FIGURE 1.17)

Introduction

PCR is a very different approach to isolation of a DNA segment. Rather than a lengthy series of manipulations involving cells, PCR is a test-tube reaction that is carried out simply by mixing together the appropriate reagents and incubating them in a thermal cycler, a piece of equipment that enable the incubation temperature to be varied over time in a preprogrammed manner. The basic steps in a PCR experiment are as follows:

1. DNA is prepared from the organism being studied and denatured by heating to 94°C.
2. A pair of oligonucleotides is added to the DNA; the sequences of these oligonucleotides enables them to anneal to either side of the gene or another DNA segment that is to be isolated, and the mixture is cooled to 50°C–60°C so that these oligonucleotides attach to their target sites.
3. A thermostable DNA polymerase is added together with a supply of deoxyribonucleotides, and the mixture is heated to the optimal temperature

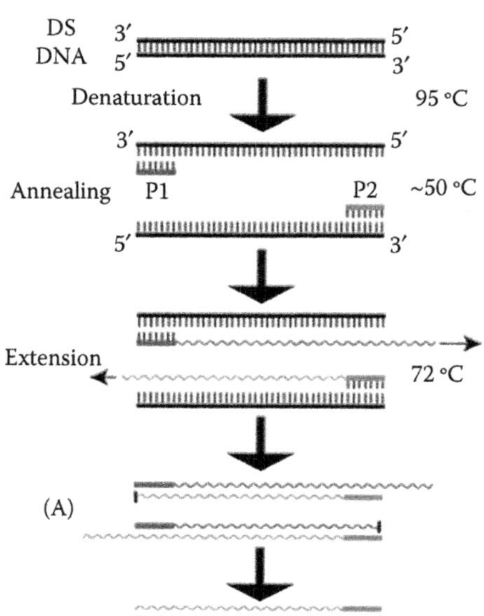

FIGURE 1.17 PCR in outline.

for DNA polymerase (the DNA polymerase I enzyme from the thermo-tolerant bacterium, *thermos aquacticus* is used).

4. The cycle of denaturation–annealing–extension is repeated 25–30 times, with each cycle doubling the number of newly-synthesized DNA molecules, copies of the DNA sequence flanked by the pair of oligonucleotides.

Choice Between Cloning and PCR

Cloning and PCR both achieve the same end: they provide samples of single, short DNA molecules derived from within a lengthy starting molecules. The clearest difference between the 2 techniques is the time required to complete the procedure. Cloning is time-consuming, especially considering that that steps illustrated in Figure 1.17 comprise just the first stage of a cloning experiment: further lengthy manipulations are needed after the clone of interest is identified from the resulting library. A competent research would expect to spend 2–3 months isolating a gene by cloning, but could achieve the same result in a single day by PCR. In view of the ease and rapidity of the PCR approach, it might seem unlikely that cloning would ever be the first-choice strategy for gene isolation. But, it often is, largely because of 2 significant limitations of PCR.

To design a PCR experiment, you need some information on the sequence of the gene that you wish to isolate. This is because the pair of oligonucleotides that anneal to either side of the DNA segment to be amplified probes the PCR. The sequences of these oligonucleotides must be complementary to their annealing sites, which means that the sequences of at least the regions bordering the gene of interest must be know. This requirement prevents the use of PCR for isolation of a gene that has never been studied before.

It is difficult to amplify long stretches of DNA by PCR, because the polymerase cannot indefinitely extend an oligonucleotide primer. PCRs up to 3 kb long are relatively easy to perform and fragments up to 10 kb long can be amplified, with practice, by standard techniques. Special procedures are needed for longer sequences, with 40 kb being the approximate limitation for even the most skilled operator.

These considerations mean that the choice between cloning and PCR can be stated as follows: If the DNA segment that you with to isolate is less than 3 kb long and you have sufficient sequence information to be able to design a suitable primer, then use PCR; if the DNA segment is longer than 3 kb and/or you cannot design primers, then use cloning.

Experimental Outline

1. Isolation of DNA.
2. Amplification of DNA.
3. Study PCR products.
4. Examine the stained products found under UV light.

Materials Required

Chemicals

- *10 × PCR buffer:* 100 m Tris- HCl (pH 8.3) containing 500 mM KCl, 15 mN $MgCl_2$, and 0.1% gelatin

- *dNTPs mix:* mixture of dATP, dCTP, dTTP, and dGTP, each at a concentration of 1.25 mM
- *Primer pair:* dissolve each primer strand in 5 mL of distilled H_2O. Primers are available commercially.
- Taq DNA polymerase
- Mineral oil
- Sterile distilled water

Glassware and Other

- Reagent bottles
- Conical flask
- Beaker
- Measuring cylinder
- PCR tubes

Equipment

- Thermocycler (PCR machine)
- Eppendorf tubes (0.5 mL capacity)
- Micropipettes

Method

1. In a sterile Eppendorf tube (0.5 μL capacity), mix the following, stepwise: sterile distilled H_2O, 5.3 μL; 10×PCR buffer, 10.0 μL; dNTP mix, 16.0 μL; primer 1, 5.0 μL; primer 2, 5.0 μL; template DNA, 10.0 μL.
2. Incubate the reaction mixture at a high temperature (94°C) for 5 min to denature the DNA completely.
3. To this, add 0.5 μL of Taq DNA polymerase and give a quick spin in a microfuge to collect the reaction mixture at the bottom.
4. Cover surface of the mixture with a drop of mineral oil to prevent evaporation of sample during repeated cycles of heating and cooling in the thermal cycler.
5. Carry out the cyclic amplification in a thermal cycler by setting the following temperatures and time periods for each step of a cycle.

Denaturation	94°C	1 min
Annealing	35°C	2 min
Polymerization	72°C	3 min

   Operate the thermal cycler for a total of 30 cycles. Depending upon the template and primer pairs, these conditions can be varied.
6. After amplification, take out the PCR products, carefully leaving behind the top mineral oil layer in the PCR tube. Store the products at –20°C.
7. Load a small fraction (5–10 μL) of PCR products on 1%–2% agarose gel along with 1 kb ladder DNA as standard and run at 100 V for 60 min. Stain the bands with ethidium bromide and visualize under UV transilluminator.

Precautions

1. Keep separate pipettes, solutions, and other materials specifically for PCR work.
2. Use all general precautions used in handling of nucleic acids.

PCR AMPLIFICATION OF SPECIFIC TARGET SEQUENCE FROM GENOMIC DNA

PCR, polymerase chain reaction, is an *in vitro* method of enzymatic synthesis of specific a DNA sequence, developed by Karry Mulis in 1983. It is a very simple and inexpensive technique for characterizing, analyzing, and synthesizing any specific piece of DNA/RNA form virtually any living organism (plant/animal/virus/bacterium). PCR is also known as *molecular photocopying*.

PCR Consist of 3 Basic Steps

1. *Denaturation:* During this step, the 2 strands melt open to form ss DNA, and all enzymatic reactions stop. This is generally carried out at 92–96°C.
2. *Annealing:* Annealing of primers to each original strand for new synthesis is carried out between 45°C and 55°C.
3. *Extension at 72°C:* The polymerase adds NTPs complementary to template at 3′ ends of the primers. Since both strands are copied during PCR, there is an exponential increase in number of copies of the genes.

These steps are repeated 20–30 times in an automated thermocycler that can heat and cool the reaction mixture in tubes within a very short time.

Amplification = 2^n
Where, n = number of cycles

Materials Required

Chemicals

- Marker
- DNA Sample

The reagents are added in PCR tubes in the following order:

• Sterile water	38 μL
• 10 × assay buffer	5 μL
• 10 mm dNTP mixture	3 μL
• Template DNA (10 ng/μL)	1 μL
• Forward primer (10 ng/μL)	1 μL
• Reverse primer (100 ng/μL)	1 μL
• Taq DNA polymerase (3 U/μL)	1 μL
Total reaction volume	50 μL

- Formamide dye (98% formamide, 4 μL dye)
- 10 m M EDTA
- Bromophenol blue
- Xylene cyanol
- 6% polyacrylamide gel (20:1 acrylamide:bis; 7.5 M urea 1× TBE buffer)

Glassware and Other

- Conical flask
- Beakers
- Measuring cylinder

Equipment

- PCR unit
- Eppendorf
- Micropipette
- Ice bucket

Method (Figure 1.18)

1. The amplification is carried out in a thermocycler for 30 cycles using the following reaction conditions:
   a. Initial denaturation is carried out for 1 min at 94°C.
   b. Denaturation is carried out for 30 s at 94°C.
   c. Annealing is carried out for 30 s at 48°C.
   d. Extension is carried out for 1 min at 72°C.
   e. Final extension is carried out for 2 min at 72°C.
2. Amplified product is separated on agarose gel.
3. The gel is visualized under UV transilluminator.

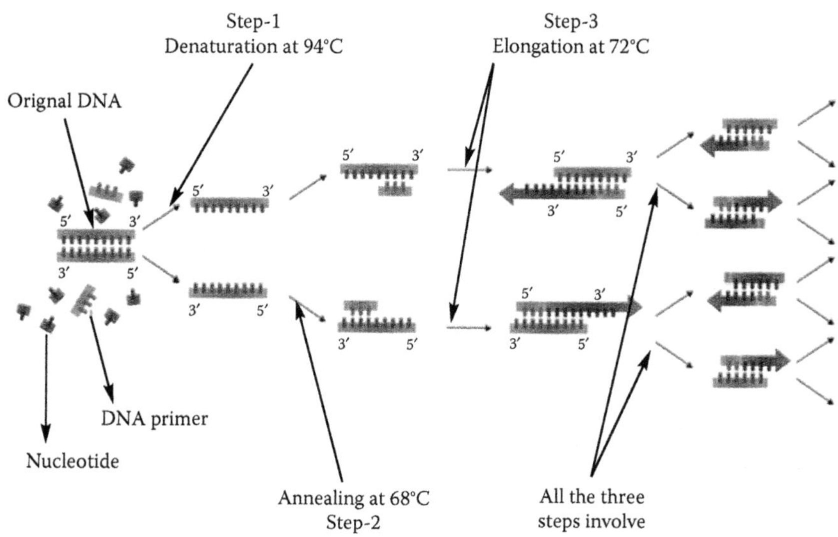

FIGURE 1.18 Steps involve in PCR amplification procedure.

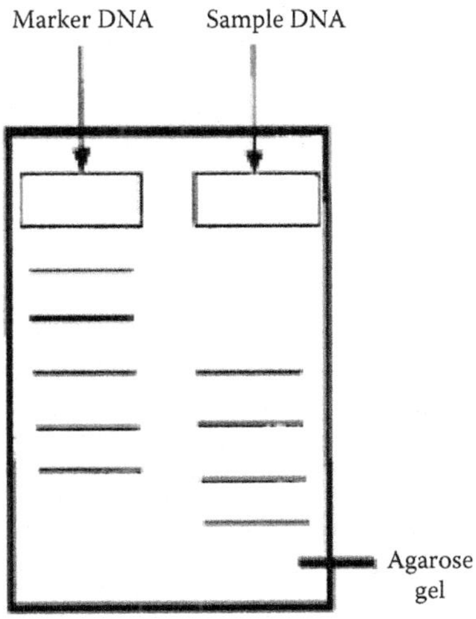

FIGURE 1.19 Gel analysis after amplification.

Gel Analysis (Figure 1.19)

1. After PCR, add an equal volume (20 μL) of formamide dye (98% formamide, 4 μL dye), 10 mM EDTA, bromophenol blue, xylene cyanol to each reaction. Heat the sample for 3 min at 90°C and immediately place on ice.
2. Pour 6% polyacrylamide gel (20:1 acrylamide:bis; 7.5 M urea 1× TBE buffer) with 0.4 mm spacers and shark tooth combs
3. Pre-electrophorese the gel at constant power for –20 min.
4. Load 3 μL of each sample on the gel.
5. Electrophorese at constant power until xylene cyanol (slower dye) is 2/3 down the length of the gel.

Precautions

1. High-purity chemicals and water should be used.
2. Ethidium bromide is carcinogenic. Gloves must be worn while handling gels in ethidium bromide solution.
3. Wear safety glasses while viewing the gel on the UV transilluminator.
4. Do not pipette any solutions by mouth. Wash your hands with hot water and soap before eating or drinking.
5. Use gloves.

ISOLATION OF TOTAL PLANT RNA FROM RICE PLANTS AND REVERSE TRANSCRIPTASE POLYMERASE CHAIN REACTION

Introduction

Reverse transcriptase PCR (RT-PCR) is the PCR amplification of c-DNA (or complementary DNA, which is complementary to mRNA), as it involves a reverse transcription step prior to amplification. The first strand c-DNA synthesis reaction is carried out in a conventional manner. But the second strand is synthesized in the first PCR cycle, which can be used to amplify first-strand cDNA (that is synthesized *in vitro* by DNA-dependent RNA polymerase). Thermostable DNA polymerases with reverse transcriptase activity have recently been described and allow RT-PCRs to be performed as a single reaction.

Its advantage as a preparative technique over conventional library-based c-DNA cloning methods include

- Speed
- Requirement for only small amounts of target material, and
- Tolerance of large amount of contaminating tRNA and rRNA, allowing whole cellular RNA to be used as the source.

Its disadvantages include the tendency to produce only part-length products and that, unlike a library, it is not a permanent resource. However, for the purpose of detecting transgene activity, RT-PCRs serve the purpose fairly well.

Nuclear Genes

Nuclear genes have a complex structure consisting of coding (exon) and noncoding (intron) regions. Consequently, transcribed mRNA must be processed prior to translation to remove the noncoding regions from its genes. mRNA splicing, the mechanism by which noncoding introns are removed from the sequence and exon ends are joined together, takes place in the cell nucleus. This splicing process results in the creation of a continuous mRNA reading frame, which encodes a full-length functional protein. The spliced molecules are then exported into the cell cytoplasm, and its coding sequence is eventually translated into protein by cytoplasm ribosomes.

Necessary Conditions for Reaction

Initial denaturation of template DNA is done at 94°C for 4 min. Then, the 34 cycles of the program scheduled as follows are done:

94°C for 1 min: Denaturation
52°C for 1 min: Primer annealing
72°C for 2 min: Primer extension

One cycle is performed at 72°C for 5 min to complete the extension of unfinished products. The product can be held at 4°C for an indefinite time (Figure 1.20).

The RT-PCR products with positive (cry1AcDNA fragment) and negative controls (untransformed rice plants) are visualized on a 1% ethidium bromide stained agarose gel along with 1 kb DNA ladder mix (MBI ferments) as the marker. The gel is photographed using Polaroid film.

Materials Required

Chemicals

- Sterile water
- Buffer 2.5 μL

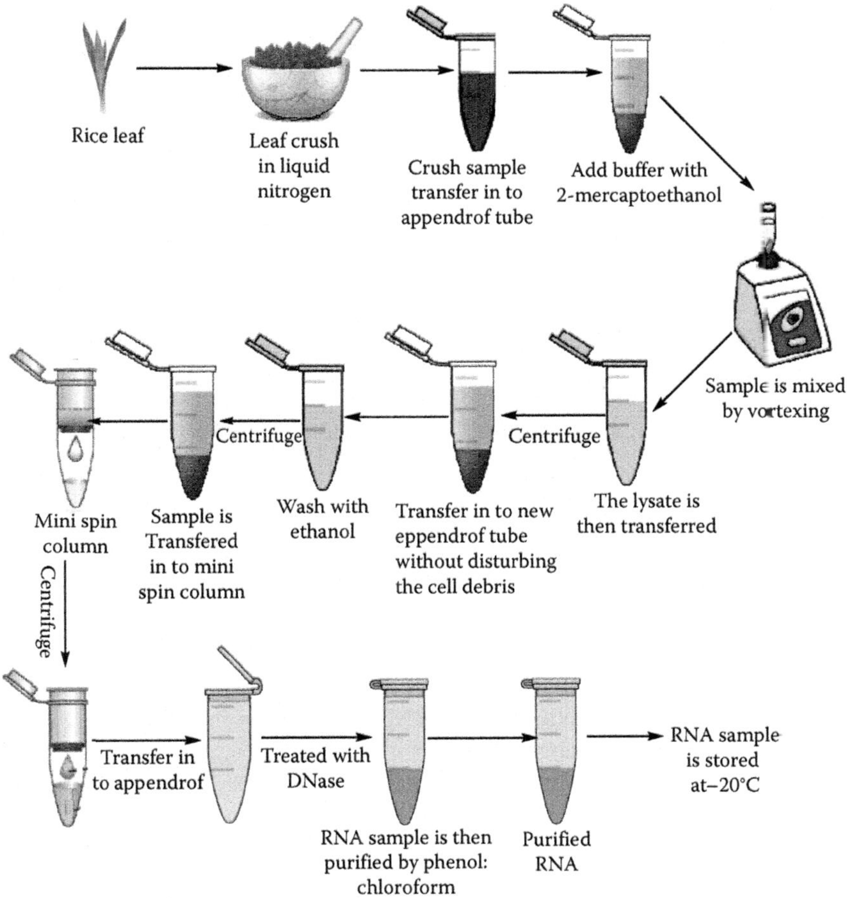

FIGURE 1.20 Isolation of total plant RNA.

- dNTPs
- Primers
- Enzymes
- Taq polymerase

Plasticware and Other

- Pestle and mortar
- Eppendorf tubes
- Spatula
- Tips

Equipment

- Thermocycler
- Pipette
- Vortex mixture

Pre-Lab Preparation

Pestle and mortar, Eppendorf tubes, spatula, tips, sterile water 18 μL, 10 × buffer 2.5 μL, dNTPs 0.5 μL (200 U) each, primers 1 μL (300 ng) each, template DNA 2 μL (100 ng), sterile water to a final volume of 25 μL, enzymes (Taq polymerase etc.).

Method

Total plant RNA is isolated from tender leaves of transformed and non-transformed (control) rice plants by the Qiagen RNase Plant Mini Kit procedure.

1. Tender leaves of rice plants are ground under liquid nitrogen in prechilled, autoclaved DEPC-treated pestle and mortar.
2. 100 mg of fine powder is taken in an Eppendorf tube and 450 μL of buffer RLT is added along with 4.5 μL of β-mercaptoethanol (10 μl/mL of RLT buffer).
3. Sample is mixed vigorously by vortexing. The lysate is then transferred to QIA shredder spin column (lilac) sitting in a 2 mL collection tube.
4. Centrifugation is done at 13,000 rpm for 2 min at room temperature.
5. The flow through fraction is transferred from QIA shredder to a new Eppendorf tube without disturbing the cell debris pellet in the collection tube.
6. 0.5 volume (usually 225 μL) of ethanol (95%–100%) is added to the clear lysate and mixed well by pipetting.
7. The sample is then applied onto an RNeasy mini spin column (pink) sitting in a 2 mL collection tube.
8. Centrifugation is done for 15 s at 13,000 rpm. The flow through is discarded and the collection tube is rescued.
9. 700 μL of buffer RWI is taken and put onto the RNeasy column. Centrifugation is done for 15 s at 13000 rpm. The flow through and collection tubes are discarded.
10. Now RNase column is transferred to a new 2 mL collection tube and 500 μL of RPE buffer (containing ethanol) is applied onto the column.
11. Again centrifugation is done for 15 s at 13,000 rpm. The flow through is discarded and the collection tube is rescued.
12. 500 μL of RPE buffer is applied onto the column and centrifuged for 2 min at 13,000 rpm.
13. The membrane of the column is dried by centrifugation at 13,000 rpm for 1 min.
14. The RNeasy mini column is then transferred to a new 1.5 mL Eppendorf tube. 30 μL of RNase-free water containing RNase-inhibitor at 0.5 units/μL is applied onto the column and centrifuged at 13,000 rpm for 1 min.
15. The flow through is again loaded on the same column and centrifugation is done at 13,000 rpm for 1 min (for better elution). The isolated RNA is then treated with DNase and kept at 28°C for 30 min. (The DNase enzyme is inactivated by adding 0.5 mM EDTA and heating it at 75°C for 10 min in a water bath).
16. The RNA sample is then purified by phenol:chloroform extraction and quantified by spectrophotometry. The RNA sample is stored at –20°C.

Reverse Transcriptase-Polymerase Chain Reaction

Reverse Transcriptase PCR is carried out using 50 μL reaction mixture containing 5 μg of total RNA, isolated from the leaves of transgenic and nontransgenic plants. Omniscript RT kit (200) from Qiagen is used for RT-PCR.

RT-Mix (for 10 Reactions)		PCR Mix (for 1 Reaction of 25 μL)	
ddH20	280 μL	ddH20	16 μL
10× buffer	50 μL	10× buffer	2.5 μL
dNTPmix	20 μL	$MgCl_2$ (25 mM)	1.5 μL
Reverse primer	20 μL	dNTP mix (2.5 mM each)	2 μL
Enzyme	10 μL	Prime rl (forward)	1 μL (20 pM)
RNA template	5ul	Primer 2 (reverse)	1 μL (20 pM)
		Taq DNA polymerase	2 μL (2 units)
		Template DNA	0.5 μL (~500 ng)

The above RT-mix is incubated at 37°C for 1 h for cDNA synthesis. After RT-reaction is over, 5 μL aliquot of mix is taken and general PCR mix (MBI Fermentas) is added in accordance with 20 μL PCR mix, along with 2 units of Taq DNA polymerase. The sample is then amplified by a thermocycler (The Perkin Elmer Thermocycler is a study machine and is so used, preferably), using the reaction conditions given below.

The primer sequences used for the Cry lAc gene are as follows:

Forward: 5′ACC ATC ATC GCA GCC GAT AGT ATT3′
Reverse: 5′TCT CCT GCA GTC CCA CTA AAG TTT3′
The T_m values of the primers are 56°C and 54°C, respectively.

Precautions

1. High purity chemicals and water should be used.
2. Ethidium bromide is carcinogenic. Gloves must be worn while handling gels in ethidium bromide solution.
3. Wear safety glasses while viewing the gel on the UV transilluminator.
4. Do not pipette any solutions by mouth. Wash your hands with hot water and soap before eating or drinking.
5. Use gloves.

ISOLATION OF GENE OF INTEREST BY PCR

Isolation of Genomic DNA (Figure 1.21)

1. For cultured cells, aspirate culture medium and briefly rinse the cells twice in PBS. Add 10 mL/60 mm culture plate (with approximately 10^5 cells) of DNA isolation buffer to the cells and allow cells to be lysed for 5 min at room temperature with shaking at 60 rpm. For tissues, grind 1 g tissue in liquid N_2 to fine powder and transfer the powder into a tube containing 15 mL of DNA isolation buffer. Allow cells to be lysed for 10 min at room temperature with shaking at 60 rpm.

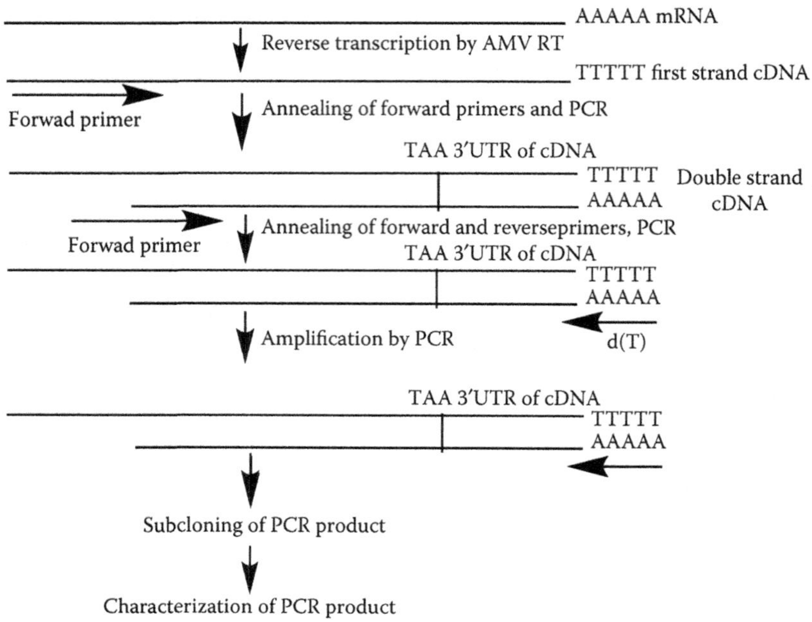

FIGURE 1.21 Diagram of Isolation of 3′—RACE PCR.

2. Incubate at 37°C for 5 h or overnight to degrade proteins, and extract genomic DNA.
3. Allow the lysate to cool to room temperature and add 3 volumes of 100% ethanol, prechilled at −20°C, to the lysate. Gently mix and allow DNA to precipitate at room temperature with slow shaking at 60 rpm. Precipitate DNA; appearance of white fibers should be visible in 20 min.
4. Taken out the DNA using a sterile glass hook or spin down the DNA. Rinse the DNA twice in 70% ethanol and partially dry the DNA for 40–60 min at room temperature to evaporate ethanol. The DNA can be directly subjected to restriction enzyme digestion without being dissolved in TE buffer or dd H_2O. We usually overlay the precipitated DNA with an appropriate volume of restriction enzyme digestion cocktail and incubate the mixture at an appropriate temperature for 12–24 h. The digested DNA is mixed with an appropriate volume of DNA loading buffer and is ready for agarose gel electrophoresis.

Materials Required

Chemicals

- *For EB (extraction buffer):* 50 mM Tris (pH 8.0), 1% CTAB, 50 mM EDTA, 1 Mm 1-10-o-phenanthroline, 0.7 M NaCl, 1% betamercraptoethanol
- 75 mM Tris-HCl (pH 6.8)
- 100 mM NaCl
- 2 mg/ml protease K (freshly added)

- 0.1% (w/v) N-lauroylsarcosine
- *10 × Sau3AI buffer:* 0.1 M Tris-HCl (pH 7.5), 1M NaCl, 70 mM $MgCl_2$
- *5× Loading buffer:* 38% (w/v) sucrose, 0.1% bromophenol blue, 67 mM EDTA
- *5 × TBE buffer:* 54 g Tris base, 27.5 g boric acid, 20 mL of 0.5 M EDTA (pH 8.0)
- *TE buffer:* 10 mM Tris-HCl (pH 8.0), 1 mM EDTA (pH 8.0), ethanol (100%, 70%)

Glassware and Other

- Pestle and mortar
- Conical flask
- Measuring cylinder
- Beakers
- Reagent bottle
- Eppendorf tubes
- Spatula
- Tips

Equipment

- Thermocycler
- Cooling centrifuge
- Water bath
- Pipette
- Vortex mixture

Partial Digestion of Genomic DNA Using Sau3AI

It is necessary to partially cut the high molecular weight of genomic DNA with a 4-base cutter, Sau 3AI, to increase the efficiency of PCR.

1. *Optimization of Partial Digestion of Genomic DNA with Sau3AI*
   To determine the amount of enzyme used to digest the high molecular weight DNA into 20–30 kb fragments, small scale reactions, or pilot experiments, should be carried out.
   a. Prepare 1 × Sau3AI buffer on ice:

10 × Sau 3AI buffer	0.2 mL
Add dd.H_2O to a final volume of	2.0 mL

   b. Perform Sau3AI dilution in 10 individual microcentrifuge tubes on ice:

Tube	Sau3AI (3 u/μL) Dilution	Dilution Factor
1	2 μL Sau 3AI + 28 μL 1 × Sau3AI buffer	1/15
2	10 μL of 1/15 dilution + 90 μL 1× Sau 3AI buffer	1/150
3	10 μL of 1/150 dilution + 10 μL 1× Sau 3AI buffer	1/300
4	10 μL of 1/150 dilution + 30 μL 1× Sau 3AI buffer	1/600

(Continued)

Tube	Sau3AI (3 u/μL) Dilution	Dilution Factor
5	10 μL of 1/150 dilution + 50 μL 1× Sau 3AI buffer	1/900
6	10 μL of 1/150 dilution + 70 μL 1× Sau 3AI buffer	1/1200
7	10 μL of 1/150 dilution + 90 μL 1× Sau 3AI buffer	1/1500
8	10 μL of 1/150 dilution + 110 μL 1× Sau 3AI buffer	1/1800
9	10 μL of 1/150 dilution + 190 μL 1× Sau 3AI buffer	1/3000
10	10 μL of 1/150 dilution + 290 μL 1× Sau 3AI buffer	1/4500

c. Perform 10 individual, small-scale digestion reactions on ice in the order listed below:

	Tube Number									
Components	1	2	3	4	5	6	7	8	9	10
Genomic DNA (1 μg/μL)	1 μL	1 μL	1 μL	1 μL	1 μL	1 μL	1 μL	1 μL	1 μL	1 μL
10 × Sau3AI buffer dd.H_2O (μl)	5 μL	5 μL	5 μL	5 μL	5 μL	5 μL	5 μL	5 μL	5 μL	5 μL
dd.H_2O (μl)	39	39	39	39	39	39	39	39	39	39
Sau3AI dilution in the same order as in (b) above	5 μL	5 μL	5 μL	5 μL	5 μL	5 μL	5 μL	5 μL	5 μL	5 μL
Final Volume (μl)	50	50	50	50	50	50	50	50	50	50

Note: The final volume Sau 3AI concentration used in tube numbers 1 through 10 should be 1, 0.1, 0.05, 0.025, 0.015, 0.0125, 0.01, 0.0085, 0.005, and 0.0035 unit/μg DNA, respectively.

d. Incubate the reactions at the same time at 37°C for 30 min. Place the tubes on ice and add 2 μL of 0.2 M EDTA buffer (pH 8.0) to each tube to stop the reaction.
e. During the reactions, prepare a large size of 0.4% agarose gel in 1 × TBE buffer.
f. Add 10 μL of 5 × DNA loading buffer to each of the 10 tubes containing the digested DNA prepared in step (d).
g. Load 30 μL of each sample to the wells in the order of 1–10. Load DNA markers (e.g., 1 μg DNA Hind III markers) to the left or the right well of the gel to estimate the sizes of digested DNA.
h. Carry out electrophoresis of the gel at 2–5 V/cm until the bromophenol blue reaches the bottom of the gel.
i. Photograph the gel under UV light and find the well that shows the maximum intensity of fluorescence in the desired DNA size range of 20–30 kb.

2. *Large-scale preparation of partially digested genomic DNA*
   a. Based on the optimal conditions established above, carry out a large-scale digestion of 50 μg of high molecular weight genomic DNA using half units of Sau3AI/μg DNA that produced the maximum intensity of

fluorescence in the DNA size range of 20–30 kb. The DNA concentration, time, and temperature should be exactly the same as those used for the small-scale digestion. For instance, if tube number 7 (0.01 unit of Sau3AI/μg DNA) in the small-scale digestion of the DNA shows a maximum intensity of fluorescence in the size range of 20–30 kb, the large scale digestion of the same DNA can be carried out as follows.

Genomic DNA (1 μg/μl)	50 μL
10 × Sau 3AI buffer	250 μL
dd.H_2O	1.95 mL
Diluted Sau 3AI (0.005 u/μg) prepared as step 1 (b), tube 9	250 μL
Final volume of	2.5 mL

b. Incubate the reaction at 37°C in a water bath for 30 min. Stop the reaction by adding 20 μL of 0.2 M EDTA buffer (pH 8.0), and place the tube on ice until use.
c. Add 2–2.5 volumes of chilled 100% ethanol to precipitate the DNA at −70°C for 30 min.
d. Centrifuge at 12,000 g for 10 min at room temperature. Carefully decant the supernatant and briefly rinse the DNA pellet with 5 mL of 70% ethanol. Dry the pellet under vacuum for 8 min. Dissolve the DNA in 50 μL of TE buffer. Take 5 μL of the sample to measure the concentration at A_{260} nm and store the DNA sample at −20°C before use.

Design and Synthesis of Specific Forward and Reverse Primers

i. *Amplification and Isolation of Exon and Intron Sequences*
A pair of forward and reverse primers can be designed based on the 5′-UTR or 3′-UTR sequences of a known cDNA. The forward primer can be designed in the 5′-UTR region in the 5′→3′ direction with 20–30 bases, which should be complementary to the first or (−) strand of the cDNA template. The reverse primer is designed in the 3′-UTR region in the 5′→3′ direction with 20–30 bases, which should be complementary to the second or (+) strand of cDNA template. Alternatively, forward and reverse primers can be designed based on the very N-terminal and C-terminal amino acid sequences if the cDNA sequence is not known.

ii. *Amplification and Isolation of Promoter Sequence*
To amplify the promoter sequence of interest, a reverse primer can be designed according to the 5′-end sequence of known cDNA or based on the N-terminal amino-acid sequence. This primer will be employed to anneal with a DNA strand of the denatured genomic DNA fragments. The primer facilitates the synthesis of a new strand, which is complementary to the template and includes the promoter sequence. After addition of oligo (dC) to the 3′ end of the new strand, an oligo (dG) anchor primer or forward primer in the cloning site of the vector will facilitate the synthesis of the DNA strand that is complementary to the new strand. As a result, major PCR products, including the promoter sequence, will be obtained.

Amplification of Specific DNA Fragments by PCR

1. Prepare a PCR cocktail in a 0.5 mL microcentrifuge tube.

Forward Primer (20–26 mer, 10–30 ng/μL)	5–8 pmol (30–40 ng), depending on the size of primer
Reverse Primer (20–26 mer, 10–30 ng/μL)	5–8 pmol (30–40 ng), depending on the size of primer
Digested genomic DNA	2–3 μg
10× amplification buffer 4 μL	4 μL
dNTPs (2.5 mM each)	4 μL
Taq or Tth DNA polymerase 5–10 units	5–10 Unit
Add dd.H_2O to a final volume of	40 μL

Note: The DNA polymerase should be of high fidelity and long expand.

2. Carefully overlay the mixture with 30 μL of mineral oil to prevent evaporation of the samples during PCR amplification. Place the tubes in a thermal cycler and perform PCR.

	Cycling (30 cycles)			
Predenaturation	**Denaturation**	**Annealing**	**Extension**	**Last**
94°C, 3 min	94°C, 1 min	60°C, 1 min	70°C, 1.5 min	4°C, forever

Purification of PCR Product by Agarose Gels

1. Load the amplified PCR mixture into a 1%–1.4% agarose gel, depending on the sizes of the PCR products, and carry out electrophoresis.
2. When electrophoresis is complete, quickly locate the DNA band of interest by illuminating the gel on a long wavelength (>300) UV transilluminator. Quickly slice out the band of interest using a sharp, clean razor blade.
3. To enhance the yield of DNA, trim away extra agarose gel outside the band, and cut the gel slice into tiny pieces with a razor blade.
4. Transfer the fine slices into a 1.5 mL microcentrifuge tube (Eppendorf).
   a. If one does not need to elute DNA out of the gel slices, the slices do not need to be further sliced into tiny pieces. They can be directly placed in a tube.
   b. At this point, there are 2 options for eluting DNA from the agarose gel pieces. The first option it to immediately carry out high-speed centrifugation. The second option is to elute the DNA as high yield as possible.
5. Centrifuge at 12,000–14,000 g or at the highest speed using an Eppendorf centrifuge for 15 min at room temperature.

Note: With high-speed centrifugation, the agarose matrix is compressed and/or even partially destroyed by the strong force of centrifugation. The DNA molecules contained in the matrix are released in to the supernatant fluid.

6. After centrifugation, carefully transfer the supernatant fluid containing the DNA into a clean microcentrifuge tube. The DNA can be used directly for ligation, cloning, labelling, and restriction enzyme digestion without ethanol precipitation. Store the DNA solution at 4°C or −20°C until use.

Note:

a. In order to confirm that the DNA is released from the gel pieces, the tube containing the fluid can be briefly illuminated with long-wavelength UV light after centrifugation. An orange-red fluid color indicates the presence of DNA in the fluid.
b. The supernatant fluid should be immediately transferred from the agarose pellet; within minutes, the temporarily compressed agarose pellet may swell, absorbing the supernatant fluid.

Precautions

1. High-purity chemicals and water should be used.
2. Ethidium bromide is carcinogenic. Gloves must be worn while handling gels in ethidium bromide solution.
3. Wear safety glasses while viewing the gel on the UV transilluminator.
4. Do not pipette any solutions by mouth. Wash your hands with hot water and soap before eating or drinking.
5. Use gloves.

ISOLATION OF pUC18 PLASMID FROM TOP10 pUC18 *E. COLI* CELLS

A plasmid is an extrachromosomal DNA present in most of bacteria and in some yeast. Most plasmids are circular, and they vary in size and in numbers. Plasmids are modified and used as vectors (carriers) in rDNA technology. In this experiment, one such modified plasmid pUC18 is isolated from an *E. coli* culture by the alkaline lysis method. This method is based on the principle that exposure of bacterial suspensions to the strongly anionic detergent at high pH opens the cell wall, denatures chromosomal DNA and proteins, and releases plasmid DNA into the supernatant. Although the alkaline solution completely disturbs chromosomal DNA, the circular plasmids of DNA are unable to separate from each other because they are topologically intertwined. During lysis, bacterial proteins, broken cell walls, and denatured chromosomal DNA become enmeshed in large complexes that are coated with dodecyl sulphate. These complexes are efficiently precipitated when the solution with sodium ions are replaced by a solution with potassium ions. The denatured materials are then removed by centrifugation. Alkaline lysis is a flexible method that works well with all strains of *E. coli* and with bacterial cultures ranging in size from 1–500 mL. The plasmid DNA obtained by this method is devoid of nuclear DNA (Figure 1.22).

Materials Required

Chemicals

- LB plate with best pUC18 colonies
- LB broth with 100 mg/L Amp
- *Solution I:* 50 mM glucose, 25 mM Tris-HCl, 10 mM EDTA
- *Solution II:* 1% SDS, 0.2 N NaOH (pH 12.0)

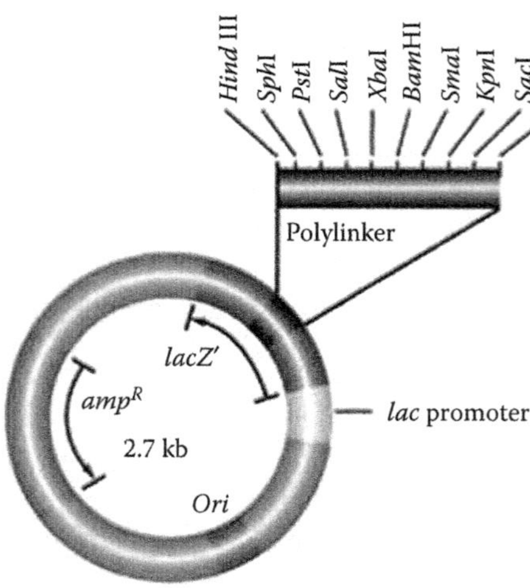

FIGURE 1.22 Structure of pUC18 vector.

- *Solution III:* 3 M sodium acetate (pH 5.0)
- TE Buffer: 10 mM Tris-HCl 10 mM EDTA
- RNase (1 mg/mL)
- Phenol:chloroform:isoamyl alcohol (25:24:1)
- Absolute ethanol
- 70% ethanol
- 1% agarose gel

Glassware and Other

- 1.5-mL microcentrifuge tubes
- Reagent bottles
- Conical flask
- Beakers
- Measuring cylinder
- Knife or razor blade

Equipment

- Microcentrifuge
- Micropipette
- Vortex unit
- Electrophoresis apparatus
- Incubator shaker
- Deep fridge

Method

1. Inoculate a loopful of pUC18 cells in 5 mL of LB broth with 100 mg/L Amp and kept in a 37°C shaker for 16 h.

2. Transfer about 1.5 mL of culture to centrifuge tubes.
3. Centrifuge for 5 min at 6,000 rpm at 4°C and remove the supernatant.
4. Add 100 μL of solution I to the pellet and resuspend by vortexing.
5. Add 200 μL of freshly prepared solution II, mix by inversion, and incubate on ice for 3 min. Complete lysis is ensured by a clear and viscous solution.
6. Add 150 μL of ice cold solution III and incubate on ice for 5 min.
7. Centrifuge at 15,000 rpm for 5 min at 4°C and carefully transfer the supernatant to a new tube.
8. Add an equal volume of isopropanol to the supernatant, mix gently, and centrifuge at 15,000 rpm for 10 min at 4°C.
9. Discard the supernatant and wash the pellet with 70% ethanol.
10. Dry the pellet and resuspend in 50 μL TE.
11. Add about 10 μL of RNase, incubate at 37°C for 1 hr, purify by phenol chloroform extraction.
12. Load 5 μL of the plasmid DNA in a 1% agarose gel. Store the remaining samples in –20°C for future use.

Precautions

1. High-purity chemicals and water should be used.
2. Ethidium bromide is carcinogenic. Gloves must be worn while handling gels in ethidium bromide solution.
3. Wear safety glasses while viewing the gel on the UV transilluminator.
4. Do not pipette any solutions by mouth. Wash your hands with hot water and soap before eating or drinking.
5. Use gloves.

TRANSFORMATION OF THE DESIRED BACTERIAL STRAIN WITH PLASMID DNA

Bacterial transformation is a process by which external DNA from the surrounding medium is taken up. It is done to introduce a foreign plasmid DNA and to use the transformed bacteria to amplify the foreign DNA. Bacterial cells can be artificially induced to take up DNA by treating them with $CaCI_2$. Such artificially induced bacterial cells are competent cells. To electroporate DNA into cells, washed *E. coli* are mixed with the DNA to be transformed and then pipetted into a plastic cuvette containing electrodes. A short electric pulse, about 2400 volts/cm, is applied to the cells, causing smalls holes in the membrane through which the DNA enters. The cells are then incubated with broth as above before planting. For chemical transformation, there is no need to pretreat the DNA. For electro-poration, the DNA must be free of all salts so the ligations are first precipitated with alcohol before they are used.

In the course of developing competence, receptors of some kind are either formed or activated on the cell wall and are responsible for initial binding of the DNA. The cells are exposed to a temperature of 42°C (heat shock) for a short time. Immediate chilling on ice ensures closure of pores. These cells are screened for transformants/recombinants. Transformants are selected based on the selection marker, generally an antibiotic-resistance gene. Screening of recombinants is done on the basis of insertional inactivation. Insertion of DNA disrupts the gene, and, as a result, the inactivated gene is no longer functional and its phenotype is not displayed by the recombinant bacterial cell. The objectives are

1. Prepare competent cells.
2. Perform transformation.
3. Learn α-complementation.

Materials Required

Chemicals

- pUC18 (20 ng/μL) *E. coli* strain
- LB agar
- Ampicillin
- 0.1 M $CaCl_2$
- X-Gal (5-bromo-4-chloro-3-indolyl-β-D-galactopyranoside)
- IPTG (isopropyl-β-D-thiogalactopyranoside)
- Distilled water
- Tryptone
- Yeast extract
- NaCl
- Agar

Glassware and Other

- Oak ridge tubes (50 mL)
- Conical flask
- Petri plates
- Spreader
- Pipettes
- Tips
- Cuvette (of 1 cm path length)
- Micropipettes

Equipment

- Refrigerated centrifuge
- Incubator, 37°C shaker
- Spectrophotometer
- Water bath

Pre-Lab Preparation

1. *Preparation of the media:*
   Tryptone 10 g
   Yeast extract 5 g
   NaCl 10 g

Agar 15 g
Distilled water 1,000 mL
Adjust the pH to 7.0 with 5N NaOH, if required, before autoclaving
Add 1.5% LB agar and autoclave

2. *Ampicillin preparation:* Dissolve 100 g of ampicillin in 1 mL of sterile water to get a stock concentration of 100 mg/mL. Store at 4°C for 2 wk. Use the antibiotic within this period.
3. *Ampicillin LB medium:* Add ampicillin to LB broth or agar at a final concentration of 10 μL/mL, when the temperature of the medium is around 40°C–45°C.
4. *Preparation of LB Amp plates with X-Gal and IPTG:* To the ampicillin LB agar plates, add 40 μL each of X-Gal and IPTG for every 20 mL of LB agar. Mix well and pour media into required number of plates.

Method (Figure 1.23)

Preparation of Competent Cells

Day 1: Revival of bacterial strain on LB plates from glycerol stock or stab by streaking.

Day 2: Inoculate a single colony into 5 mL of LB medium and incubate at 37°C (in a gyratory shaker) overnight for 12–16 h.

Day 3:

1. Inoculate 1 mL of overnight culture into 100 mL LB medium taken in 500 mL conical flask and incubate at 37°C in a shaker. Grow until the OD_{600}

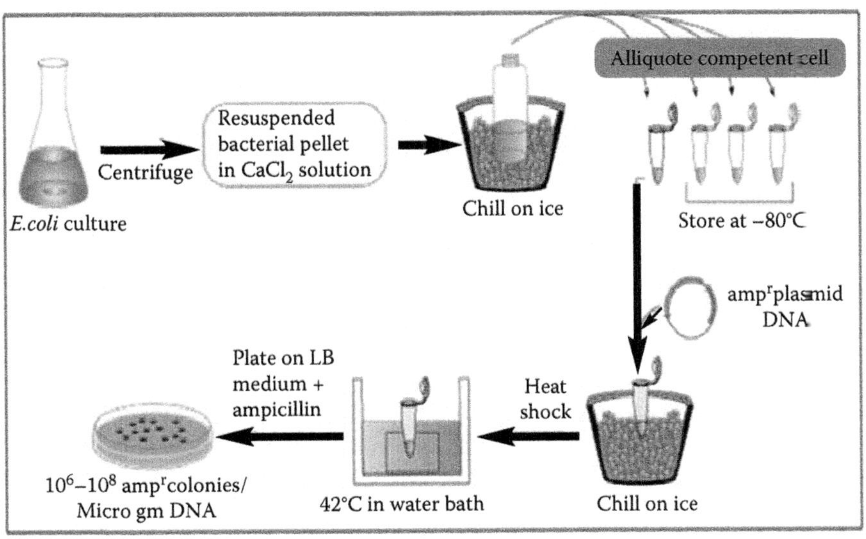

FIGURE 1.23 Transformation of the desired bacterial strain with plasmid DNA.

reaches 0.25–0.30; this takes about 2–3 h. When desired OD is reached, chill the culture flask on ice for 10–20 min.
2. Transfer the cell culture into centrifuge tubes in a laminar hood and centrifuge for 10 min at 6,000 rpm at 4°C to pellet down the cells.
3. Discard the supernatant and aseptically add 1.0 mL of cold 0.1 M $CaCl_2$ solution. Place the tube on ice. Resuspend the cell pellet gently in the solution using a pipette.
4. Keep the tube with the cell suspension on ice for 30 min.
5. Centrifuge at 6,000 rpm for 8 min, preferably at 4°C.
6. Discard the supernatant and resuspend gently in 0.6 mL of cold 0.1 M $CaCI_2$
7. Aliquot 100 μL of competent cells into 6 prechilled vials.
8. Competent cells are ready for immediate use.

Transformation

1. Add 100 ng of the plasmid DNA to a aliquot of 100 μL of competent cells. Gently tap and incubate on ice for 20 min (for DNA to bind to the cells).
2. Heat shock the cells by placing the vial in 42°C water bath for 2 min, then immediately place the tube on ice for 5 min.
3. Aseptically add 1 mL of LB broth to the vials and incubate at 37°C in a shaker for 1 h.
4. Label 3 LB agar plates with X-Gal and IPTG as a, b, c. Pipette 100 μL of LB on each plate. Add 25, 50, 100 μL of transformed cells to plates a, b, and c, respectively. Mix well and spread thoroughly using a spreader.
5. Plate 100 μL of competent cells that have not been transformed to check for any contamination. Label this as control plate/nontransformed plate.
6. Incubate the plates at 37°C overnight

Precautions

1. Carry out the procedure under aseptic conditions.
2. Prepare competent cells within 3 d of reviving the strain.
3. Precool tubes, 0.1 M $CaCl_2$ solution, and centrifuge tubes.
4. Storage of cells may result in poor or no transformants.
5. X-Gal is light-sensitive so cover with aluminium foil.
6. On transforming competent cells with pUC18, antibiotic resistance is conferred on the host as this plasmid carries the gene for amp resistance. As a result, only transformed cells grow on LB ampicillin.
7. These transformed cells appear blue on LB-Amp-X-Gal-IPTG plates due to X.-complementation, that is, active β-galactosidase produced cleaves X-Gal to give the blue color on IPTG induction.
8. Using the calcium chloride method for cell preparation, the expected transformation efficiency on transforming 100 ng of pUC18 is approximately 10^5/μL of DNA. Efficiency lower than this could be due to improper conditions during preparation of competent cells. So, always keep the cells a 4°C.

PURIFYING pUC18/HIND III/EcoRI DIGEST BY GEL ELUTION

INTRODUCTION

DNA can be recovered from agarose gel through different methods, such as electroelution, tresqueezes DNA kits, and using low-melting-point agarose gel. In this experiment, low-melting-point agarose gel is used for elution. In this method, a special grade of agarose that melts at low temperature without significant deterioration in the strength of the hardened gel is used. This grade of agarose is also free of inhibitors and nucleases. The fragments of DNA are first separated by electrophoresis through an agarose gel of the appropriate concentration and located by staining with ethidium bromide. The DNA is recovered by melting the agarose, extracting by phenol:chloroform, and precipitating the DNA. The protocol works best for DNA fragments ranging in size from 0.5–5.0 Kb.

MATERIALS REQUIRED

Chemicals

- DNA source: pUC18/Hind III/ EcoRI
- 1% low-melting-point agarose gel
- Phenol:chloroform (1:1, v/v)
- Absolute ethanol
- 70% ethanol
- TE buffer
- 1% agarose gel

Glassware and Other

- Measuring cylinder
- Conical flask
- Beaker
- Centrifuge tubes
- Scalpel

Equipment

- Electrophoresis apparatus
- Centrifuge
- Gel documentation system
- Deep fridge

Method

1. Load the entire volume of pUC18/Hind III/EcoRI digest in a 1% low-melting-point agarose gel, electrophorese, and visualize the bands under UV.
2. Using a sharp scalpel, cut down the DNA band referring the pUC18 digest, and place in 1.5 mL centrifuge tubes.
3. Chop the gel pieces and kept in −80°C overnight.
4. Thaw the frozen gel pieces and add about 200 μL phenol:chloroform; vortex for 3 min.

5. Centrifuge at 15,000 rpm for 10 min and transfer the upper phase into a new tube.
6. Add an equal volume of ice-cold ethanol and keep at –20°C for 1 h.
7. Centrifuge at 15,000 rpm for 10 min and discard the supernatant.
8. Wash the pellet with 70% ethanol and dry the pellet.
9. Resuspend in 20 μL TE buffer and load 2 μL of the purified samples in 1% agarose gel, and run electrophoresis.
10. Store the remaining samples in –20°C for future use.

Precautions

1. High-purity chemicals and water should be used.
2. Ethidium bromide is carcinogenic. Gloves must be worn while handling gels in ethidium bromide solution.
3. Wear safety glasses while viewing the gel on the UV transilluminator.
4. Do not pipette any solutions by mouth. Wash your hands with hot water and soap before eating or drinking.
5. Use gloves.

DEPHOSPHORYLATION OF RESTRICTION-ENZYME-DIGESTED VECTOR pUC18

Introduction

Alkaline phosphatase removes 5′P from ends of DNA or RNA. Dephosphorylated vector DNA is thus prevented from self-annealing prior to the insertion of foreign DNA fragments.

Materials Required

Chemicals

- pUC18 linearized with EcoRI
- Bacterial alkaline phosphatase
- Tris-saturated phenol
- Chloroform
- Isoamyl alcohol
- *Phenol-chloroform solution:* Mix equal volumes of Tris-saturated phenol and chloroform

Glassware and Other

- Reagent bottle
- Conical flask
- Measuring cylinder
- Micro tips

Equipment

- Incubator
- Micropipette
- Microcentrifuge

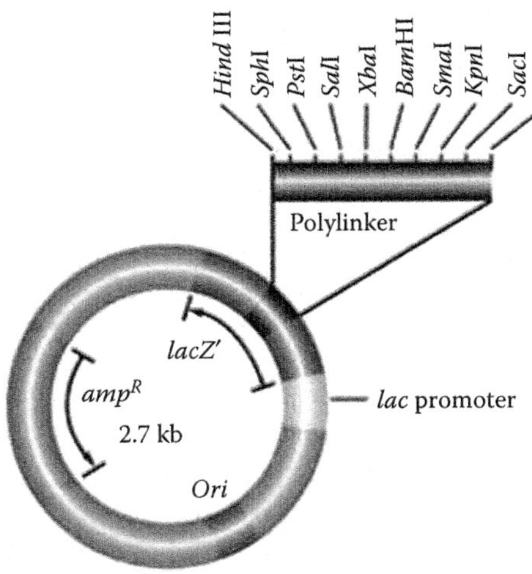

FIGURE 1.24 pUC18 vector.

Method (Figure 1.24)

1. Add 2–3 μg of linearized vector DNA (pUC18 cut with EcoRI) to a 1.5 mL tube kept on ice.
2. Add bacterial alkaline phosphatase at a concentration 1U/μg DNA to be dephosphorylated.
3. Incubate the reaction mixture at 60°C for 1 h.
4. Add an equal volume of phenol-chloroform solution and spin in a microfuge at full speed for 5 min.
5. Collect the aqueous layer and add an equal volume of chloroform:isoamyl alcohol.
6. Precipitate the DNA and store at 4°C until further use.

Precautions

1. High-purity chemicals and water should be used.
2. Ethidium bromide is carcinogenic. Gloves must be worn while handling gels in ethidium bromide solution.
3. Wear safety glasses while viewing the gel on the UV transilluminator.
4. Do not pipette any solutions by mouth. Wash your hands with hot water and soap before eating or drinking.
5. Use gloves.

RESTRICTION DIGESTION OF pUC 18 AND A-DNA

A restriction enzyme is an enzyme that cuts double-stranded DNA at specific sites known as restriction recognition sequences, Most of the restriction recognition sequences are palindromic and vary in length from 4–8 nucleotides. Restriction enzyme makes 2 incisions, once through each sugar-phosphate backbone (i.e., each strand) of the DNA double helix. Some restriction enzymes cut the double-stranded

DNA in 2 different positions and generate ends that are staggered, with 5′- or 3′-protruding terminal nucleotides; others cut at the same position and produce blunt ends. Discovery of restriction enzymes led to the development of recombinant DNA technology, and restriction enzymes are routinely used for DNA modification and manipulation in laboratories. In this experiment, pUC 18 and A-DNA are double digested with EcoRI and Hind III, and the restriction recognition sites for these enzymes are given below:

EcoRI: G/AATTCATTAAIC
Hind III: *A*/AGCTTTTCGAI A

Materials Required

Chemicals

- DNA source: pUC18 and λ DNA
- Double distilled water
- Hind III
- 10 mM Tris-HCl
- 50 mM NaCI,
- 10 mM $MgCl_2$
- 1 mM DTT
- EcoRI
- 50 mM Tris-HCl
- 100 mM NaCI
- 10 mM $MgCl_2$
- 5 mM mercaptoethanol
- Hind III and EcoRI restriction enzymes
- 0.5 M EDTA
- 1% agarose gel

Glassware and Other

- 1.5-mL microcentrifuge tubes
- Reagent bottle
- Conical flask
- Pipette
- Measuring cylinder

Equipment

- Electrophoresis apparatus
- Microcentrifuge

Method

A clean 1.5-mL microcentrifuge is taken and 2 reactions were set according to the table pUC18 digestion A DNA digestion pUC18 plasmid (lug/Ill) Hind III assay buffer (10 X) Hind III (10 U/μL)

TABLE 1.1
Restriction Digestion pUC8

pUC18 digestion		λ DNA digestion	
pUC18 plasmid (1 μg/μl)	1	λ DNA	1
Hind III assay buffer (10×)	2	Hind III assay buffer (10×)	2
Hind III (10 U/μl)	0.5	Hind III (10 U/μl)	0.5
EcoRI (10 U/μl)	0.5	EcoRI (10 U/μl)	0.5
Sterile distilled water	16	Sterile distilled water	16
Total	20 μl		20 μl

EcoRI (10 U/μL) with sterile distilled water.

DNA Hind III assay buffer (10 × Hind III (10 U/μL) EcoR I (10 U/μL) with sterile distilled water (Table 1.1).

Note: For double digestion, Hind III buffer is used in which both EcoRI and Hind III will work because of the low salt concentration.

1. Mix the added reagents gently and centrifuge briefly.
2. Keep the tubes at 37°C for 4 h.
3. Stop the enzyme reaction by adding 1 μL of 0.5 M EDTA.
4. Load 5 μL of the samples in a 1% agarose gel and store the remaining samples in –20°C for future use.

Precautions

1. High-purity chemicals and water should be used.
2. Do not pipette any solutions by mouth. Wash your hands with hot water and soap before eating or drinking.
3. Use gloves.

ENZYME LINKED IMMUNOSORBENT ASSAY (ELISA)

Enzyme linked immunosorbent assay (ELISA) is an immunological technique used to detect proteins and is based on the ability of low-molecular-weight antibodies to couple with enzymes/secondary antibodies and conjugate with an enzyme (e.g., alkaline phosphatase), which has a visual detection by reaction with an appropriate substrate (e.g., PNPP).

Using this principle, ELISA was initially developed in 1971 independently by 2 groups (Engvall and Perlmann; Weeman and Schuurs).

At least 3 different ELISA techniques are available. In all 3, a polystyrene or polyvinylchloride microtiter plate is used, which has wells to provide for a solid phase for the immune reaction.

DAS-ELISA

In the double antibody sandwich (DAS) ELISA, the wells are coated with antibody (or immunoglobulins, Ig) prepared from antisera. The test sample is then added to allow trapping of antigen by coated antibody in the well. This is followed by the addition of enzyme-labelled Ig, which will attach to the trapped antigen. A substrate for the enzyme is now added for producing a color reaction. Peroxidase (OD 405) or alkaline phosphatase (OD 405) is used, which allows quantitative measurement through a spectrophotometric device (ELISA reader with automated photometer).

DAC-ELISA

In this direct antigen coating (DAC) ELISA, plant extracts prepared in the carbonate buffer are applied directly to the wells. In the second step, diluted unfractionated antiserum is added. Ig attached to antigen is detected by the addition of enzyme-Ig conjugates.

PAC-ELISA

This protein-A coating (PAC) ELISA technique involves 4 steps. The wells are coated with high-protein dilutions of processed antisera and then the test samples and, finally antisera (cross absorbed with healthy plant components, to minimize nonspecific reactions), are added.

Materials Required

Chemicals

- Na_2CO_3
- $NaHCO_3$
- NaCl
- Na_2HPO_4
- KH_2PO_4
- KCl
- Tween-20
- Bovine serum albumin
- p-nitrophenyl phosphate disodium salt (PNPP)
- NaOH

Pre-Lab Preparation

Procedure for DAC-ELISA (Transgenic Rice)

Solutions

Coating buffer

Na_2CO_3	1.59 g
$NaHCO_3$	2.93 g

Add distilled water to make 1 L.

PBS (phosphate buffer saline, pH 7.4)

	1×	5×
NaCl	8 g	40 g
Na_2HPO_4	1.5 g	7.5 g
KH_2PO_4	0.2 g	Ig
KCl	0.2 g	Ig

Add distilled water to make 1 L.

ELISA-wash buffer (PBS-T)
Add 0.5 mL Tween-20 to 1 L PBS.

Blocking solution
Add 5 g of bovine serum albumin (BSA) to 1 L PBS-T.

Protein Isolation from Rice Plants

1. With an autoclaved pestle and mortar, take fresh leaves (2–3 g) from transgenic and nontransgenic rice plants and grind with liquid nitrogen into a fine powder.
2. Allowed the powder to thaw and then add 3–4 mL of coating buffer.
3. Mix the powdered material thoroughly to allow better extraction of protein in the coating buffer and leave the solution for some time to allow the particulate material to settle.
4. Suck (by cut tips) the supernatant and dispense in Eppendorf tubes.
5. Centrifuge the tubes at maximum speed (15,000 rpm) for 15 min and use the supernatant as a crude protein sample.

DAC-ELISA

1. Coat the polystyrene ELISA plates with 200 μL (~150 μg) of crude plant protein, diluted in coating buffer. Make 2 replicates of each sample (crude protein). Use coating buffer as blanks, nontransformed plants as negative controls, and pure Bt protein (cry 1Ac) as the positive control.
2. Allow the samples to stand for 1 h at 37°C; coated plates can be stored for at least 2 wk at 4°C.
3. Empty the plates with a quick jerk, so as not to cross contaminate the wells.
4. Wash the plates 3 times with 200 μL of ELISA wash buffer (PBS+ 0.05% Tween-20).
5. After washing, add 200 μL of blocking solution (PBS+5% BSA) to each well. Incubate the plates at 37°C for 2 h or at 4°C overnight.
6. Again, wash the plates 3 times, as above.
7. Add 200 μL of the primary antibody (1:1000 dilution with blocking solution) to the wells, and incubate the plates at 37°C for 1 hr.
8. Wash the plates 3 times with ELISA wash buffer, emptying the plate each time with a quick jerk.
9. Add 200 μL of the secondary antibody enzyme conjugate (Goat-anti-rabbit IgG-AP) (1:10,000 dilution in PBS containing 5% BSA) to each well.

10. After incubation of plates for 2 hr at 37°C, wash the plates 3 times with wash buffer.
11. Add 200 μL of substrate, PNPP (p-nitrophenyl phosphate disodium salt) tablets, and incubate the plates at 37°C for 1 h.
12. Add 50 μL of stopping solution (1 M NaOH) to each well to stop the reaction of the substrate, and read the plates at 405 nm using the ELISA reader (preferably within 0.5-1h of the addition of the stopping solution).

Precautions

1. High-purity chemicals and water should be used.
2. Do not pipette any solutions by mouth. Wash your hands with hot water and soap before eating or drinking.
3. Use gloves.

ELECTROPHORETIC SEPARATION OF PLASMA PROTEIN

In this experiment, proteins obtained from human plasma by salt precipitation are separated by electrophoresis.

Materials Required

Chemicals

- 1% alcohlic solution of bromophenol blue
- Sodium nitrite solution 12%
- Normal saline solution (0.9% NaCl)
- Barbitone buffer 0.07 M (pH 6.8)
- Acetic acid solution

Glassware

- Beaker
- Measuring cylinder
- Reagent bottle
- Conical flask

Equipment

- Electrophoretic unit
- Centrifuge
- Microwave oven
- Power pack

Method

Fractionation of Human Plasma

1. Mix 9.5 mL sodium nitrite solution in 0.5 mL collected human plasma.
2. Allow to stand at room temperature for 10–15 min.
3. Collect the precipitate by centrifugation at 3000 g for 10 min and store. At this concentration of salt, fibrinogen is precipitated.

4. Decant the supernatant, measure the volume and add to it a sufficient amount of sodium nitrite so that the final concentration is 15%. At this stage, the α-globulins precipitate out.
5. Collect the precipitate by centrifugation as before and store.
6. The γ-globulins in the supernatant are precipitated out by raising the salt concentration up to 20%, and the precipitate is collected. Save the supernatant also.
7. Dissolve the precipitates in 0.5 mL saline (0.9% NaCl).

Electrophoresis of Samples

1. Fill the tanks with barbitone buffer (0.07 M, pH 6.8) to equal levels.
2. Moisten a filter paper strip (Whatman 1 mm) with the buffer. Drain off the excess fluid by slightly folding in between blotting papers, and lay it on a solid support (a glass plate) between the 2 tanks; the 2 ends should dip in the buffer solution.
3. Alternately, paper wicks can be used to bring about contact between the buffer and the paper.
4. Carefully streak 10 μL of unfractionated plasma across the breadth, midway between the tanks.
5. Cover the set-up, turn on the power supply, and adjust the voltage to 36 V (4 V/cm).
6. Leave the set up undisturbed for about 6 h, and, at the end of the period, turn off the power supply.
7. Remove the strip and dry by placing in an oven for a few min.
8. Repeat for various fractions obtained from plasma.
9. In some instruments, it is not possible to use 3 or 4 strips at the same time; in that case, the voltage has to be increased correspondingly.

Detection of Protein Bands

1. Soak the dry strips in a 1% alcoholic solution of bromophenol blue to stain the protein.
2. After soaking of 15 min, remove the strips and soak in another tray containing 0.5% acetic acid solution. The acid treatment removes the excess dye, which remains unabsorbed to the protein.
3. Continue the acid washing until a clear background is obtained with blue bands of proteins.
4. Give a quick wash with water to remove acid and dry the strips in air.
5. Compare the bands obtained with the whole plasma and those from a fractionated samples and comment on the results.

SODIUM DODECYL SULPHATE-POLYACRYLAMIDE GEL ELECTROPHORESIS

SDS-gel electrophoresis is widely used for the analysis and characterization of protein samples. It is useful in molecular weight determination relative to estimation of purity and amino acid sequencing of proteins.

Materials Required

Chemicals

Monomer solution

Acrylamide	97.5 g
Bis	2.5 g
Distilled water	q.s. to 200 mL; store at 4°C in the dark

Running gel buffer

Tris	36.6 g
Distilled water	q.s. to 200 mL; adjust to pH 8.8 with HCl

Stacking gel buffer

Tris	6.0 g
Distilled water	q.s. to 200 mL adjust to pH 6.8 with HCl

10% SDS

SDS	50 g
Distilled water	q.s. to 500 mL

Initiator

Ammonium per sulphate	0.75 g
Distilled water	q.s. to 50 mL

Running gel overlay

Tris	25 mL
SDS	1.0 mL solution
Distilled water	q.s. to 200 mL

2 × Treatment buffer

Tris	2.5 mL solution
SDS Solution	4.0 mL
Glycerol	2.0 mL
2-mercaptoethanol	1.0 mL
Distilled water	10 mL

Tank buffer

Tris	12 g
Glycine	57.6 g
SDS	40 mL solution
Distilled water	q.s. to 4.0 L

Stain buffer

Coomassie blue	2.0 g
Distilled water	q.s. to 200 mL
Stir and filter	

Glassware and Other

- Beakers
- Microtips
- Measuring cylinder
- Conical flask

Equipment

- Pipette
- Electrophoretic unit

Pre-Lab Preparation

Preparation of the Separating Gel

1. In a sidearm vacuum flask, mix 40 mL of separating gel solution witih the help of a magnetic stirrer.
2. Apply a vacuum to the flask for several min.
3. Add the TEMED and ammonium persulphate.
4. Swirl the flask to mix.
5. Pipette the solution into the sandwich to about 4.0 cm from the top.
6. With the help of a smooth running syringe, layer the slab with water.
7. Formation of a very sharp water gel interface will indicate the completeness of gelling.
8. Remove the water layer by tilting casting stand.
9. Rinse the surface of the gel once with distilled water.
10. Add about 1.0 mL of running gel overlay solution.
11. Allow the gel to set for several h.
12. Stir by using a small magnetic stirrer.

S.No.	Composition of SDS polyacrylamide separating gels				
1.	Acrylamide stock (%)	4.0	6.0	10.0	12.0
2.	Running gel buffer (mL)	15.0	15.0	15.0	15.0
3.	Water (mL)	19.7	17.7	15.7	11.7
4.	10% SDS (mL)	0.4	0.4	0.4	0.4
5.	1.5% ammonium persulphate (mL)	0.9	0.9	0.9	0.9
6.	TEMED (mL)	0.02	0.02	0.02	0.02

Preparation of the Stacking Gel

1. Pour the liquid onto the surface of the gel.
2. Degas the stacking gel solution as described before.
3. Add the ammonium persulphate and TEMED.

4. Gently swirl the flask to mix.
5. Add 1–2 mL of stacking gel solution to each sandwich to rinse the surface of the gel.
6. Rock the casting stand and pour off the liquid.
7. Fill each sandwich with stacking solution.
8. Insert a comb into each sandwich. Take care not to trap any bubbles below the teeth of the combs.
9. Allow the gel to set for at least 1/2 h.
10. Combine equal parts of protein sample and treatment buffer in a test tube.
11. Put the tube in a boiling water bath for 90 s.
12. Remove the sample and put it on ice until ready to use.
13. This treated sample can be put in the freezer for future runs.

S.No.	Composition of stacking gel		
1.	Final acrylamide concentration (%)	3.0	6.0
2.	Acrylamide stock (mL)	0.6	1.0
3.	Stacking gel buffer (mL)	1.25	1.25
4.	Water (mL)	7.55	7.15
5.	10% SDS (mL)	0.1	0.1
6.	1.5% ammonium persulphate (mL)	0.5	0.5
7.	TEMED (mL)	0.005	0.005

Loading and Running the Gels

1. Slowly remove the combs from the gels. Be careful to pull the comb straight up to avoid disturbing the well dividers.
2. Rinse each well with distilled water. Carefully invert the casting stand to drain the wells.
3. Fill each well with tank buffer.
4. Using a Hamilton syringe, under layer the sample in each well.
5. Add a spinner to the lower buffer chamber.
6. Place the chamber on a magnetic stirrer.
7. When the lower bubble is circulated, the temperature of the buffer remains uniform. This is important, because uneven heating distorts the banding pattern of the gel.
8. Put a drop of 0.1% phenol red in the upper buffer chamber.
9. Set the power supply to constant voltage.
10. Turn the power on and adjust the voltage so that proteins stack better.
11. Run the separating gel at 100–150 V.
12. When the dye reaches the bottom, turn the power supply off and disconnect the power cables.

Staining and Destaining the Gels

Composition of staining solution

Coomassie blue	62.5 mL stain stock
Methanol	250 mL
Acetic Acid	50 mL

Composition of destaining solution

Acetic acid	700 mL
Methanol	500 mL
Water	10 liter

Method

1. Disassemble the sandwiches.
2. Put the gel into stain.
3. Gently shake the gels for 4–8 h.
4. Remove the gels and put them in destaining solution.
5. Shake for 1 h.
6. Use destaining solution for further destaining.

PRODUCTION OF ANTIBODIES FROM LABORATORY ANIMALS

Primary antigenic stimulation of an immunologically mature animal is followed, after an induction period of several days, by slight to moderate liberation of antibody into the circulating blood; the antibody titre is then decrease rapidly. Secondary antigenic stimulation after a suitable interval promptly incites a steep (logarithmic) rise in antibody titer to a higher level, which is maintained for a longer duration than that of a primary response. A series of injections of antigen induce a stepwise increase of antibodies in the circulating blood, but eventually a maximum is reached beyond which further injection produces no increase in antibody titer.

Antibodies are formed in cells of the lymphoid series, which appear to develop into plasma cells when subjected to suitable antigenic stimulation. *In vivo* as well as *in vitro* experiments indicate that antibodies are synthesized *de novo* from free amino acids rather than from more complex precursors. The antibody-producing cells are situated in various organs, notably the spleen and lymph nodes. The former is particularly active following intravenous or intraperitoneal injection of antigen, the latter after subcutaneous or other local injection. Antibody produced in a given animal may vary in their serological properties. These variations may be attributed to the participation of different organs in antibody formation and to the multiplicity of determinant sites against which antibodies are formed. There are 3 kinds of antibodies: cell bound, which formed first and are responsible for delayed hypersensitivity; a circulating type with marked affinity for tissue cells; and classic circulating antibody.

There are basically 2 hypotheses that have been proposed as mechanisms of antibody formation. According to the template hypothesis, the antigen serves as a pattern against which antibody globulin is synthesized or molded to produce a molecule-processing area with a reverse structural image of determinant sites of the antigen, or else it modifies the DNA or ribonucleoprotein of globin-producing cells so that they thereafter form globulin. The template hypothesis accounts satisfactorily for specificity but does not explain the difference between primary and secondary response and immunologic tolerance. The selective hypothesis postulates that certain body cells are selected to produce antibody by virtue of an inherent or acquired specific reactivity with the antigen and are stimulated to do so by contact with antigen. The selective hypothesss accounts for most of the observed features of the nature of antibodies and antibody production.

Preparation of Immunizing Material

Protein Solution

Solutions containing about 1% protein are satisfactory for inoculation into laboratory animals. Serum diluted 1:7 with saline contains about this concentration of protein. Dissolve egg albumin and other dried proteins by adding a little saline and mixing with a glass rod to make a thick paste; dilute the paste gradually. The solution should be stirred instead of shaken, which creates troublesome foam. Remove undissolved particles by centrifugation or by filtration through paper. Add a preservative if desired.

Use of Adjuvants

Adjuvants are used to reduce the number of injections required and often increase the antibody response. Aluminum hydroxide and paraffin oil are sometimes incorporated into material used for animals inoculation for this purpose.

Alum Precipitation

Prepare 50 mL of bacterin or other immunizing agent and its preservative in double strength. Add 2.33 mL of 10% $AlCl_3$ and sufficient 20% NaOH to bring the reaction to pH 7.0; dilute to 100 mL with saline.

Oil in Water

Prepare the immunizing material and preservative in double strength. Sterilize good quality light paraffin oil and Arlacel A by autoclaving. Mix 8.5 volume of the paraffin oil with 1.5 volume of Arlacel A, using a Waring blender or by aspirating it repeatedly into syringe without a needle. Add 10 volumes of antigen and mix as before. Test the emulsion by allowing a drop to fall from an applicator stick onto the surface of water. If the drop remains perfectly formed and does not spread over the surface, the emulsion is ready for use.

An additional adjuvant effect is sometime provided by incorporation of killed acid-fast bacteria. Dried *Mycobacterium tuberculosis* or *M. bytyricum* that have been killed by heat are added to the paraffin oil at a concentration of 10 mg/100 mL and thoroughly mixed by grinding in a mortar. Aracle A and antigen are then added as described previously.

Inoculation and Bleeding of Animals

Rabbit

For antisera production, usually young adult rabbits of 2–4 kg body weight are used. Inoculations may be done by intravenous, intraperitoneal, or subcutaneous routes. The intravenous route is commonly used for this purpose. For the production of antibacterial sera in rabbits, the schedule of inoculation of killed suspensions of gram-negative rod bacteria consists of following intravenous injections detailed in Table 1.2.

TABLE 1.2
Intravenous Injections for the Production of Antibacterial Sera in Rabbits

S.No.	Day	Inoculum Size (mL)
1	1	0.1
2	4	0.3
3	8	0.5
4	11	1.0
5	15	2.0

Serum of high titer value is usually secured 3–5 d after the last injection. For living bacteria, closely spaced intravenous injections are satisfactory. Stock saline suspension having any visible change in consistency are tested employing the antigen agglutination test. Fresh stock suspensions are prepared as indicated in Table 1.3 and refrigerated when not in use. An anti-erythrocyte effect is produced by a series of twice weekly intravenous injections of 1 mL of 50% suspension of red blood cells. Animal serum is then titrated 3 to 4 times after the fourth injection, and if the titer is not satisfactory, additional injections are conducted.

Another injection routine consists of 4 daily intravenous injections of 1 mL of 10% red blood cells per kg body weight followed by 6 injections every second day. Peak titer is usually found about 2 wk after the first injection and is often maintained for several days.

Protein solution, 1%, is injected intravenously once or twice a wk. The number of injections is determined by the results of trial titrations of the animal's serum. After the first 2 wk, injections are given following a rest period of a wk or more by intraperitoneal route to reduce the likelihood of anaphylactic shock. One or 2 0.5 mL subcutaneous injections of antigens emulsified in paraffin oil are usually followed, which produce antisera of high titer. A trial titration is performed 3–4 wk after the first injection, and if the antibody titer is not sufficiently high, a second injection is given (Table 1.3).

TABLE 1.3
Preparation of Stock Solution

S.No.	Day	Inoculum Size	Dilution
1.	1	0.1	Fresh stock suspension 1:1000
2.	2	0.2	Suspension 1:100
3.	3	0.3	Suspension 1:100
4.	4	0.1	Fresh stock solution undiluted
5.	5	0.2	Suspension undiluted
6.	6	0.3	Suspension undiluted
7.	11	0.5	Fresh stock solution undiluted
8.	16	0.5	Trial titration, if titration is satisfactory, bleed from the heat otherwise, inject 0.5 mL fresh undiluted stock suspension and titrate after an additional 5 days.

Intravenous Injection

The marginal ear vein of the rabbit, located at the outer edge of the dorsal side of the ear, is readily accessible. One ear is reserved for injection and the other for bleeding. The first injection of a series is always made as near the tip of the ear as possible, succeeding injections being made closer toward the animal's head, so that scar tissue will not prevent injected material from entering the circulation.

- Hold the animal on the lap facing the operator and shave the skin over the vein. Rub vigorously with 70% ethanol. Hold the ear with the left hand so that the middle finger supports the area to be injected. Insert the inoculating needle, bevel up, in the direction of blood flow through the skin and into the vein at a very acute angel to the vein so that it does not pass completely through.
- Inject the inoculum slowly. When the injection is made correctly, the inoculum can be seen passing toward the heart as it partially replaces the blood. If the needle is not within the vein, the antigen will produce a blanched, raised area in the neighboring tissue. Gentle massage is used to force the material out of the needle puncture, and injection should be made at another site.
- Firmly apply cotton moistened with alcohol after the inoculation is completed and withdraw the needle.

Intraperitoneal Injection

- Clip the hair in the median abdominal line and disinfect with alcohol. Pinch up a fold of skin and peritoneum between the thumb and forefinger, and insert the needle into the ridge of the skin and through the fold of peritoneum; release the peritoneum and skin to make a successful inoculation.

Subcutaneous Injection

- Clip the hair and shave the hair on the side of the back and disinfect with alcohol. Pinch up a fold of skin between the thumb and forefinger, and insert the needle into the ridge of the skin; release the skin and make the injection.
- Wash the area with alcohol.

Bleeding from the Ear Vein

The marginal vein of the ear is used to collect a small amount of blood. The first bleeding of the series is made near the base of the ear and succeeding punctures are made toward the distal end from the head.

- Hold the animal on the lap facing away from the operator and rub the ear vigorously to increase the blood circulation. A drop of xylol can also be placed on the tip of the ear to produce mild inflammation. Xylol must be removed by washing 3 times with alcohol before the animal is returned to the cage.
- Shave the area and place an artery clamp over the vein proximal to the site to be punctured.
- Hold the ear so that middle finger of the left hand supports the site of puncture and make a short cut with a sharp razor blade. Collect the blood

in a centrifuge tube while keeping the other hand at the site of puncture; remove the tube and move the artery clamp distal to the puncture and hold dry cotton firmly over the puncture until bleeding stops.
- Remove the clot by centrifugation and collect the serum carefully with a pipette.

Bleeding from the Heart

A large quantity of blood can be collected from the heart or jugular vein. Cardiac bleeding is usually simpler and can be repeatedly performed on the same animal if proper technique is used.

- Clip the area over the sternum and disinfect with alcohol and tincture of iodine.
- Insert the needle between 2 ribs in the area of maximum pulsation, usually about midway of the sternum and slightly to the left and advancing the needle in a straight line toward the right shoulder. Blood will appear in the needle as soon as the needle punctures the heart. 50 mL may safely be collected from an average rabbit.
- If finding the heart is difficult, withdraw the needle and reinsert in another direction. Twisting the needle within the pericardial cavity can tear the heart and cause immediate death.
- Collect the blood in a centrifuge tube, and remove the clot by centrifuge. Collect the serum with a pipette.
- Serum is stored in glass bottles in the refrigerator and can be preserved with methiolate or phenol.

Mice

- Intraperitoneal injection is usually employed with mice. Hold the mouse by its tail with the right hand and grasp it firmly between the ear with the thumb and forefinger of the left hand and turn it over.
- Hold the tail with the help of the little finger of the left hand. Hold the animal's head down so that the intestine falls forward.
- Make the injection in the posterior region of the abdomen.

PURIFICATION OF ENZYMES FOR IMMOBILIZATION

The solubility of protein depends on, among other things, the salt concentration in the solution. At low concentrations, the presence of salt stabilizes the various charged groups on a protein molecule, thus attracting protein into the solution and enhancing the solubility of the protein. Further, as the salt concentration is increased, a point of maximum protein solubility is usually reached. Further increase in the salt concentration implies that there is less and less water available to solubilize protein. Finally, protein starts to precipitate when there are not sufficient water molecules to interact with protein molecules. This phenomenon of protein precipitation in the presence of excess salt is known as salting out. Many types of salts have been employed to effect protein separation and purification through salting out. Of these salts, ammonium sulphate has been the most widely used chemical because it has high solubility and is relatively inexpensive.

As enzymes are proteins, enzyme purification can also be carried out by using the same principle; however, some attention must be paid to the consideration of permanent loss of activity due to denaturation under adverse conditions. There are 2 major salting out methods:

By increasing salt concentrations: By slow addition of a saturated salt solution or powdered salt crystals to the protein mixture, the salt concentration of the mixture is raised. The precipitated protein is collected and categorized according to the concentration of the salt solution at which it is formed. This partial collection of the separated product is called fractionation.

By decreasing salt concentrations: Proteins are first precipitated with a concentrated salt solution. Then a series of cold (near 0°C) ammonium sulphate solutions of decreasing concentrations are employed to extract selectively the protein components that are the most soluble at higher ammonium sulphate concentrations. The extracted protein is recrystallized and thus recovered by gradually warming the cold solution to room temperature. This method has the added advantages that the extraction media may be buffered or stabilizing agents may be added to retain the maximum enzyme activity. The efficiency of recovery typically ranges from 30% to 90%, depending on the protein.

Materials Required

- Protein solution (200 mL, 1% hemoglobin)
- Fungal α-amylase (1%)
- Protease (1%)
- Saturated $(NH_4)_2SO_4$ solution
- Add 750 g of ammonium sulphate to 1000 mL of water in a beaker or flask. Simply stir the solution at room temperature with a magnetic stirrer for 15 min or until saturation. Gently decant the clear supernatant solution after the undissolved solids settle on the bottom of the flask.

Equipment

- Test tube
- Graduated cylinder
- Pipettes
- Balance
- Centrifuge
- Filtration device

Method

Purification of Hemoglobin

1. Record the absorbance of the hemoglobin solution at the wavelength of 577 nm. This measurement is to be used in the calculation of the recovery of the protein.

2. Pipette 4 mL of the hemoglobin solution into a test tube.
3. While stirring, add the saturated ammonium sulphate solution, dropwise, to the protein solution using a burette until precipitates start to form. It is critical to avoid spatial nonuniformity in the salt concentration during the addition of the salt solution. Localized concentration hot spots will prematurely initiate the precipitation of other proteins and inadvertently affect the purity of the protein crystals.
4. Record the volume of the saturated ammonium sulphate solution needed to cause precipitation. One should keep in mind that protein precipitation is not instantaneous and may require 15–20 min to equilibrate.
5. Centrifuge the mixture at 10,000 g for 15 min.
6. Collect the precipitate by carefully discarding as much supernatant as possible.
7. Reconstitute the pelleted hemoglobin by resuspending the precipitate and bringing the volume up to 4 mL with water. For effective recovery of protein, water should be added in small aliquots.
8. Measure the absorbance of the reconstituted hemoglobin solution with a spectrophotometer.

Purification of Fungal α-Amylase

1. Instead of hemoglobin solution, use 4 mL of 1% fungal α-amylase.
2. Salt-out the enzyme with a saturated ammonium sulphate solution as described previously.
3. Record the volume of the saturated salt solution added.
4. Collect the protein precipitate. Avoid the dilution of the enzyme solution too much. Filtration through a syringe filter unit may be conveniently employed if the crystals are not too small and if the collected crystals can be easily washed off the filter paper. Usually, the quantitative analysis of the activity of the enzyme collected with the filtration method is not as accurate as the centrifugation method.
5. After resuspending the proteins, analyze the enzyme activity.

Purification of Protease

- Instead of hemoglobin solution, use 4 mL of saturated protease. Note that protease may not be totally soluble, and supernatant can be obtained by centrifugation.

Notes: To assure maximum yield and to avoid unnecessary denaturation of the enzymes, most of the protein purification work is usually carried out at low temperature, that is, between 0 and 40°C. However, it is simply far more convenient to work in a regular laboratory room as opposed to a cold room.

The recovery of protein can have significant economic implications. Because a fixed fraction of the original protein stays soluble in the solution, the recovery of protein is often not near 100% and indicates that there may be problems associated with the assay method.

Enzyme Immobilization in Polyacrylamide Gel

Three easily available media for enzyme immobilization are polyacrylamide, calcium alginate, and gelatin. All these gels are easy to form and require a simple set of equipment. To immobilize the enzyme, enzymes are well mixed with monomers/polymers and crosslinking agents in a solution. The solution is then exposed to polymerization promoters to start the process of gel formation. The solution is poured into a mold to achieve the desired shape. A gel block may be cut into smaller cubes to increase the surface area. To form spherical beads, the unpolymerized solution is forced through a set of nozzles. In such a set preparation, adjusting the back pressure would easily control the size. The resulting beads may be further hardened to enhance structural integrity. In summary, the efficiency of an immobilization process can be measured with the following parameters. Most important of all is the percentage of the enzyme initially retained in gel matrices. Second, the enzyme activity must be intact during and after immobilization, and thirdly, the enzymes must be released by diffusing back into the substrate solution at a later time. Another important parameter is the uniformity of the enzyme dispersion in the gel matrix.

Polyacrylamide is the most widely used matrix for entrapping enzymes. It has the advantage that it is nonionic. The consequence is that the properties of the enzymes are only minimally modified in the presence of the gel matrix. At the same time, the diffusion of the charged substrate and products is not affected. However, dimethylaminopropionitrile, the polymerization initiator, is highly toxic and must be handled with great care. The requirement to purge the monomer solution with nitrogen is also troublesome, although not totally crippling. This technique is based on the polymerization of acrylamide with N,N′-methylenebisacrylamide as the crosslinking agent. The degree of crosslinking, thus, can be partly controlled by adjusting the ratio of acrylamide to Bis used.

Equipment

- Beakers
- Pipettes
- Balance
- Graduated cylinder
- Syringe
- Needle

Materials Required

Chemicals

Buffered Monomer Solution

- 0.1 mM EDTA
- 0.1 M Tris-HCl
- % N,N′-methylenebisacrylamide
- 20% acrylamide
- Adjust the pH to 7.0

Washing Solution

- 0.5 M NaCl
- 0.1 mM EDTA
- 0.1 M Tris-HCl
- Adjust the pH to 7.0

Dimethylaminopropionitrile (polymerization initiator)
Potassium persulphate solution, 1% (polymerization catalyst)
Nitrogen gas cylinder
Enzyme

Procedure

1. Prepare buffered monomer solution by adding 1.1 g of Bis and 20 g of acrylamide to 100 mL of buffered solution (pH 7.0) of 0.1 mM EDTA and 0.1 M Tris HCl in a beaker. (The pH of the buffer should be adjusted to match the optimum value of the enzyme to be entrapped.)
2. To 10 mL of the buffered monomer solution of the above step, add enzyme powders (approximately 0.1 mL of 7.5 fungal amylase or an equivalent concentrated enzyme solution) and mix.
3. For 20 min, purge the dissolved oxygen in the solution that can interfere with the polymerization process with nitrogen. This step is critical in achieving a high degree of crosslinking. Add 0.1 mL of dimethylaminopropionitrile and mix. Add 1.0 mL of freshly prepared 10% potassium persulphate solution to initiate polymerization.
4. Now is the time to pour the solution into a mold if one does not desire the gel to form in the original beaker. Leave the solution undisturbed and approximately 10–30 min later, the gel will form. Hardening can be accelerated by using more dimethylaminopropionitrile.
5. Cut the resulting gel into small cubes of approximately 3 mm per side. Alternatively, if smaller pieces are desired, the gel can be forced through a syringe fitted with a fine needle.
6. Gently wash the free enzyme off the gel surface in 10 mL of the washing solution. Repeat the washing processes 2 additional times.

Note: The above methods of enzyme immobilization by gel entrapment can be directly applied to live cells with minor modifications. For example, dimethylaminopropionitrile used in forming the polyacrylamide gel may not be employed because of its toxicity to viable cells. The monomers of acrylamide are also somewhat toxic to cells. On the other hand, cells can be immobilized with much less degree of crosslinking due to their much larger size.

Enzyme Immobilization in Alginate Gel

Alginate, commercially available as alginic acid sodium salt, commonly called sodium alginate, is a linear polysaccharide normally isolated from many strains of marine brown seaweed and algae, thus the name alginate. The copolymer consists of 2uronic acids: d-mannuronic acid (M) and L-guluronic acid (G). Because it is the skeletal component of the algae, it has the property of being strong and yet flexible. Alginic acid can be either water soluble or insoluble depending on the type of associated salt. The salts of sodium, other alkali metals, and ammonia are soluble, whereas the salts of polyvalent cations, for example, calcium, are water insoluble, with the exception of magnesium. Polyvalent cations bind to the polymer whenever there are 2 neighboring guluronic acid residues. Thus, polyvalent cations are responsible for the crosslinking of both different polymer molecules and different parts of the same polymer chain. The process of gelation, simply the exchange of calcium ions for sodium ions, is carried out under relatively mild conditions. Because the method is based on the alginate, the molecular permeability does not depend on the immobilization conditions. Rather, the pore size is controlled by the choice of the starting material.

$$2\text{Na (alginate)} + \text{Ca}^{++} \rightarrow \text{Ca (alginate)}_2 + 2\text{Na}^+$$

The ionically linked gel structure is thermostable over the range of 0°C–100°C; therefore, heating will not liquefy the gel. However, the gel can be easily redissolved by immersing the alginate gel in a solution containing a high concentration of sodium, potassium, or magnesium. Maintaining a ratio of sodium:calcium smaller or equal to 25:1 will help avoid gel destabilization. In fact, it is recommended to include 3 mM calcium ions in the substrate medium. On the other hand, citrate or phosphate pH buffers cannot be effectively used without destabilizing the alginate gel. Alginate is currently widely used in food, pharmaceutical, textile, and paper products. The properties of alginate utilized in these products are thickening, stabilizing, gel-forming, and film-forming. Alginate polymers isolated from different alginate sources vary in properties. Different algae, or for that matter different parts of the same algae, yield alginate of different monomer composition and arrangement. There may be sections of homopolymeric blocks of only 1 type of monomer (-M-M-M-) (-G-G-G-), or there may be sections of alternating monomers (-M-G-M-G-M-). Different types of alginate are selected for each application on the basis of the molecular weight and the relative composition of manuronic and guluronic acids. For example, the thickening function depends mainly on the molecular weight of the polymer; whereas, gelation (affinity for cation) is closely related to the guluronic acid content. Thus, high guluronic acid content results in a strong gel.

Materials Required

Chemicals

- Sodium alginate
- Sodium salt
- Calcium chloride ($CaCl_2$)
- Enzyme

Equipment

- Beakers
- Graduate cylinder

- Balance
- Pipettes
- Syringe

Method

- Dissolve 30 g of sodium alginate in 1 L to make a 3% w/v solution.
- Mix approximately 0.015 g of enzyme with 10 mL of 3% w/v, depending on the desired hardness.
- The beads are formed by dripping the polymer solution from a height of approximately 20 cm into an excess (100 mL) of stirred 0.2 M $CaCl_2$ solution with a syringe and a needle at room temperature.
- The bead size can be controlled by pump pressure and the needle gauge. A typical hypodermic needle produces beads of 0.5–2 mm in diameter. Other shapes can be obtained by using a mold with a wall that is permeable to calcium ions. Leave the beads in the calcium solution to cure for 0.5–3 h.

Notes: Sodium alginate solution is best prepared by addition the powder to agitated water, rather than vice versa, to avoid the formation of clumps. Prolonged stirring may be necessary to achieve the complete dissolution of sodium alginate. After sodium alginate is completely dissolved, leave the solution undisturbed for 30 minutes to eliminate the air bubbles that can later be entrapped and cause the beads to float.

2 Regulative Techniques

RESTRICTION FRAGMENT LENGTH POLYMORPHISM (RFLP)

This phenomenon was first described for a mutant strain of adenovirus by Grodzicker et al. (1974).

Experimental Outline

RFLP is environmentally independent. RFLP technically includes the following steps:

1. DNA isolation.
2. Production of DNA fragments using restriction enzyme(s).
3. The enzyme is used in a range of 4 units/mg of DNA, and the reaction can be proceeded for about 8 h (or preferably) longer. Digestion of the DNA will be carried out at optimum conditions depending upon the enzyme as specified by the respective company.
4. Choice of the enzymes depends on ability to detect the polymorphism (e.g., EcoRI, EcoRV, Hind III etc.).
5. DNA fragment separation on agarose gel.
6. Transfer of DNA fragments onto nylon/nitrocellulose membrane.
7. Visualization of DNA fragments using labelled probes.
8. Analysis of the results. Figure 2.1 shows the steps of DNA analysis by RFLP.

Materials and Pre-Lab Preparations

TBE buffer	10×
Tris base	108 g
Boric acid	55 g
40% sucrose	4 g
0.25% xylene cyanol	25 mL
0.25% BPB	25 mg
5 L of loading/25 μL sample	(1:5 ratio)

Method

1. Isolate DNA (as per the method outlined in previous experiments).
2. Choose appropriate restriction enzyme.
3. Total volume of the reaction mixture is 25 mL.
4. Calculate DNA stock required for 10 mg DNA to be used.
5. 4 units of enzyme/mg of DNA is to be used.
6. All the components are added in the following order.

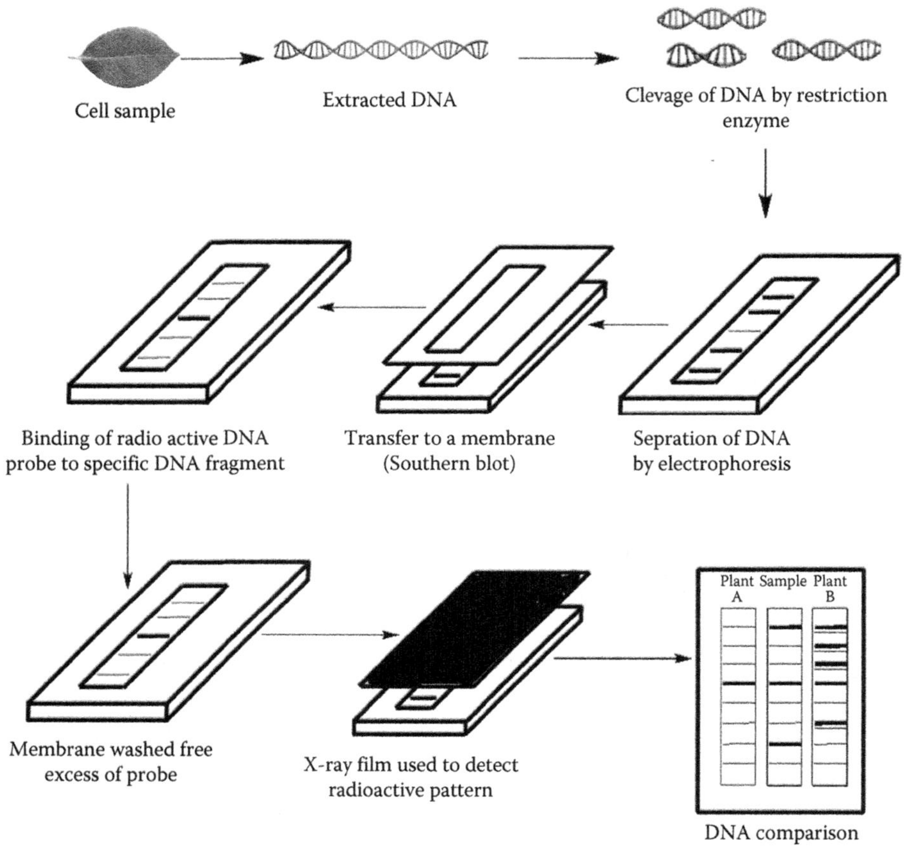

FIGURE 2.1 DNA analysis of RFLP.

7. Spermidine, to inhibit the ability of polysaccharides, which hinder the digestion:
   a. Water
   b. Buffer (2.5 μL)
   c. DNA
   d. Enzyme
8. Mix well by shaking and incubate at 37°C in an incubator.
9. Analyse the samples on 0.8% agarose gel.

DNA Fragment Separation on Agarose Gel

1. Seal the sides of the gel casting plate with cellophane tape.
2. Select a comb that fits into the grooves of the gel plate.
3. Add 0.8 g agarose for 100 mL TBE buffer.

4. Heat to dissolve the agarose and allow to cool.
5. Add 5 mL (stock of 10 mg/mL) of 0.5 × TBE buffer into the gel and remove the comb gently.
6. Remove the cellophane tape and transfer the gel plate to the electrophoresis unit.
7. Add a sufficient quantity of 0.5 × TBE buffer to cover the gel completely.
8. Restricted DNA is loaded into wells by mixing it with 1/10 volume of gel loading buffer (bromophenol blue). A molecular weight marker (e.g., Hind III digest) can be used.
9. Apply 30 V and allow to run until the bromophenol blue reaches 1 cm away from the end of the gel.
10. In a dark room (with red safety light), place the membrane in a film cassette with the DNA-containing surface facing upward, and place a sheet of X-ray film on top. Carefully close the cassette and keep at 70° for 2–3 days.
11. Develop the film in the dark room (with red safety light) for 3 min in X-ray developer; rinse briefly with deionized water. Light for 3 min in X-ray fixer. Rinse with deionized water for 5 min and let the film air-dry.
12. Cut the nitrocellulose/nylon membrane to the same size as the gel and wet it in 20 × SSC:2 Whatman paper (15 × 35 cm).

If using Hybord^{TM+} N^+ it can be transferred by 0.4 nM NaOH, and there is no need to denaturate and renaturate; after washing by 2 × SSC, membranes can be used immediately for hybridization.

Gels can be denaturated before transfer by 4% HCl for 20 min and neutralized by 0.4 M NaOH for 20 min.

Materials and Pre-Lab Preparations

1. Denaturation solution:
   1.5 M NaCl and 0.5 M NaOH
   20 g of NaOH/L
   87 g of NaCl/L
2. Neutralizing solution:
   1.5 M NaCl and 0.5 M Tris-HCl (pH 7.5)
   1 M Tris = 121:14 g in L
   0.5 M Tris = 60.57 g in 1 L
3. 20 × SSC:
   3 M NaCl
   0.3 M sodium citrate (pH 7.2)
   NaCl: 175.3 g in 1 L
   Sodium citrate: 83.23 g in 1 L

Hybridization

Temperature 65°C

Church buffer is used

$Na_2HPO_4{\cdot}2H_2O$	31.8 g/L
$NaH_2PO_4{\cdot}H_2O$	7.4 g/L
SDS	70 g/L
BSA	10.0 g/L
EDTA	370 mg/L

Washing 2 × SSC + 0.1% SDS at 65°C for 15 min.

Southern Blotting Procedure

1. After electrophoresis, incubate the gel for 15 min in 0.5 M HCl to depurinate the DNA (shaking).
2. Incubate the gel twice for 30 min (2 times in 20 × SSC).
3. Incubate the gel twice in neutralizing solution for 20 min.
4. Cut the nitrocellulose/nylon membrane to the same size as the gel and wet it in 20 × SSC. 2 Whatman papers for the bridge (15 × 35 cm).
5. The gel is put upside down on the first 3 Whatman papers using 2 glass plates.
6. Make sure that the tissues above the Whatman papers are dry.
7. Capillary blot to the membrane over input with 20 × SSC.
8. After blotting, the membrane is placed on UV light for the DNA face down. Then the membrane is shortly washed in 2 × SSC and placed in Saran Wrap (thus can be stored for a few days at 4°C).

Probe Labelling

Probe labelling is carried out to produce the radioactively labeled DNA. This is dependent on DNA polymerase 1 (from *E. coli*) to add nucleotide residues to the 3′ hydroxyl terminus (which can be created), when one strand of double-stranded DNA molecule is nicked. By replacing the pre-existing nucleotides with radioactive nucleotides, it is possible to prepare p^{32}-labeled DNA. The radioactively labelled DNA can be used to detect the presence of homologous sequences.

Method

1. Add 8 μL of sterile distilled water to approximately 25–50 ng of DNA.
2. Boil the probe for 5–10 min to denature the DNA and keep it on ice immediately.
3. Add 1 μL each of dGTP, dTTP, dATP and 5 mL of dCTP P^{32}.
4. Add 2 μL of buffer and add 1 μL of Klenow enzyme (enzyme should be added at the end).
5. Incubate at 37°C for 30 min.
6. Check the reaction for proper labelling with Geiger Muller (GM) counter.
7. Precipitate the DNA by adding 3 μL of sodium acetate and 100 mL of ethyl alcohol (absolute) and incubate at 70°C for 30 min.
8. Centrifuge and dissolve the precipitate in 100 μL of hybridization buffer (church buffer).
9. Boil the probe and add to the filter (sec for dGTP, dCTP, etc.).

Note: (1) Concentration of washing solution and time vary depending on if the probe used is homologous or heterologous. (2) After washing, the filter is exposed to autoradiography to observe the signals.

Autoradiography

Radioactive nucleic acids and proteins can be detected by autoradiography. Because it is more sensitive and gives higher resolution, it can be used extensively.

Materials

1. X-ray film developer

Water at 52°C	500 mL
Metol	1 g
Sodium sulphite	75 g
Hydroquinone	9 g
Sodium carbonate (monohydrate)	25 g
Potassium bromide	5 g

Dissolve in the order listed above and bring the volume up to 1 L with cold water.

2. Fixer

Sodium thiosulphate	80 g
Potassium metabisulphate	5 g

Bring the volume up to 250 mL with water and filter.

Method

1. Remove the film after exposing and develop it in developer for 3–5 min in dark.
2. Wash the film in stop bath for 1 min and fix the film in fixer for 5 min in dark.
3. Wash the film under running water for 30 min and dry it at room temperature.
4. Avoid reloading the moist or cold cassette. P^{32} samples are covered from contamination with penetrating radioactivity.

Results

Observe the polymorphic bands on the X-ray film and document the variations.

Precautions

1. Avoid pipetting with the mouth.
2. Place the contaminated and disposable items into plastic radioactive trash and mark as radioactive waste.
3. Check your hands for radiation after the completion of the work.

AMPLIFIED FRAGMENT LENGTH POLYMORPHISM

Amplified fragment length polymorphism (AFLP) technology is a technique for fingerprinting genomic DNA. DNA fingerprinting is used to visualize DNA polymorphisms between individuals. The fingerprints may be used as a tool for determining the identity of a specific DNA sample or to assess the relatedness among individual. Fingerprints are also used as the source for genetic makers to generate linkage maps and to identify molecular markers linked to phenotypic traits of agricultural importance.

AFLP technology is based on selective amplification of a subset of genomic restriction fragments using PCR. DNA is digested with restriction endonucleases, and double-stranded DNA adapters are ligated to the ends of the DNA fragment to generate template DNA for amplification. Thus, the sequence of adapters and the adjacent restriction site serve as primer binding sites for subsequent amplification of the restriction fragment by PCR. Selective nucleotides extending into the restriction fragment are added to the 3′ end of the PCR primers such that only a subset of the restriction fragments are recognized. Only restriction fragments in which the nucleotides flanking the restriction site match the selective nucleotide are amplified. The subset of amplified fragments is then analyzed by denaturing polyacrylamide gel electrophoresis to generate the fingerprint. When the DNA fingerprints of related samples are compared, common bands, as well as different bands, are observed. Polymorphisms detected in DNA fingerprints obtained by restriction cleavage can result from alterations in the DNA sequence, including mutation abolishing or creating a restriction site, and insertions, deletions, or inversion between 2 restriction sites. The DNA polymorphisms identified using AFLP are typically inherited in Mendelian fashion and may therefore be used for typing and mapping of genetic loci.

Summary of the AFLP Analysis System

The AFLP analysis system 1 has been designed for use with plants having genomes ranging in size from 5×10^8 to 6×10^9 bp including tomato, corn, soybean, cucumber, lettuce, barley cotton, oilseed rape, potato, sunflower, pepper, and brassica. The AFLP technique involves 3 major steps (Figure 2.2):

1. Restriction endonuclease digestion of the DNA and ligation of adapters.
2. Amplification of the restriction fragments.
3. Gel analysis of the amplified fragments.
4. Typically, 50–100 restriction fragments are coamplified in each AFLP reaction and detected by denaturing gel electrophoresis. This technique is, therefore, extremely powerful for the identification of DNA polymorphisms.

Restriction Endonucleases Digestion

To prepare an AFLP template, genomic DNA is isolated and digested with 2 restriction endonucleases simultaneously. This step generates the required substrate for ligation and subsequent amplification.

The restriction fragments for amplification are generated by 2 restriction endonucleases, EcoR1 and Mse1. EcoR1 has a 6-bp recognition site, and Mse1 has a 4-bp recognition site. When used together, these enzymes generate small DNA fragments that will amplify well and are in the optimal size range (<1 kb) for separation

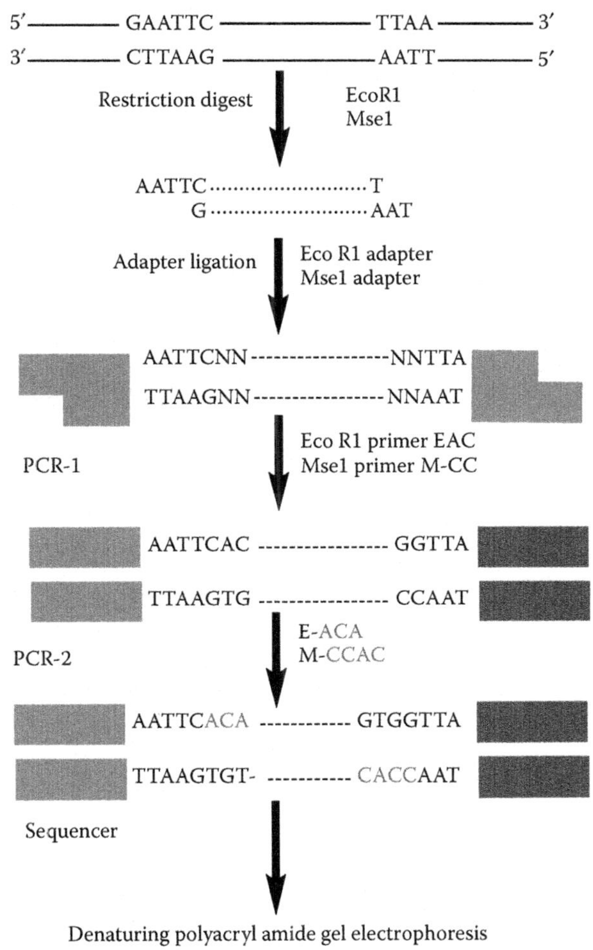

FIGURE 2.2 Amplified fragment length polymorphism (AFLP).

on denaturing polyacrylamide gels. Due to primer design and amplification strategy, theses EcoR1/Msel fragments are preferentially amplified (rather than EcoR1/EcoR1 or Mse1/Mse1 fragments).

Ligation of Adapter

Following heat inactivation of the restriction endonucleases, the genomic DNA fragments are ligated to EcoR1 and Mse1 adapters to generate template DNA for amplification. These common adapter sequences that flank variable genomic DNA sequences serve as primer binding sites on these restriction fragments. Using this strategy, it is possible to amplify many DNA fragments without having prior sequence knowledge.

Amplification Reaction

PCR is performed in 2 consecutive reactions. In the first reaction, called preamplification, genomic DNAs are amplified with AFLP primers, each having 1 selective nucleotide. The PCR products of the preamplification reaction are diluted and used as a template for the selective amplification using 2 AFLP primers, each containing 3 selective nucleotides. (The EcoR1 selective primer is 32 P or 33 P-labelled before amplification). This 2-step amplification strategy results in consistently cleaner and more reproducible fingerprints with the added benefit of generating enough template DNA for thousands of AFLP reactions.

The most important factor in determining the number of restriction fragments amplified in a single AFLP reaction is the number of selective nucleotide in the selective primers. The selective primers in the AFLP Analysis System 1 contain 3 selective nucleotides. In practice, using the AFLP Analysis System 1 with plants having genomes ranging in size from 5×10^8 to 6×10^9 bp, the number of fragments amplified per sample per primer pair averages 50, but ranges from as low as 1 up to 100, depending on the sequence context of the selective nucleotides and the complexity of the genome.

A second factor in determining the number of amplified fragments is the C and G composition of the selective nucleotide. In general, the more Cs and Gs as selective nucleotides in the amplification primers, the fewer DNA fragments amplified. Also, the smaller the genome being analyzed, the fewer fragments amplified and the simpler the fingerprint.

Method

Additional Materials Required

The following items are required for use with the AFLP Analysis System 1, but they are not included in the system:

- Taq DNA polymerase
- 1.5-mL microcentrifuge tubes
- 0.2–0.5-mL thin-walled microcentrifuge tubes (depending on thermal cycler)
- Automatic pipettes capable of dispensing 1–20 μL and 20–200 μL
- Autoclaved, aerosol-resistant tips for autotrophic pipettes
- Microcentrifuge capable of generating a relative centrifugal force of 14,000 × g
- 37°C, 70°C, 90°C water bath or programmable thermal cycler
- Denaturing polyacrylamide gel reagents and apparatus
- X-ray film
- Film cassette
- [33 P or 32 P] ATP (for radioactive detection)
- AFLP nonradioactive probe
- Silicone oil (if thermal cycler requires)

Pre-Lab Preparation

For best results, high quality, intact DNA must be used as starting material for AFLP. It is essential to prevent incomplete restriction digestion of genomic DNA, which may result in the incorrect identification of polymorphisms.

Restriction Digestion of Genomic DNA

Add the following to a 1.5-mL microcentrifuge tube:

Component	Control	Sample
i. 5× reaction buffer	5 μL	5 μL
ii. Tomato control DNA (100 ng/μL)	2.5 μL	–
iii. Sample DNA (250 ng in <18 μL)	–	<18 μL
iv. EcoR1/Mse1	2 μL	2 μL
v. Distilled water	15.5 μL	25 μL
Total volume	25 μL	25 μL

Mix gently and collect the reaction by brief centrifugation. Incubate the mixture 2 h at 37°C. Incubate the mixture for 15 min at 70°C to inactivate the restriction endonucleases. Place tube on ice and collect content by brief centrifugation.

Ligation of Adapter

Add the following to the digested DNA.

Component	Volume
i. Adapter ligation solution	24 μL
ii. T_4DNA ligase	1 μL

Mix gently at room temperature, centrifuge briefly to collect contents, and incubate at 20 ± 2°C for 2 h.

Perform 1:10 dilution of the ligation mixture as follows:

- Take 10 μL of the reaction mixture and transfer to 1.5-mL microcentrifuge tube.
- Add 90 μL TE buffer and mix well.

The unused portion of the reaction mixture may be stored at 20°C. Stop pre-amplification reactions.

Add the following to a 0.2 or 0.5 mL thin-walled microcentrifuge tube:

Component	Volume
i. Diluted template DNA	5 μL
ii. Preamp primer mix	40 μL
iii. 10× PCR buffer plus Mg	5 μL
iv. Taq DNA polymerase (1 unit/μL)	1 μL
Total volume	51 μL

Mix gently and centrifuge briefly to collect reaction. Overlay with 2–3 drops of silicon oil, if required. Perform 20 cycles at:

94°C for 30 s
86°C for 60 s
72°C for 60 s
Soak temperature: 4°C

Perform a 1:50 dilution as follows: transfer 3 μL to a 1.5-mL microcentrifuge tube containing 147 μL TE buffer. Thus is sufficient for 30 selective AFLP amplifications. If necessary, a new dilution can be made of the preamplification reaction to give an additional template for the selective AFLP amplification. Both unused diluted and undiluted reactions can be stored –20°C.

Primer Labelling

Primer labelling is performed by phosphorylating the 5′ end of the EcoRI primers with ATP and T4 kinase. The reaction products are less prone to degradation because of auto radiolysis; please refer to Table 2.1 for information regarding choice of primers.

Note: This procedure generates enough labelled primer for 100 selective AFLP amplifications.

a. Add the following components to a 1.5-mL microcentrifuge tube:

Component	Volume (μL)
i. EcoR1primer	18
ii. 5× kinase buffer	10
iii. ATP (3000 Ci/mmol)	20
iv. T4 kinase	2
Total volume	50

b. Add the following component to a 1.5 mL microcentrifuge tube:

Component	Volume (μL)
i. EcoR1 printer	18
ii. Distilled water	10
iii. 5× kinase buffer	10
iv. ATP (3000 Ci/mmol)	20
v. T4 kinase	2
Total volume	60

a. Mix gently and centrifuge briefly to collect content of tube. Incubate reaction at 37°C For 1 h.
b. Heat inactivate the enzyme at 70°C for 10 min. Centrifuge briefly to collect reaction contents.

TABLE 2.1
Primer Pair Selection

BARLEY:

	M-CAA	M-CAC	M-CAG	M-CAT	M-CTC	M-CTG	M-CTG	M-CTT
E-AAC	+	+	+	+	+	+	+	+
E-AAG	+	+			+	+	+	+
E-ACA	+	+	+	+	+	+	+	+
E-ACC	+	+	+	+	+	+	+	+
E-ACG	+	+	+	+	+	+	+	+
E-ACT	+	+		+	+	+	+	+
E-AGC	+	+	+	+	+		+	+
E-AGG	+	+	+	+	+	+	+	+

LETTUCE:

	M-CAA	M-CAC	M-CAG	M-CAT	M-CTC	M-CTC	M-CTG	M-CTT
E-AAC	+	+	+	+	+	+	+	+
E-AAG	+	+	+		+	+		+
E-ACA	+	+	+	+	+	+	+	+
E-ACC	+	+	+	+	+	+	+	+
E-ACG	+	+	+		+	+	+	
E-ACT	+		+	+	+	+	+	+
E-AGC	+	+	+	+	+	+	+	+
E-AGG	+	+	+		+	+	+	+

MAIZE:

	M-CAA	M-CAC	M-CAG	M-CAT	M-CTC	M-CTG	M-CTG	M-CTT
E-AAC		+	+	+	+	+	+	+
E-AAG	+	+	+	+		+	+	+
E-ACA		+	+	+	+	+	+	+
E-ACC	+	+	+	+	+	+	+	+
E-ACG	+	+	+	+	+		+	
E-ACT	+		+		+	+		
E-AGC	+	+	+	+	+	+	+	
E-AGG	+	+	+	+	+	+	+	+

OILSEED RAPE:

	M-CAA	M-CAC	M-CAG	M-CAT	M-CTC	M-CTG	M-CTG	M-CTT
E-AAC	+		+	+	+			+
E-AAG	+	+		+		+		+
E-ACA								+
E-ACC	+	+	+			+		
E-ACG								
E-ACT				+		+		+
E-AGC								
E-AGG		+	+	+	+	+		

SUNFLOWER:

	M-CAA	M-CAC	M-CAG	M-CAT	M-CTC	M-CTG	M-CTG	M-CTT
E-AAC	+		+	+	+	+	+	+
E-AAG		+	+	+		+	+	+
E-ACA		+	+			+	+	+
E-ACC	+	+	+	+	+	+	+	+
E-ACG	+		+	+		+	+	+
E-ACT	+	+	+	+	+	+	+	+
E-AGC	+	+	+	+	+		+	+
E-AGG	+	+	+	+	+		+	+

SUGAR BEET:

	M-CAA	M-CAC	M-CAG	M-CAT	M-CTC	M-CTG	M-CTG	M-CTT
E-AAC	+	+	+	+	+	+	+	+
E-AAG	+	+	+	+	+	+	+	
E-ACA	+	+	+	+	+	+	+	+
E-ACC	nd	nd	nd	nd	nd	nd	nd	nd
E-ACG	nd	nd	nd	nd	nd	nd	nd	nd
E-ACT			+	+	+	+	+	+
E-AGC	nd	nd	nd	nd	nd	nd	nd	nd
E-AGG	nd	nd	nd	nd	nd	nd	nd	nd

+: Recommended primer pair
n.d: No data
Blank: Primer pair not recommended

List of Primers

EcoR I primers	Mse I primers
primer E-AAC	primer M-CAA
primer E-AAG	primer M-CAC
primer E-ACA	primer M-CAG
primer E-ACT	primer M-CAT
primer E-ACC	primer M-CTA
primer E-ACG	primer M-CTC
primer E-AGC	primer M-CTG
primer E-AGG	primer M-CTT

Selective AFLP Amplification

a. For each primer pair, add the following component to a 1.5-mL microcentrifuge tube and label it "Mix 1."

Component	Volume (μL)
i. Labelled EcoR1 primer	5
ii. Mse1 primer	45
Total volume	50

b. Add the following to another 1.5-mL microcentrifuge tube and label it "Mix 2."

Component	Volume (μL)
i. Distilled water	79
ii. 10 × PCR buffer plus Mg	20
iii. Taq DNA polymerase (5 units/μL)	1
Total volume	100

c. Each AFLP amplification is assembled by combining the following in a 0.2 or 0.5 mL thin-walled microcentrifuge tube:

Component	Volume (μL)
i. Pre-amplified tomato DNA undiluted	5
ii. Mix 1 (primers/dNTPs)	5
iii. Mix 2 (taq DNA polymerase/buffer)	10
Total volume	20

Gel Analysis

1. After PCR, add an equal volume (20 μL) of formamide dye (98% formamide, 4 μL dye), 10 mM EDTA, bromophenol blue, and xylene cyanol to each reaction. Heat the samples for 3 min at 90°C and immediately place on ice.
2. Pour a 6% polyacrylamide gel (20:1 acrylamide/bis solution; 7.5 M urea 1× TBE buffer) with 0.4 mm spacers and sharktooth combs
3. Pre-electrophorese the gel at constant power for 20 min.
4. Load 3 μL of each samples on the gel.
5. Electrophorese at constant power until xylene cyanol (slower dye) is 2/3 down the length of the gel.
6. Dry the gel.
7. Expose the gel to X-ray film at −8°C overnight.

DNA BLOTTING TECHNIQUES

A mixture of DNA (or RNA) and protein can be separated by gel electrophoresis when separate discrete bands can be stained and visualized on the gels. However, to confirm the identities of these bands or to find similarity of 1 or more of these bands with known or available molecular profiles requires additional testing. It is possible to hydrolyze these bands with labeled probes which facilitates hybridization. The bands are often transferred to a nitrocellulose membrane through a technique known as "blotting." The "DNA bands" are thus blotted. It is called "Southern blotting" when the DNA fragment is transferred from the gel into solid immobilizing media such as nitrocellulose paper, either electrophoretically or by capillary action. This blotting technique was first discovered by Southern in 1975 and is hence called Southern blotting (Figure 2.3).

After the transfer of separated DNA onto nitrocellulose paper, a variety of analytic tests may be carried out, such as

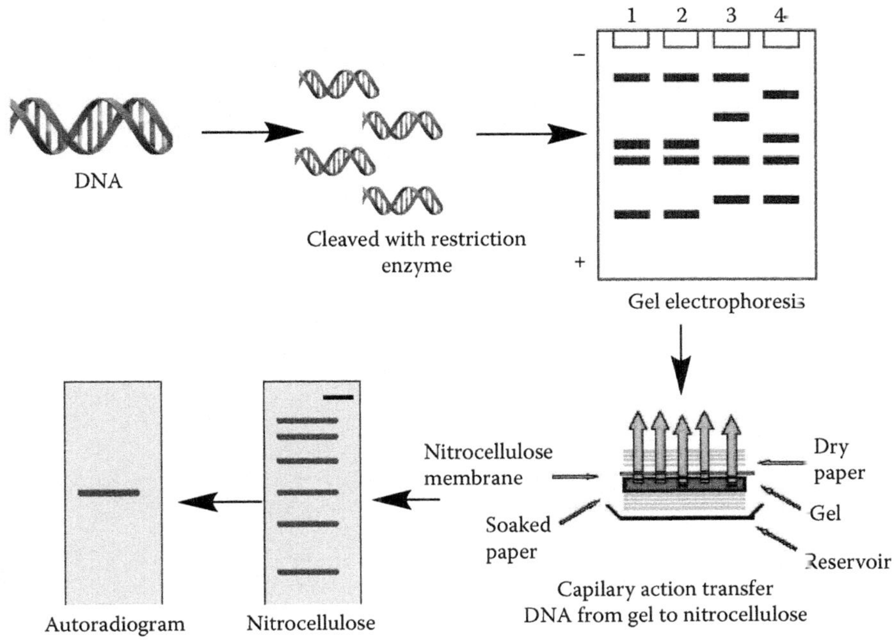

FIGURE 2.3 Stages of Southern blotting.

1. Hybridization with labelled nucleic acid probes.
2. Determination with labelled antibodies.
3. Demonstration of spelotic staining procedure.
4. Autoradiograph; there are many advantages with blot paper of thew original gel or rapid staining destaining.
5. Autoradiography can be easily carried out on the original gel.
6. Samples in the preparation gel may be located repeatedly by briefly blotting the gel.
7. Low concentration of samples can be detected as they are concentrated at the surface of the blotting paper and are detected by low specific probes.
8. The blot is convenient for a permanent record of the gel run. In this experiment, the initial electrophoreses gel, which had been denatured and transferred out of the gel onto the nitrocellulose membrane, has greater affinity, thus making a replica of gel patterns; small fragments of DNA are transferred almost quantitatively from 0.7% to 1% agarose gel within an h, and large fragments are transferred slowly and less effectively.

Precautions

- The glassware used should be washed thoroughly.
- Use gloves when you are transferring the DNA bands from the agarose gels to the nitrocellulose membrane.

Experimental Outline

1. Transfer of the DNA molecule from the agarose gel to the nitrocellulose membrane.
2. Transfer can be carried out either by capillary action or by electroblotting.

Materials

1. Nitrocellulose membrane
2. Whatman no. 1 filter papers
3. Sodium chloride
4. Sodium citrate
5. Tissue paper
6. Weight tray
7. Distilled water
8. NaOH
9. Electrophoresed agarose
10. Tris
11. Gel DNA digested with Pcl-sv-CVB
12. HCl
13. NaCl
14. Viral DNA
15. λ EcoR1 digest, Hind III digest, EcoR1 double digest

Pre-Lab Preparations

1. Denaturation solutions (1.5 Mm NaCl + 0.5 M NaOH): weigh 8.7 g NaCl and 20 g NaOH and dissolve in distilled water and bring up to a volume of 1 L and autoclave.
2. Neutralizing solution (Tris – NaCl pH 7.4) 1 mM Tris + 1.5 M NaCl.
3. Transfer solution: 20 × SSL (pH 7.0): sodium citrate (0.3 M) weigh 44.1 g of salt and dissolve in distilled water and bring up to a volume of 500 mL.
4. Sodium chloride (3 M): weigh 8.5 g and dissolve in distilled water and bring up to a volume of 500 mL.
5. Saline sodium cobalt (SSC).
6. Saline sodium cobalt 1 (pH 7).
7. Sodium cobalt 0.3 M (82.2 g).
8. Sodium chloride (3M) 17 μg in 1 L of water; store in a refrigerator (20 × SSC).

Method

Take the electrophoresed agarose gel and transfer to a large dish containing denaturation solution and soak for 15 min. Subsequently, transfer the gel to the neutralizing solution for 30 min, then to fresh solution for 15 min. In the middle of the tray, the agarose gel electrogram is kept on this support with large pieces of

Whatman no. 1. The tray is filled with 20 × SSC transfer solution and a wick was connected. Excise a piece of nitrocellulose membrane 1.5 times greater than the gel and gently immerse in water and then 20 × SSC solution.

Transfer the agarose gel onto the wick, wet the surface with 20 × SSC solution; cover the rest of the area with paraffin having a rectangular hole. Afterward, gently place nitrocellulose paper on the gel with a pipette to exclude air bubbles between the gel and nitrocellulose membrane sheets of Whatman no.1 paper, having the exact size of membrane over layered on the membrane (15.8 cm thick paper towel stacked on the top of the filter paper). Later, place a glass plate on the top of the paper towels and plae a 500 g weight on it. Leave the whole set undisturbed for 3 h. Afterward, remove the stock and wash the membrane in 6 × SSC solution for 15 min and observe the DNA bands under UV light.

Results

The bands of the sample and markers, which are transferred from gel to nitrocellulose membrane, will be observed making a replica of the gel pattern.

Discussion

A mixture of DNA or RNA or protein can be separated by gel electrophoresis and separate bands can be stained and visualized in the gel. However, to confirm the identities of these bands or to find similarity of 1 or more of these bands with a known and available molecular probes requires additional testing. It is possible to hydrolyze these bands with labelled probes, which facilitates hybridization. The bands are often transferred to a nitrocellulose membrane through a technique known as "blotting." When the DNA bands are thus blotted, it is called Southern blotting.

RNA SEPARATION OF SPECIES BY ION EXCHANGE COLUMN CHROMATOGRAPHY ON METHYLATED ALBUMIN KIESSELGUHER COLUMNS (MAK)

Nucleic acids bind to an anion exchanger due to their negatively charged phosphate groups and these can then selectively be eluted by changing the ionic strength of the elution buffer. The method is commonly used for fractionation of different types of RNA. Kiesselguher acts as an inert support material for albumin. The negative charges on albumin are masked by methylation while the positive charges remain exposed and are thus available for binding nucleic acids, which are negatively charged.

Experimental Outline

Preparation of methylated albumin and Kiesselguher column, and elution of the column with different molarities of NaCl.

Materials

- Gradient maker
- Spectrophotometer
- Glass column
- Albumin
- Kiesselguher
- HCl
- KOH
- NaCl
- RNA

Pre-Lab Preparations

1. *Methylated albumin*: Prepare by suspending 5 g albumin in 500 mL absolute methyl alcohol and 4.2 mL 12 N HCl. The protein dissolves and eventually precipitates again. Allow the mixture to stand in the dark for 3 d or more with occasional shaking. Collect the precipitate in 250 mL centrifuge bottles; wash twice with methyl alcohol and twice with anhydrous ether in a centrifuge. Evaporate most of the ether in air and then in a vacuum over KOH until a dry powder is obtained; store in desiccators over KOH. Failure to remove promptly the residual acid lowers the basicity of the final product. The material is readily soluble in water and is stable both in solution and in dry form. Reproducible batches can easily be made. Make 1% solution of dry powder in water.
2. NaCl solutions (0.1, 0.4, and 2 M).
3. Nucleic acid sample containing approximately 500 μg RNA in 1–2 mL of 0.4 M NaCl solution.

Method

1. Suspend 20 g of Kiesselguher in 100 mL of 0.1 M NaCl (pH 7.0) in a beaker and boil the suspension briefly to expel air. Cover the beaker, cool the solution and then add 5 mL of 1% (w/v) solution of methylated albumin and stir well. Wash the material thoroughly with 0.4 M NaCl and finally suspend it in 0.4 M NaCl label it as MAK suspension.
2. Boil and then cool a suspension of 6 g Kieselguher in 1 mL of 0.1 M NaCl. Allow it to cool and then pour it down a column with a base plugged with glass wool. When Kiesselguher settles down, add 10 mL of MAK suspension (prepared in step 1) to form a second layer. The third layer in the column is obtained by suspension of 1 g Kiesselguher in 10 mL of 0.4 M NaCl; wash the column with 0.4 M NaCl.
3. Load the nucleic acid sample on the surface of the ion exchanger and elute the column with a 0.4–2 M linear gradient of NaCl solution. Collect 5 mL fractions and record absorbance of each fraction.
4. Pool fractions showing absorbance at 260 nm of the individual peaks and calculate the amount of nucleic acid in the pooled fractions.

Note: The nucleic acids are eluted in the following order: rRNA, tRNA, and finally mRNA.

PREPARATION OF RADIO-LABELLED PROBE BY RANDOM PRIMER METHOD

The random primer method is one of the techniques employed for preparing a labelled DNA probe. The original DNA is first denatured by boiling to obtain a single-stranded DNA template. Synthetic hexanucleotide primers of random sequence are added and these anneal with the template (Figure 2.4). The Klenow enzymes then extend the primers and synthesize a radioactively labelled complementary strand due to incorporation of ^{32}p dCTP from the reaction mixture into the elongating chain.

Experimental Outlines

Denature the DNA in the Eppendorf tubes for obtaining single-stranded DNA. Incorporate the denatured single-stranded DNA with a ^{32}pd CTP in the presence of Klenow enzyme.

Materials

- Boiling water bath
- Water bath which maintains constant temperature (37°C)
- Eppendorf tubes (1.5 mL)
- Micropipettes (0–10 and 100 μL)
- DNA probe
- Random primer labelling kit known enzyme
- DNA digest
- ^{32}pd CTP (1 μ Ci/μ)

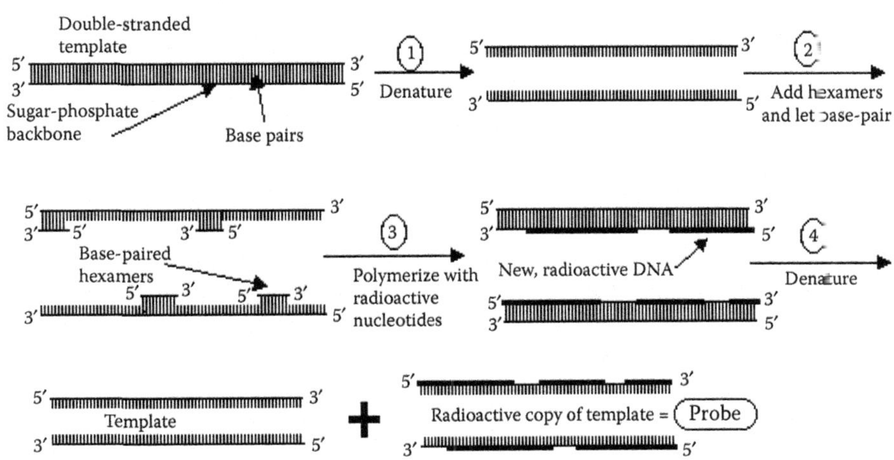

FIGURE 2.4 Preparation of radiolabelled probe by random primer method.

- TE buffer
- SDS

Pre-Lab Preparations

1. DNA probe: plasmid DNA containing plant DNA insert (5 ng/μL).
2. Random primer labelling kit. These kits are available commercially and usually contain the following components: random primer mix (10 constitutes of random primers in a buffer and mixture of dNTP (dATP, dGTP, dCTP, and dTTP at 100 mM concentration each).
3. λ DNA Hind III digest: available commercially or prepare as described elsewhere in this manual.
4. TE buffer (pH 8.0): add 0.4 mL or 0.25 M EDTA solution to 1.0 mL of 1 M Tris-HCl buffer (pH 8.0). Bring volume up to 100 mL with distilled water.
5. SDS (10%): dissolve 10 g of SDS in 100 mL of water.

Method

1. Take 5 μL of probe DNA (25 mg) in an Eppendorf tube with a hole poked in the cap; add 20 μL water and 8.0 μL TE buffer.
2. Place the Eppendorf tube in a boiling water bath in a floater for 5 min to denature DNA. After this, immediately place the tube on ice.
3. Add the following reagents, while the tube is on ice (or follow the instructions supplied by the manufacturer of the labelling kit): dNTP mix (6 μL), random primer mix (5 μL), Klenow enzyme (1 μL), and dCTP (5 μL).
4. Incubate the tubes at 37°C for 60 min.
5. Stop the reaction with 2.5 μL of 10% SDS.
6. Add 500 μL of TE buffer and keep in a boiling water bath for 5 min.
7. Cool in ice for 5 min.
8. Label λ DNA (Hind III digest) in a similar way.

Precautions

1. To keep DNA in denatured form, immediately place the Eppendorf tube on ice without any undue delay after boiling.
2. Radioactive substances are hazardous to health. Hence, necessary precautions such as use of gloves and protective clothing must be observed while handling these materials. Never handle radioactive materials with bare hands.
3. Carry out radioactive work only behind radioactive shield or screen.
4. All glassware/plasticware used for work with radioisotopes should be clearly marked with a radiation symbol.
5. Always wash your hands and monitor yourself before leaving a radioactive area.
6. Dispose of radioactive wastes in approved containers provided for the purpose.
7. Use all the general precautions for handling nucleic acids as mentioned previously.

DETECTION OF A SPECIFIC FRAGMENT OF PLANT DNA FROM RESTRICTION DIGEST BY HYBRIDIZATION WITH A LABELLED PROBE

Complementary strands of DNA hybridize with each other under suitable conditions of temperature and ionic strength. After blotting, the membrane is treated with radioactively labelled probe, which hybridizes with the specific plant DNA fragment bearing sequence homology with the probe, and the radioactive spot on the membrane is detected by autoradiography (Figure 2.5). Prior to hybridization, the membrane is prehybridized with salmon-sperm DNA to block the nonspecific binding sites where the probe may otherwise get bound spuriously.

Experimental Outline

Hybridization of complementary strands of DNA under suitable conditions of temperature and ionic strength.

Materials

- X-ray cassette
- X-ray film
- Plastic bags and boxes
- Boiling water bath

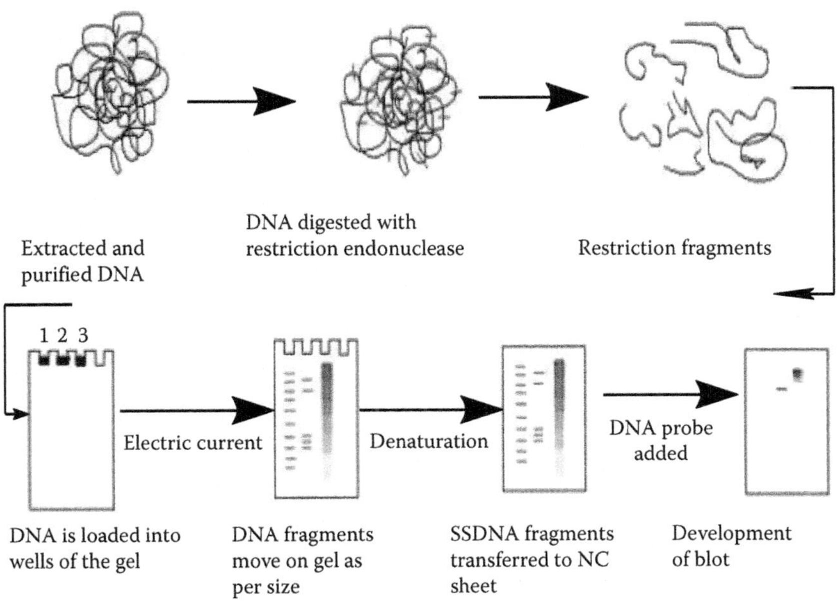

FIGURE 2.5 Plant DNA from restriction digest by hybridization with a labelled probe.

- Shaking water bath at 65°C
- Ultrafreezer, (–70°C)
- Geiger Muller Counter
- HSB (5×)
- Carrier DNA
- Denhart's III
- TES solution
- 10 × SSC
- Prehybridization solution

Pre-Lab Preparations

1. *Carrier's DNA*: Add 100 mL distilled water to 0.45 g salmon-sperm DNA and autoclave. Store in refrigerator.
2. *Denhardt's III*: Dissolve 2 g gelatin, 2 g ficoll, 2 g polyvinylpyrolidine, 10 g SDS, and 5 g tetrasodium pyrophosphate in water and bring up the volume to 100 mL.
3. *TES solution*: 10 mM Tris-HCl containing 1 mM EDTA and 0.1% SDS.
4. *10 × SSC*:1.5 M NADP and 0.15 M sodium citrate. Make a 0.5 × 1 × 2 × SSC solution by diluting the above solution accordingly.
5. *Prehybridization solution*

5 × MSB	8.0 mL
Denhardt's III	4.0 mL
Salmon sperm DNA (freshly denatured by boiling for 5 in)	4.0 mL
Sterile distilled H_2O	4.0 mL
Mix well and incubate at 68°C for 15 min	

6. *Hybridization buffer*

5 × MSB	4.0 mL
Denhardt's III	2.0 mL
25% dextran DNA	8.0 mL
Salmon-sperm DNA (freshly denatured by boiling for 5 min)	0.5 mL
Sterile distilled H_2O	5.5 mL

7. *Nylon membrane containing DNA fragments*: Labelled probe (as in the earlier experiment).

Method

The experimental procedure consists of 2 distinct phases.

1. Hybridization, during which nylon membranes with blotted DNA are incubated with labelled probe.
2. Autoradiography for location of any radioactive zones on the membranes.

Various steps involved in performing hybridization and autoradiography are described as follow:

Hybridization

1. Place the nylon membrane containing the immobilized DNA fragments in a plastic bag, add 20 mL prehybridization solution, and seal the bag without leaving any entrapped air bubbles.
2. Incubate the bag at 65°C in a water bath with gentle shaking for at least 2 h.
3. Denature the probe by incubating in a boiling water bath for 5 min and then quickly chill it on ice.
4. Cut open the plastic bag and drain prehybridization solution. Add labelled probe (freshly denatured by boiling for 5 min) in 20 mL hybridization solution and pour the the mixture into the plastic bag. Seal the bag, taking utmost care that no air bubbles are entrapped.
5. Incubate at 65°C overnight with gentle shaking for hybridization to take place.
6. The next day, cut open the bag, remove the membrane with forceps, and place it in a box with a lid. Dispose of the plastic bag and the solution in radioactive waste reservoirs.
7. Wash the membrane successively with 200 mL of the following buffers in a vigorously shaking water bath:

First wash	2 × SSC, 0.1% SDS 20 min at room temperature
Second wash	1 × SSC, 0.1% SDS 20 min at 65°C
Third wash	0.5 × SSC, 0.1% SDS 20 min at 65°C

Autoradiography

1. For autoradiography, dry the washed membrane between folds of filter paper and then wrap in Saran Wrap. Measure the strength of the radioactivity signal with a hand Geiger Muller counter.
2. In a dark room (with red safety light), place the membrane in a film cassette with the DNA-containing surface facing upward and place a sheet of X-ray film on top of it. Carefully close the cassette and keep at –70° for 2–3 d.
3. Develop the film in the dark room (with red safety light) for 3 min in X-ray developer, rinse briefly with deionized water, and fix for 3 min in X-ray fixer. Rinse with deionized water for 5 min and let the film air-dry.

Precautions

1. Radioactive substances are hazardous to health. Hence, necessary precautions such as use of gloves and protective clothing must be observed while handling these materials. Never handle radioactive material with bare hands.
2. Carry out radioactive work only behind a radioactive shield or screen.
3. All glassware/plasticware used for work with radioisotopes should be clearly marked with a radiation symbol.

4. Always wash your hands and monitor yourself before leaving a radioactive area.
5. Dispose of radioactive waste in approved containers provided for the purpose.
6. Use all the general precautions for handling nucleic acids as mentioned previously.

IN VITRO TRANSCRIPTIONS

Plasmids are used in *in vitro* transcription. Many plasmids and phagemid vectors that contain various combinations of bacteriophage promoters and polycloning sites are commercially available (e.g., the PGEN series and the PSP series). Some of these vectors also encode the lacZ complementing fragment, which allows selection by color of recombination on plates containing X-Gal (5-bromo-4-chloro-3-indolyl-β-D-galactopyranoside). The choice among these vectors is largely a matter of personal preference. However, if transcripts of both strands of the template are required, it is better to use a vector carrying 2different bacteriophage promoters than to clone the template in 2 orientations in a vector carrying a single promoter. It is also important to consider the disposition of restriction sites within the template DNA and downstream from it. The 5′ terminus of the transcript is fixed by the bacteriophage promoter but the 3′ terminus is defined by the downstream site of cleavage by the restriction enzyme. By using different restriction enzymes, RNAs of various lengths can be synthesized from a series of linear templates generated from the same polylinker plasmid.

The DNA sequence to be transcribed is cloned into a restriction site in a polycloning region that is flanked by promoters for 2 bacteriophage polymerases (usually T_7 and SP6). The DNA is then linearized with the restriction enzyme that cleaves downstream from the region to be transcribed. This cleavage prevents synthesis of multimeric transcripts from circular template DNA. Transcription begins from the promoter proximal to the insert when the appropriate polymerase and rNTPs are added in a suitable buffer. The RNA products that are complementary to 1 strand of the template are equal in length to the distance between the active promoter and the end of the liner DNA distal to the target sequence. At the end of the reaction, the template is removed by digestion with RNase-free DNase. Unincorporated nucleotides and oligodeoxynucleotides produced by digestion with DNase can be removed by chromatography through Sephadex G-75.

However, the presence of plasmid sequences in RNA probes can increase the level of background hybridization to levels that are unacceptable if the probe is to be used to screen a plasmid or cosmid library.

During generation of template DNAs, complete cleavage of superhelical plasmid DNA by a restriction enzyme is essential. Small amounts of circular plasmid DNAs will dramatically reduce yields by producing multimeric transcripts. Restriction enzymes that generate blunt or 5′ protruding termini produce the best linear templates. However, both of these types of termini yield RNA products that show heterogeneity at their 3′ ends. Transcription of templates with 3′ protruding termini results in the synthesis of significant amounts of RNA molecules that are aberrantly initiated at the termini of the templates and thus in the production of double-stranded RNA molecules. Restriction enzymes that generate protruding 3′ termini should therefore be avoided.

In addition to plasmids, some bacteriophage and cosmid vectors also contain bacteriophage promoters, usually arranged in opposite orientations on either side of the cloning site for foreign DNA. When a recombinant constructed in a vector of this type is digested with the restriction enzyme that cleaves many times within the foreign DNA, a large number of fragments are generated, one of which contains a particular bacteriophage promoter and the foreign sequences that lie immediately adjacent to it. If the fragments do not carry portending 3′ termini, only the fragment bearing the bacteriophage promoter serves as a template for *in vitro* transcription. The resulting radiolabelled RNA, which is complementary to sequences located at 1 end of the original segment foreign DNA, can then be used as a probe to isolate overlapping clones from genomic DNA or cDNA libraries. These vectors greatly simplify the task of "walking" from 1 recombination clone to another along the chromosome. The following are 2 highly efficient methods available to generate strand-specific RNA probes.

1. The relevant DNA fragment may be cloned or subcloned into specialized plasmids that contain promoters for bacteriophage-encoded DNA-dependent RNA polymerases. The recombinant plasmids are a source of double-stranded templates that can be transcribed *in vitro* into single-stranded RNAs of defined length and strand specificity.
2. The DNA fragment to be transcribed may be amplified in PCRs with primers whose 5′ ends encode synthetic promoters for bacteriophage-encoded DNA-dependent RNA polymerase. Following purification, the products of the PCRs are used as double-stranded templates for *in vitro* transcription reactions.

In both cases, the synthesis of RNA is remarkably efficient. When *in vitro* transcription reactions are saturated with ribonucleoside triphosphates (rNTPS), the templates can be transcribed many times, yielding a mass of RNA that exceeds the weight of the template several fold. In addition, because bacteriophage-encoded DNA-dependent RNA polymerase functions efficiently *in vitro* in the presence of relatively low concentrations of rNTPs (1–20 μM), full-length probes of high specific activity can be synthesized relatively inexpensively. Finally, RNA probes can be freed from template DNA by treating the reaction products with RNase-free DNase probes that usually do not need to be purified by gel electrophoresis (Figure 2.6). However, when using RNA probes of high specific activity to detect rare mRNA transcripts, background hybridization can be kept to a minimum using probes purified by denaturing gel electrophoresis.

In the protocol, we describe procedures for synthesizing RNA probes of high specific activity from plasmids containing promoters for bacteriophage-encoded RNA polymerases.

Experimental Outline

The DNA used as a template in the *in vitro* transcription reaction need not be highly purified. Crude mini preparations work well. The essential requirement is that the

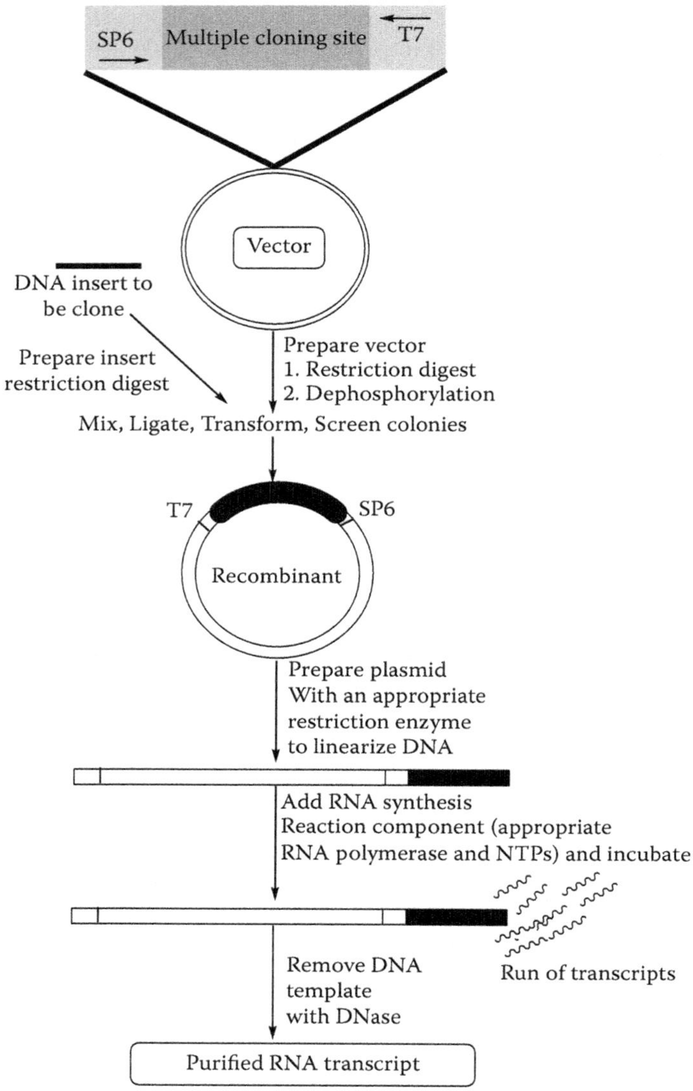

FIGURE 2.6 Schematic representation of *in vitro* transcription.

template be free of RNase, a criterion that can usually be fulfilled by extracting the preparation of plasmid DNA twice with phenol:chloroform. However, if RNase was added to the plasmid at a late stage in the purification process, that is, after deproteinization, it should be removed by treatment with proteinase K, as follows:

1. To the plasmid DNA preparation, add 0.1 volume of 10 × proteinase K buffer (100 Mm Tris-HCl (pH 8.0)/500 mM NaCl), 0.1 volume of 5% (w/v) SDS, and proteinase K (20 mg/mL to a final concentration of 100 μg/mL).
2. Incubate the reaction for 1 h at 37°C.

3. Extract the DNA with phenol:chloroform, and recover the DNA by standard precipitation with ethanol.
4. Resuspend the DNA in RNase-free TE (pH 7.6) at a concentration of >100 μg/mL.

Materials

- Ammonium acetate (10 N)
- Bovine serum albumin (2 mg/mL, fraction V, Sigma)
- Dithiothreitol (1 N)
- Ethanol
- Phenol:chloroform (1:1 v/v)
- Placental RNase inhibitor (20 units/μL)
- Sodium acetate (3 M, pH 5.2)
- 10× transcription buffer
  - 400 mM Tris-Cl (pH 7.5 at 37°C)
  - 60 mM $MgCl_2$
  - 20 mM spermindine HCl
  - 50 mM NaCI

Sterilize the 10× buffer by filtration and then store it in small aliquots at −20°C. Discard each aliquot after use.

- Agarose gel (0.8%–1.0%).
- Nucleic acids and oligonucleotides: rNTP solution containing rATP, rCTP, rCTP, each at 5 mM. and rGTP (0.5 mM).
- Template DNA: The DNA fragment to be transcribed should be cloned into 1 of the commercially available plasmids containing bacteriophage RNA polymerase promoters on both sides of the polycloning sequence (e.g., PGEM from Promega or P blue scripts from Strategene).
- Radioactive compound: (^{32}P) rGTP (10 mCi/mL, sp. act. 400–3000 mCi/mmol) to minimize problems caused by radiolysis of the precursor. It is best, whenever possible, to prepare radiolabelled probes on the day the (32P) dNTP arrives in the laboratory.

Enzymes and Buffers

- Appropriate restriction enzymes.
- DNA-dependent RNA polymerase of bacteriophage T_3, T_7, or SP_6. These enzymes are available from several companies and are usually supplied at concentrations of 10–20 units/μL.
- Most manufacturers also supply a 10× transcription buffer that has presumably been optimized for their particular preparation of the DNA-dependent RNA polymerase. The generic 10× transcription buffer can be used if the manufacturers buffer is in short supply.

- RNase-free pancreatic DNase 1 (1 mg/mL) This enzyme is available from several manufacturers (e.g., RQl RNAse-free DNase I from Promega). Alternatively, contaminating RNase can be removed from standard preparation of DNase 1 by affinity chromatography an agarose 5′-(4-aminophenyl phosphoryl)-uridine-2′(3′)-phosphate.

Equipment

- Microfuge tubes (0.5 mL)
- Sephadex G-50 spin column, equilibrated with 10 mM Tris-Cl (pH 7.5)
- Water bath at 40°C

Method

1. Prepare 5 pmol of linear template DNA by digestion of superhelical plasmid DNA with a suitable restriction enzyme. Analyze an aliquot (100 ng) of the digested DNA by agarose gel electrophoresis. If necessary, add more restriction enzyme and continue incubation until there is no trace of the undigested DNA.

Note: Approximately 2 mg of a plasmid 3 kb in length is 1 pmol.

Important: It is essential that plasmid DNA templates be cleaved to completion, because trace amounts of closed, circular plasmid DNA result in the generation of extremely long transcripts that include plasmid sequences. Because of their length, these transcripts can account for a substantial proportion of the incorporated radiolabelled rNTP.

2. If restriction enzymes, such as PSTI or SSTI, that generate protruding 3′ termini must be used, treat the digested DNA with bacteriophage T_4 DNA polymerase in the presence of all 4 dNTPs to remove the 3′ protrusion.

Note: protruding 3′ termini create an opportunity for the DNA-dependent RNA polymerase to transfer to the complementary strand of the template and to generate long u-turn transcripts with extensive secondary structure.

3. Purify the template DNA by extraction with phenol:chloroform and strand precipitate with ethanol. Dissolve the DNA in H_2O at a concentration of 100 nm (i.e., 200 μg/mL) for a 3 kb plasmid).
4. Warm the first 6 components listed below to room temperature. In a sterile 0.5 mL microfuge tube, mix in the components in the following order at room temperature:

Template DNA	0.2 mole (400 mg for 3 kb plasmid)
RNase-free H_2O	106 μL
100 mM dithiothreitol	2 μL
5 mM rNTP solution	2 μL
10× transcription buffer	2 μL
2 mg/mL bovine serum albumin	1 μL
10 m Ci/mL (^{32}P) or GTP (sp. act. 400–300 Ci/mmol)	5 μL

Mix the components of the mixture by gentle tapping the outside of the tube. Then add:

Placental RNase inhibitor (10 units)	1 μL
Bacteriophage DNA-dependent RNA polymerase	1 μL

The components are added in the order shown and at room temperature to avoid the possibility that the template DNA may be precipitated by the high concentration of spermidine in the transcription buffer.

Mix the reagents by gently tapping the outside of the tube. Centrifuge the tube for 1–2 s to transfer all of the liquid to the bottom. Incubate the reaction for 1–2 h at 37°C (bacteriophages T_3 and T_7 DNA-dependent RNA polymerases) or 40°C (bacteriophages SP_6 DNA-dependent RNA polymerase).

Note: the reaction may be scaled from 20 to 5 μL to accumulate more dilute reagents.

When the reaction is carried out as described above, 80%–90% of the radiolabel will be incorporated into RNA. The yield of RNA will be 20 ng (sp. act. 4.7×10^9 dpm/mg) when the specific activity of the (^{32}p) GTP is 3000 Ci/mole and 150 ng (sp. act. 6.2×10^8 dpm/μg) when the specific activity of the precursor is 400 ci/mmole.

5. If full-length transcripts are desired, add 2 μL of 0.5 mM rGTP, and incubate the reaction mixture for an additional 60 min at the temperature appropriate for the polymerase.
6. Terminate the *in vitro* transcription reaction by adding 1 μL of 1 mg/mL RNase-free pancreatic DNase 1 to the reaction tube. Mix the reagents by gentle tapping the outside of the tube. Incubate the reaction mixture for 15 min at 37°C.
7. Add 100 μL of RNase-free H_2O and purify the RNA by extraction with phenol chloroform.

Note: If the probe will be used in experiments where length is important (e.g., RNase protection), purify the radiolabelled RNA by polyacrylamide gel electrophoresis. The aim of this extra step is to eliminate truncated radiolabelled molecules from the preparation.

8. Transfer the aqueous phase to a fresh microfuge tube, and separate the radiolabelled RNA from undesired small RNAs and rNTPs by 1 of the 3 methods.

To Purify RNA by Ethanol Precipitation

- Add 30 μL of 10 M ammonium acetate to the aqueous phase. Mix and then add 250 μL of ice-cold ethanol to the tube. After storage for 30 min on ice, collect the RNA by centrifugation at maximum speed for 10 min at 4°C in a microfuge.
- Remove as much of the ethanol as possible by gentle aspiration and leave the open tube on the bench for a few minutes to allow the last visible traces of ethanol to evaporate. Dissolve the RNA in 100 μL of RNase-free H_2O.
- Add 2 volumes of ice-cold ethanol to the tube and store the RNA at 70°C until needed.

To recover the RNA, transfer an aliquot of the ethanolic solution to a fresh microfuge tube. Add 0.25 volume of 10 M ammonium acetate, mix, and then store the tube for at least 15 min at 20°C. Centrifuge the solution at maximum speed for 10 min at 40°C in a microfuge. Remove the ethanol by aspiration and dissolve the RNA in the desired volume of the appropriate RNase-free buffer.

Troubleshooting

1. *No RNA is synthesized.* The most common cause of an apparent lack of RNA synthesis is contamination of tubes or reagents with RNase. This contamination can be avoided by taking the precautions.

   Less frequently, failure to synthesize RNA is a consequence of precipitation of the DNA template by the spermidine in the 10× transcription buffer. Make sure that the components of the reaction are assembled at room temperature and in the stated order. If necessary, the presence of soluble template can be confirmed by analyzing an aliquot of the reaction by electrophoresis through an agarose gel.
2. *Transcription is from the wrong strand of DNA.* Usually, >99.8% of transcripts synthesized *in vitro* by bacteriophage DNA-dependent RNA polymerases are derived from the correct DNA strand (Melton et al. 1984). However, this high degree of specificity is only achieved provided templates are both linear and lacking protruding 3′ termini. The template can be contaminated with superhelical plasmid DNA chains on both strands of the DNA. The presence of a protruding 3′ terminus downstream from the bacteriophage promoter leads to the synthesis of transcripts complementary to the full length of the wrong strand of DNA (i.e., 4-turn transcripts). Both of these problems can be avoided by careful preparation of the template (please see steps 1 and 2).

   Synthesized transcripts are shorter than the desired length. Synthesis of abbreviated transcripts can be due to the chance occurrence in the template of sequences that terminate transcription by the particular DNA-dependent RNA polymerase being used. Another contributing factor can be a limiting concentration of precursor (usually the radiolabelled rNTPs).

   The first of these obstacles may be resolved by constructing a new plasmid in which transcription of the desired sequence is driven by the polymerase from a different bacteriophage. Transcription terminator sequences are not always recognized by DNA-dependent RNA polymerases.

   Although the strengths of transcription termination signals vary greatly, all but the strongest of them can be overcome, at least in part, by increasing the concentration of the limiting rNTP. In most cases, it is impractical to increase the reaction, because the improvement in yield of full-length product is gained at the expense of reducing the specific activity of the probe. In this case, a number of addition steps can be taken:

   a. Lower the temperature of incubation of the transcription reaction to 30°C (Krieg and Melton 1987).
   b. Transcribe the sequence to the minimum. In this way, it may be possible to eliminate the termination sequence from the clone.

c. Purify the desired product by electrophoresis through a polyacrylamide or agarose gel. The transcription reactions are so efficient that it is often possible to purify sufficient quantities of the desired RNA, even if it is only a relatively minor proportion of the total RNA synthesized in the reaction.

IN VITRO TRANSLATION OF mRNA

Messenger RNA molecules are read and translated into protein by complex RNA-protein particles termed ribosomes, which are 70S or 80S, depending on their sedimentation coefficient. Prokaryotic cells have 70S ribosomes while those in the eukaryotic cytoplasm are 70 S. Ribosomes are composed of 2 subunits that are held apart by ribosomal binding proteins until translation proceeds. There are sites on the ribosomes for the binding of 1 mRNA and 2 tRNA molecules and the translation process occurs in 3 stages:

1. *Initiation*, involving the assembly of the ribosome subunits and the binding of the mRNA.
2. *Elongation*, in which specific amino acids are used to form polypeptides, this being directed by the codon sequence in the mRNA.
3. *Termination*, which involves the disassembly of the components of translation following the production of a polypeptide.

Transfer RNA molecules are also essential for translation. Each of these are covalently linked to a specific amino acid, forming an aminoacyl tRNA, and each has an exposed triplet of bases that is complementary to the codon for that amino acid. This exposed triplet is known as the anticodon, and allows the tRNA to act as an "adaptor" molecule, bringing together a codon and its corresponding amino acid. The process of linking an amino acid to its specific tRNA is termed charging and is carried out by the enzyme aminoacyl tRNA synthetase.

In prokaryotic cells, the ribosome binds to the 5′ end of the mRNA at a sequence known as the ribosome-binding site or sometimes termed the Shine-Dalgarno sequence, after those who discovered of the sequence. In eukaryotes, the situation is similar, although less well understood. It is thought to involve a so-called Kozak sequence. The ribosome appears to scan the mRNA for this sequence before translation. Following translation initiation, the ribosome moves toward the 3′ end of the mRNA, allowing an aminoacyl tRNA molecule to base-pair with each successive codon, thereby carrying in amino acids in the correct order for protein synthesis.

There are 2 sites for tRNA molecules in the ribosome (the A site and the P site) and when these sites are occupied, directed by the sequence of codons in the mRNA, the ribosome allows the formation of a peptide bond between the amino acids. The process is also under the control of the enzyme peptidyl transferase. When a ribosome encounters a termination codon (UAA, UGA, or UAG), a release factor binds to

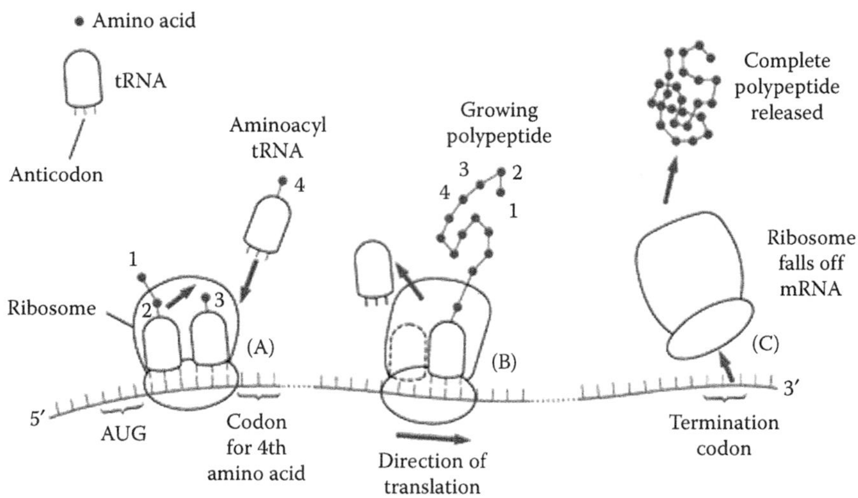

FIGURE 2.7 Translation ribosome A has moved only a short way from the 5′ end of the mRNA and has built up a dipeptide (on 1 tRNA) that is about to be transferred onto the third amino acid (still attached to tRNA). Ribosome B has moved much further along the mRNA and has built up an oligopeptide that has just been transferred onto the most-recent aminoacyl tRNA. The resulting free tRNA leaves the ribosome and will receive another amino acid. The ribosome moves toward the 3′ end of the mRNA by a distance of 3 nucleotides, so that the next codon can be aligned with its corresponding aminoacyl tRNA on the ribosome. Ribosome C has reached a termination codon, has released the completed polypeptide, and has fallen off the mRNA.

the complex and translation stops, the polypeptide and its corresponding mRNA are released, and the ribosome divides into its 2 subunits (Figure 2.7). A myriad of accessory initiation and elongation protein factors are involved in this process; in eukaryotic cells, the polypeptide may then be subjected to post-translation cations such as glycosylation and, by virtue of specific amino sequences, may be directed to specific cellular compartments of the cell.

Because the mRNA base sequence is read in triplets, an error of 1 or 2 nucleotides in the positioning of the ribosome will result in the synthesis of an incorrect polypeptide. Thus, it is essential for the correct reading frame to be used during translation. This is ensured in prokaryotes by base-pairing between the Shine-Dalgarno sequence and a complementary sequence of 1 of the ribosome's rRNAs, thus establishing the correct starting point for movement of the ribosome along the mRNA. However, if a mutation such as a deletion/insertion takes place within the coding sequence it will also cause a shift of the reading frame and result in an aberrant polypeptide.

Experimental Outline

The availability of a cloned gene makes it possible to synthesize the encoded protein by *in vitro* transcription and translation (Figure 2.8). Such *in vitro* synthesized proteins are extremely useful for a wide variety of purposes, including the analysis

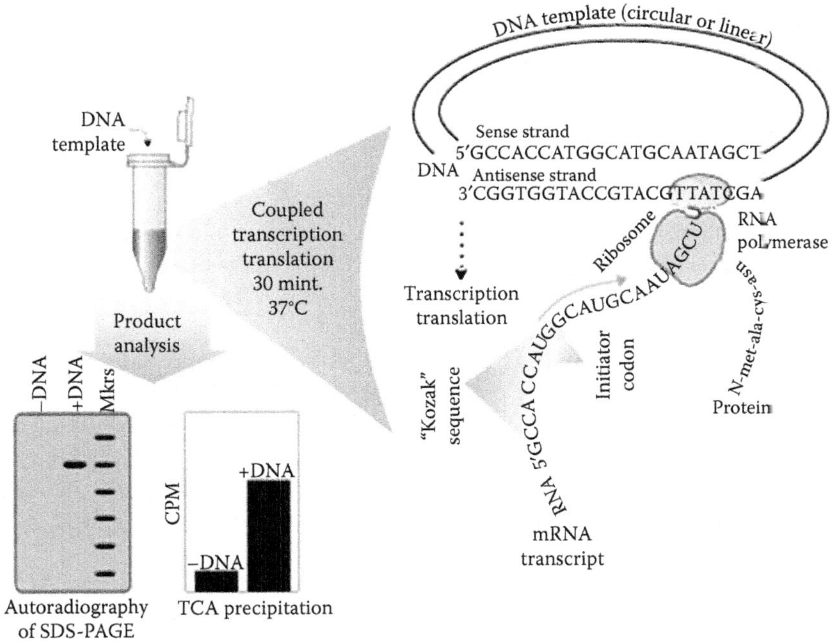

FIGURE 2.8 *In vitro* translation process.

of DNA–protein interactions. The basis of the method is to clone the protein-coding sequences into a vector containing a promoter for SP6 or T7 RNA polymerase, to produce messenger RNA by transcribing the DNA template with this enzyme, and to synthesize the desired protein (often as a ^{35}S-labelled species) by translation of this mRNA *in vitro*. A major advantage of this method is that any desired mutant protein can be generated simply by altering the DNA template.

Materials

- Plasmid DNA containing SP6 or T7 promoter
- DNA-containing cloned gene or cDNA encoding protein of interest
- Appropriate restriction endonucleases
- TE buffer 5×
- Ribonucleoside triphosphate mix 10×
- SP6/T7 RNA polymerase buffer
- 10 mM spermidine (for SP6 RNA polymerase only)
- Ribonuclease inhibitor
- SP6 or T7 RNA polymerase
- Buffered phenol
- Isobutanol
- 10 M ammonium acetate
- 100% ethanol
- *In vitro* translation kit (wheat germ extract or reticulocyte lysate)

- ^{35}S-labelled methionine (1400 Ci/mmol)
- 0.1 M NaOH
- 10% trichloroacetic acid (TCA)
- EN^3 HANCE (Du Pont NEN)

Procedure

1. Subclone protein-coding DNA sequences of interest into a plasmid vector that contains a promoter for SP6 or T7 RNA polymerase (e.g., pSP64) at a site downstream of the promoter.

Note: The protein-coding sequences must be contiguous—that is, uninterrupted by introns. It is also crucial that the protein-coding sequences be cloned in the correct orientation downstream of the bacteriophage promoter such that the correct initiation codon is the first AUG in the RNA to be synthesized. It is best if the AUG initiation codon is relatively close (25–100 bases) to the 5′ end of the RNA; this may require additional DNA manipulations of the protein-coding sequences such as nested deletions (oligonucleotide-directed mutagenesis)

2. Prepare plasmid DNA containing protein-coding sequences downstream of the bacteriophage promoter (step 1) by CsCl/ethidium bromide centrifugation or PEG precipitation.
3. Cleave 10 μg DNA with a restriction endonuclease that cuts just downstream of the termination codon (ideally 50–200 bp) and does not cut within the protein-coding region. Remove a small aliquot, and check the efficiency of cleavage by agarose gel electrophoresis.
4. Purify the DNA by phenol extraction and ethanol precipitation. Resuspend in 50 μL TE buffer.

Prepare the mRNA by in vitro Transcription

5. Set up the following 25 μL reaction mixture at room temperature (to avoid precipitation of the DNA template by spermidine):
   - 8 μL H_2O
   - 5 μL DNA (total 1 μL)
   - 5 μL 5× ribonucleoside triphosphate mix
   - 2.5 μL 10× SP6/T7 RNA polymerase buffer
   - 2.5 μL 10 mM spermidine (for SP6 RNA polymerase only)
   - 1 μL RNase in (30 to 60 U)
   - 1 μL SP6 or T7 RNA polymerase (5–20 U)
   - Incubate 60 min at 40°C for reactions with T7 RNA polymerase, omit the spermidine and add an additional 2.5 μL water.
6. Add 25 μL buffered phenol, vortex, and extract immediately. Transfer the aqueous phase to a new microcentrifuge tube, extract twice with isobutanol, and add 6 μL of 10 M ammonium acetate and 70 μL ethanol. Ethanol precipitates the RNA; wash once with ethanol.

7. Resuspend the RNA in 24 μL TE buffer, add 6 μL of 10 M ammonium acetate and 70 μL ethanol, reprecipitate, and wash once with ethanol. Resuspend RNA in 10 μL TE buffer.

Prepare the Protein by *in vitro* Translation

8. Add 1–10 μL RNA to the appropriate reagents in an *in vitro* translation kit and follow the directions of the manufacturer. Add 15 μCi of [^{35}S]methionine (1,400 Ci/mmol) to radioactively label the protein. Reactions are typically performed in 30-μL volumes at room temperature for 30–60 min. After the reaction is complete, store at 0–4°C.

Note: The amount of RNA to be added depends on the efficiency of the *in vitro* transcription reaction. In general, 1–2 μL should be enough, but if necessary, up to 10 μL can be used.

9. Remove 1 μL of the translation products and add to 50 μL of 0.1 M NaOH; incubate 15 min at 37°C. Add 1 mL of 10% TCA and incubate an additional 15 min on ice.
10. Collect the precipitated protein on glass fiber filters and quantitate the incorporated [^{35}S] methionine by scintillation counting.
11. Remove 3 μL of the translation reaction and analyze the products by 1-dimensional SDS–polyacrylamide gel electrophoresis, including lanes for molecular weight standards. Visualize the [^{35}S]proteins by fluorography with EN^3 HANCE and autoradiography.

Reagents and Solutions

5× ribonucleoside triphosphate mix
5 mM each ATP, UTP, CTP
5 mM diguanosine triphosphate (G-5′ ppp5′-G) TP
0.5 mM GTP

RETICULOCYTES AND THEIR USE IN THE STUDY OF PROTEIN SYNTHESIS

Reticulocytes are immature red blood cells which contain the normal protein synthesizing machinery typical of eukaryotic cells. They are produced in bone marrow and appear in the blood in large numbers in response to phenylhydrazine-induced anemia. The intact cells will readily incorporate amino acids from a suitable medium and, if the amino acids are tagged with a radioactive atom, protein synthesis can be monitored. As with most eukaryotic cells, the mRNA molecules found in reticulocytes are relatively stable compared with their bacterial counterparts. Thus, this cell, whose protein synthesizing machinery is mainly geared to the synthesis of large amounts of globin, yields a lysate which contains globin mRNA (species for α and β chains are present) in addition to ribosomes, tRNA, and other factors Also present, but in small amount are mRNA species for proteins other than globin. Again, protein synthesis may be studied by adding radioactive amino acids to the lysate. The translation process can therefore be studied *in vitro* using the natural endogenous messenger rather than employing exogenous templates.

The reticulocyte lysate system is readily available and has also been widely used for assessing the template activity of exogenous mRNA. However, in this case, the very presence of globin mRNA creates a problem, as a high background incorporation of amino acids into globin occurs. Therefore, a measure of overall protein synthesis in terms of incorporation of labelled amino acids into total protein cannot be used as an activity index for added mRNA, although this problem can be overcome if product analysis is performed by antiserum precipitation or SDS electrophoresis. An alternative approach is to try and eliminate the endogenous mRNA using ribonuclease, without damaging other types of RNA. Exogenous mRNA is then added as the sole template for translation.

During maturation, *in vivo* reticulocytes lose their ability to initiate globin chains and eventually degrade all mRNA and ribosomes. Mature animal erythrocytes are therefore incapable of protein synthesis, although this ability is retained by avian erythrocytes.

Experimental Outline

1. Induction of reticulocytosis.
2. Collect blood after 9 d of induction of reticulocytosis.
3. Store the reticulocytes for the use in the next experiment.

Materials and Pre-Lab Preparation

- Rabbits
- Phenylhydrazine hydrochloride (2.5% w/v adjusted to pH 7 with dilute NaOH)

1.	Wash medium (0.14 mol/L NaCl, 0.05 mol/L KCl, 5 mmol/L Mg acetate)	250 mL
2.	Sterile glassware and centrifuge tubes	–
3.	Syringes and needles for blood collections	10

Method

Reticulocytosis is induced in rabbits by subcutaneous injections of a w/v solution of phenylhydrazine hydrochloride solution using the following protocol: day 1, 1.0 mL; day 3, 1.0 mL; day 4, 1.0 mL; day 5, 0.5 mL.

Collect 5–10 mL of blood from an ear on day 9, when a reticulocyte content of more than 80% should be obtained. Centrifuge the blood on a bench centrifuge for 7 min at 4°C. Remove the supernatant with a Pasteur pipette (autoclaved), taking care to remove any buffy coat, which appears at the interface between the packed cells and the plasma. Resuspend the cells in the wash medium with a glass rod and centrifuge again. Repeat the washing procedure twice or more then give a final spin for 15 min.

Discard the supernatant and store the cells on ice until required. If a lysate is required, then retain 0.5 mL of the cells and lyse the remainder as detailed later.

Precautions

- Wear lab coat.
- Select healthy rabbits.
- Use sterile injection needles for blood collection.

PROTEIN SYNTHESIS IN INTACT RABBIT RETICULOCYTES

The value and use of reticulocytes is discussed under the introduction of previous experiment.

Experimental Outline

1. Study the protein synthesis by using radiolabelled leucine using rabbit reticulocytes.
2. Determine the radioactivity in the protein precipitated with 5% of TCA.

Materials and Pre-Lab Preparations

1.	Rabbit reticulocytes from previous experiment	10 mL
2.	Ferrous ammonium sulphate (0.105% w/v, prepare fresh)	–
3.	Glass fiber discs (Whatman SF/A 2.1 cm)	50 mL
4.	Scintillation acid (5% w/v)	–
5.	Trichloroacetic acid (5% w/v)	–
6.	Trichloroacetic acid (10% w/v)	250 mL
7.	Water baths at 80°C and 37°C (shaking)	250 mL
8.	Ethanol:acetone (1:1)	5
9.	Ovens at 50°C	200 mL
10.	Scintillation fluid	5
11.	Liquid scintillation counter	–
12.	Saline solution (0.13 mol/L NaCl, 0.005 mol/L KCl sterile)	100 mL
13.	Amino acid mixture, prepared by weighing out the following amounts of each amino acid and dissolving in 100 mL of the saline solution	100 mL

Alanine	23.0 mg
Glycine	50.0 mg
Methionine	6.0 mg
Arginine	16.0 mg
Histidine	63.0 mg
phenylalanine	33.0 mg
Aspartic acid	48.0 mg
Isoleucine	5.0 mg
Proline	20.0 mg
Cysteine	6.0 mg
Leucine	33.0 mg
Serine	22.0 mg
Glutamine	50.0 mg

(*Continued*)

Lysine HCl	41.0 mg
Threonine	25.0 mg
Tryptophan	8.0 mg
Tyrosine	19.0 mg
Valine	46.0 mg

14. Master mix 2.70 mL

Master mix	2.70 mL
Amino acid mixture	0.15 mL
$MgCl_2$ (0.25 mol/L in 10 w/v glucose)	1.45 mL
Tris-HCl (0.15 mol/L, pH 7.6)	1.10 mL
Trisodium citrate (0.01 mol/L)	1.60 mL
Total volume	7.00 mL

15. Radioactive master mix. (Dissolve leucine in the master mix to give a final concentration of 1 mmol/L and add 20 μL of ^{14}C leucine with an activity of 50 μCi/mL).

Method

Protein synthesis: Combine 0.25 mL of washed reticulocytes with 0.75 μL of radioactive mix and 3.75 μL/mL of freshly prepared ferrous ammonium sulphate solution and incubate the mixture at 37°C in a shaking water bath. Remove duplicate 50 μL aliquots at 0, 10, 20, 30, 40, and 60 min (gently mix before sampling), then spot each sample on to a glass fiber disc, which is then placed in a scintillation vial. Cover the discs with 4 mL of 5% w/v TCA. When all samples are collected, wash the discs as follows:

1. 5 mL 10% w/v TCA for 10 min
2. 5 mL 10% w/v TCA at 80°C for 10 min
3. 5 mL ethanol: acetone for 5 min

Please note that the supernatants will be radioactive, so remove with a Pasteur pipette into an appropriate counter.

Dry in an oven at 50°C for 10–15 min, add scintillation fluid, and count the radioactivity in a scintillation counter.

Calculations

1. Determine the specific activity of ^{14}C leucine in the incubation mixture.
2. Plot the counts per minute as a function of time and comment on the shape of the graph.
3. Determine the rate of globin synthesis (pmol/min packed cells) assuming 17 leucine residues per globin chain.

Note: 1 μ Ci $= 2.22 \times 10^6$ disintegrations per minute (dpm).

Precautions

1. Wear plastic flexible gloves.
2. Handle the radiolabelled amino acids carefully.
3. Dispose of the radiolabelled wastes at a place far away from the laboratory.

METABOLIC LABELLING OF PROTEINS AND IMMUNOPRECIPITATION

Plants and animals have developed mechanisms to withstand stress conditions such as high and low temperatures. Many of the adaptive mechanisms are a consequence of stress perception and are likely to be mediated. Stress-induced expression leads to the synthesis of specific stress-responsive genes both in plants and animals. Protein turnover is a common mechanism during the developmental program of both plants and animals. Protein synthesis can be studied normally by using labelled amino acids. In this experiment, methionine is used to trace the synthesis of a new protein.

EXPERIMENTAL OUTLINE

1. Incubation of either plant or animal material in the radiolabelled amino acid.
2. Extraction of protein with cold amino acid followed by a buffer.
3. Measurement of protein fraction by scintillation counting system.
4. Subject a protein fraction to electrophoresis followed by fluorography.

MATERIALS AND OTHER

- Germinating pea seedlings are best to study
- Tris-HCl
- SDS
- PMSF (phenzzene methosulphate)
- MCE (mercaptoethone)
- NNN′N′-methylene bisacrylamide
- TEMED
- Ammonium persulphate
- ^{35}S methionine
- Scintillation solution PPO (2,5-diphenyloxazole)
- Bromophenol blue
- X-ray sheet (Kodak)

EQUIPMENT

- Deep freezer
- Homogenizer
- High-speed centrifuge
- Electrophoretic unit
- Seed germinator

METHOD

1. Germinate pea (pea) *(Pisum sativum)* on moist filter paper in a tray at 30°C in a seed germinator. After 48 h of germination, remove the seedlings and wash thoroughly. Incubate the seedlings with 200 μCi of ^{35}S-methionine with specific activity 500 mCi/mm with activity of 100 μCi/mL for 3 h.

2. After pretreatment with ^{35}S methionine, wash the seedlings with cold methionine followed by water and extraction buffer.
3. Homogenize in 150 mm Tris-HCl buffer (pH 8.0), centrifuge the homogenate at 10,000 g for 10 min. Use the supernatant for further studies.
4. Take an aliquot of supernatant and proteins with cold acetone in the ratio of (1:5), measure the radioactivity of the precipitated protein with the scintillation system, using a scintillation liquid to quantify the protein yield.
5. Subject another aliquot of protein for SDS-PAGE. Stain with either Coomassie blue or used silver stain for locating the protein bands and determining the molecular weights of individual proteins with the help of molecular markers.
6. Treat the gels with PPO (2,5 diphenyloxazole) and dry the gels and expose to Kodak X-ray film at –70°C for 48 h and develop the film in dark.
7. View the newly synthesized protein bands.
8. Carryout immunoprecipitation of these proteins using western blots.

Western Blots

1. Separate proteins by SDS-PAGE using 12% gel transfer to a polyvinylidene difluoride (PVDF) membrane by electroblotting.
2. Carry out western blotting using polyclonal antiserum containing antibodies for a protein.
3. Separate proteins by SDS-PAGE using 12% gel transfer to a polyvinylidene difluoride (PVDF) membrane by electroblotting.
4. Detect the new protein's polyclonal antibodies specific to a particular protein and secondary antibody conjugated with alkaline phosphatase using BCIP (5-bromo-4-chloro-3-indolyl phosphate) as substrate.

Precautions

Always wear gloves while handing either polyacrylamide gels or radiolabelled isotopes. Dispose the radioactive materials into a pit far away from your laboratory and cover with sand fill.

MARKER-ASSISTED PYRAMIDING OF LEAF RUST RESISTANCE GENE IN WHEAT

For the last 5 decades, breeders have exploited intergenomic transfer to transfer resistance to pathogens into bread wheat from its wild relatives like *Agropyron* spp., *Aegilops* spp., and *Triticum* spp. Several of the leaf rust genes (*Lr genes*) have been effective in imparting resistance to the tropical races of *P. triticina* after their transfer into different cultivars. Most of the genes from cultivated germplasm are not effective against the prevalent races of leaf rust, and due to a narrow genetic base and continuously evolving pathogen races, resistant varieties become susceptible to the disease. One promising approach to overcome this problem is to develop a germplasm carrying combinations of several effective genes, especially those from wild relatives and related species. The identification of molecular markers for resistance genes can efficiently facilitate the pyramiding of major genes into a valuable background in less time, making it more cost effective.

Method

1. Label 0.5 mL PCR tubes for each genomic DNA sample.
2. Prepare the reaction mixture by adding the following constituents:

S.No.	Stock Solution	Quantity Required (μL)
1.	DNA (40–50 ng/μL)	2.0
2.	PCR buffer (10×) (without $MgCl_2$)	2.0
3.	$MgCl_2$ (15 mM)	2.0
4.	dNTPs (1 mM)	2.0
5.	Primer R (5 μM)	1.0
6.	Primer F (5 μM)	1.0
7.	Taq polymerase (5 U/μL)	0.2
8.	ddH_2O	9.8
	Total	20.0

PCR Reactions

Initial denaturation	94°C—5 min	
Denaturation	94°C—5 min	35 Cycles
Annealing	55°C—1 min	
Primer extension	72°C—2 min	
Final extension	72°C—5 min	
Storage	4°C—forever	

1. Prepare 1.5% agarose gel in 1× TAE buffer along with ethidium bromide.
2. Add 4 μL of loading dye to each tube.

STS MARKER-BASED DETECTION OF 1B/1R TRANSLOCATION

Wheat-rye translocation has been widely used by breeders all over the world because genes located on the translocated fragments of the rye chromosome determine a number of characters such as high yield, wide adaptation, and insect resistance to cereal cyst nematode. For copper utilization efficiency in wheat genotypes with 1B/1R translocation, the short arm of chromosome 1B of wheat is replaced by the short arm of the 1R chromosome of the rye, presumably as a natural consequence of a centric break fusion event between chromosomes 1B of the wheat (*Triticum aestivum*) and 1R of the rye. Several methods are known to detect the wheat-rye translocations. Molecular evidence for wheat-rye translocations is based on identification of rye chromatic sequences in wheat by genome mapping using rye-specific repetitive sequences; 1 of such methods is explained herewith.

Reagents

- Taq DNA polymerase (3 U/μL)
- dNTPs (1 mM)
- Taq DNA polymerase buffer (10×)
- Primers (50 ng)

- Agarose (1.5%)
- Ethidium bromide (2.5 μL/50 mL)
- Tris acetate EDTA buffer (50×)

Method

1. Label 0.5/0.2 mL PCR tubes for each genomic DNA sample.
2. Prepare the reaction mixture by adding the following constituents:

S.No.	Constituents	Volume/Reaction (μL)
1.	Sterile water	13.9
2.	10× PCR buffer	2.5
3.	dNTPs	1.5
4.	Primer I	2.5
5.	Primer II	2.5
6.	Taq DNA Polymerase	0.1

3. Distribute 23 μL of reaction mixture in each labelled PCR tube and add 2 μL of template DNA in each tube to make the final reaction volume 25 μL.
4. Mix gently and carry out the polymerase chain reaction in a thermal cycler with the following program:

Initial denaturation	94°C—4 min	
Denaturation	94°C—15 s	44 Cycles
Annealing	55°C—1 min	
Primer extension	72°C—1 min	
Final extension	72°C—5 min	
Storage	4°C—forever	

5. Prepare 1.5% agarose gel in 0.5 × TAE buffer along with ethidium bromide.
6. Add 2 μL of loading dye to each tube.
7. Load the samples along with a molecular weight marker (100 bp ladder) and carry out electrophoresis for 3 h.
8. View the gel under gel documentation system.
9. Primer sequences:

   Primer 1: GGAGACATCATGAAACATTTG
   Primer 2: CTGTTGTTGGGCAGAAAG

STS MARKER-BASED DETECTION OF Lr 26 GENE

The short arm of chromosome 1 (IRS) of rye (*Secale cereale*) carries several important disease resistance genes that confer resistance to rust (*Puccinia* spp.) and powdery mildew. This rye chromosome arm carries genes Lr 26, Sr 31, Yr 9, and Pm 8, conferring race-specific resistance to leaf rust (*Puccinia striiformis* f. sp. *tritici*) and powdery mildew (*Erysiphe graminis* f. sp. *tritrici*).

Reagents

1. Taq DNA polymerase (3 U/μL)
2. dNTPs (1 mM)
3. Taq DNA polymerase buffer (10×)
4. Primers (50 ng)
5. Agarose (1.5%)
6. Ethidium bromide (2.5 μL/50 mL)
7. Tris acetate EDTA buffer (50×)

Method

1. Label 0.5/0.2 mL PCR tubes for each genomic DNA sample.
2. Prepare the reaction mixture by adding the following constituents:

S.No.	Constituents	Volume/Reaction (μL)
1.	Sterile water	11.2
2.	10× PCR buffer	2.0
3.	dNTPs	1.0
4.	Primer I	0.4
5.	Primer II	0.4
6.	Taq DNA polymerase	1.0

3. Distribute 16 μL of reaction mixture in each labelled PCR tube and add 2 μL of template DNA in each tube to make the final reaction volume 18 μL.
4. Mix gently and carry out the polymerase chain reaction in a thermal cycler with the following program:

Initial denaturation	94°C—3 min.	
Denaturation	94°C—30 s	30 Cycles
Annealing	55°C—1 min	
Primer extension	72°C—70 s	
Final extension	25°C— 5 min	
Storage	4°C—forever	

5. Prepare 1.5% agarose gel in 0.5 × TAE buffer along with ethidium bromide.
6. Add 2 μL of loading dye to each tube.
7. Load the samples along with a molecular weight marker (100 bp ladder) and carry out electrophoresis for 3 h.
8. View the gel under a gel documentation system.
9. Primer sequences of Lr 26 gene:

Primer 1: CTCTGTGGATAGTTACTTGATCGA
Primer 2: CCTAGAACATGCATGGCTGTTACA

MICROSATELLITE MARKERS

Microsatellites are tandem repeats of DNA sequences of only a few base pairs (1–6 bp) in length in which the most abundant are the dinucleotide repeats. Microsatellites are also known as short tandem repeats (STRs) or simple sequence repeats (SSRs). The DNA sequences flanking SSRs are known to be conserved and they have been exploited to design suitable primers for amplification of the SSR loci using PCR. Differences in the length of the PCR-amplified product from different genotypes reveal the SSR polymorphism. The length differences are attributed to the variation in the number of repeat units at a particular SSR locus possibly caused by slippage during replication.

Advantages of Microsatellite Markers

The uniqueness and value of microsatellites arise from their multi-allelic nature, codominant transmission, case of detection by PCR, relative abundance, extensive genome coverage, and requirement of only a small amount of starting DNA. These markers can be distributed between labs as primer sequences acting as universal genetic mapping reagents. In addition, the SSR primers could also be used in related plant species provided the primer sites flanking the SSR region is conserved between those species.

Applications

1. *Marker-assisted selection in plant breeding*: The ability to distinguish between closely related individuals is important for crop species, which tend to have a narrow genetic base. To date, diversity detected with microsatellites reveal more polymorphism compared with other assay procedures.
2. *Determination of hybridity*: The codominant nature of the microsatellite primers allows the allelic contribution of each parent to be detected in sexual or somatic hybrids. Recently, microsatellites have been used to characterize and determine the hybridity of intraspecific somatic hybrids of potato.
3. *Genome mapping*: Microsatellites can provide reference (anchor) points for specific regions of the genome. Hence, creation of a microsatellite index is possible, although a high-resolution linkage map based exclusively on SSR is not possible.
4. *Gene tagging*: One of the reasons for developing microsatellites is their potential use as diagnostic markers for important traits in plant-breeding programs. Several resistant genes for soybean mosaic virus and peanut mottle virus have been identified using microsatellite primers.
5. *Cultivar identification, estimation of genetic relatedness, and germplasm conservation*: The unique mechanism responsible for generating high levels of microsatellite allele variability by "replication slippage" may reduce the usefulness of microsatellites for assessing genetic similarity between genotypes. However, this phenomenon is exploited to describe phonetic relationships between accessions that correspond to their geographical origins.

Method

Microsatellite PCR reaction mix (50 μL):

DNA (50 ng/μL)	4 μL
Microsatellite primer Forward (100 ng/μL)	2 μL
Microsatellite primer Reverse (100 ng/μL)	2 μL
1.0 mM dNTPs	4 μL
Taq DNA polymerase	0.6 μL
10× PCR buffer with $MgCl_2$	5.0 μL
Sterile double-distilled water	32.4 μL

Microsatellite PCR reaction conditions:

Initial denaturation	94°C—3 min	
Denaturation	94°C—1 min	45 Cycles
Annealing	50/55/60°C—1 min	45 Cycles
Primer extension	72°C—2 min	45 Cycles
Final extension	72°C—10 min	
Storage	4°C—forever	

INTER-SIMPLE SEQUENCE REPEAT (ISSR)

The ISSR technique is a PCR-based multilocus marker system that employs oligonucleotide primers homologous to microsatellite or SSR sequences such as (GATA) to amplify, mainly, the inter-SSR regions. Amplification products are obtained only if SSRs in opposite orientation are found within a PCR amplifiable distance, with flanking sequences matching the oligonucleotide primers. ISSR markers are multi-locus, unmapped dominant markers. Various techniques have been developed to use microsatellite sequences as PCR primers with minor variations. This has resulted in many synonyms of ISSR techniques, namely SSR-anchored PCR, anchored SSR-PCR (ASSR-PCR), microsatellite-primed PCR (MP-PCR), anchored microsatellite-primed PCR (AMP-PCR), single primer-amplification reaction (SPAR), randomly amplified microsatellites (RAM), and randomly amplified multiple polymorphism (RAMP). Distinction is required to be understood between ISSR and SIMS; STMS (sequence tagged microsatellite site) is also a PCR-based marker, in which unique DNA sequences flanking microsatellites are used to construct oligonucleotide primers. Resulting amplification products spanning the microsatellite region show rich allele divergence. However, ISSR polymorphisms are not influenced by repeat polymorphisms within the SSR.

Reaction Mixture for ISSR

St D/W	X
Buffer (10×)	2.5
dNTP (1 mM)	2.5
Spermidine	0.5
Formamide	0.5
Taq (5 u/μL)	0.16
Primer	0.4
DNA template	3.0 μL (35–30 ng)

AMPLIFICATION

Initial denaturation	94°C—5.0 min	
Denaturation	94°C—30 s	45 Cycles
Annealing	50°C—0.45 s	
Primer extension	72°C—70 s	
Final extension	25°C—5.0 min	
Storage	4°C—forever	

ISOLATION OF SPECIFIC FULL-LENGTH cDNAs BY RT-PCR METHOD

A full-length cDNA can be amplified via RT-PCR. To achieve this objective, a forward (or sense) primer and a reverse (or antisense) primer should be designed in the 5′-UTR (untranslated region) and the 3′-UTR regions of a known cDNA sequence or based on a known amino acid sequence.

ISOLATION OR RNAs BY TRIzol METHOD

1. From tissue: Harvest tissue and immediately drop it in liquid nitrogen. Store at −80°C until use.
2. Homogenize 0.3 g tissue in 1 mL of TRIzol reagent solution using a polytron at the top speed for 3 30-s bursts. Allow the cells to be lysed/solubilized for 5 min at room temperature.
3. Add 0.2 mL of chloroform per 1 mL of the Trizol reagent used. Cap the tube and vigorously shake the tube for 20 s to shear genomic DNA. Incubate at room temperature for 3 min for phase separation.
4. Centrifuge at 12,000 g for 20 min at 4°C in an Eppendorf tube or at 7,000–8,000 rpm/min for 30–40 min at 4°C using a 14 mL polypropylene tube in a swing-out rotor. Carefully transfer the top, aqueous phase to a fresh tube. Decant the bottom, phenol phase into a special container for disposal.
5. Add 1 mL of isopropanol, mix, and precipitate RNA at −20°C for 2 h.
6. Centrifuge at 12,000 g for 15 min at 4°C. Briefly rinse the total RNA pellet with 1 mL of 75% ethanol. Partially dry the RNA pellet at room temperature for 15 min.
7. Dissolve the total RNA in 20–50 μL of 0.5% SDS solution. Take 2 μL to measure the concentration with the spectrophotometer at 260 and 280 nm.

DESIGN AND SYNTHESIS OF SPECIFIC FORWARD AND REVERSE PRIMERS

A pair of forward (sense, or upstream) and reverse (antisense or downstream) primers should be designed based on the 5′-UTR or 3′-UTR sequences of a known cDNA to be isolated (Figures 2.9 through 2.11). These sequences can be found in published cDNA data on the same species or possible from different organisms. Specifically, the forward primer can be designed from the 5′-UTR in the 5′→3′ direction with 20–30 bases, which should be complementary to the first or (−) strand of the cDNA template. The reverse

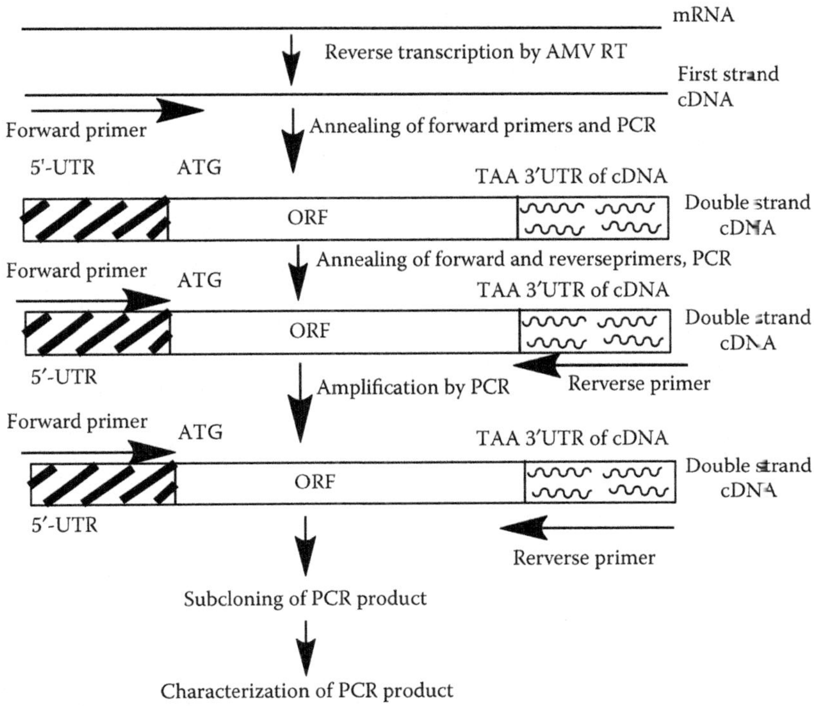

FIGURE 2.9 Diagram of isolation of full-length cDNA by RT-PCR.

primer is designed from the 3′-UTR region in the 5′→3′ direction with 20–30 bases, which should be complementary to the second or (+) strand of the cDNA template.

Forward and reverse primers can be designed based on the very N-terminal and C-terminal amino acid sequences in case the cDNA sequence is not available. As a result, if all goes well, a cDNA including the entire open reading frame (ORF) can be isolated with ease. For example, assuming that the N-terminal and C-terminal amino acid sequences are NDPNG and DPCEW, respectively, the forward and reverse primers can be accordingly designed as 5′-AAC(T)GAT(C)CCIAA (C)TGGI-3′ and 3′-GGTGAGCGTCCCTAG-5′. In case only the N-terminal amino acid sequence is available, a reverse primer may be designed as oligo(dT).

It is very important that each pair of primers is carefully designed to anneal with 2DNA strands. If 2 primers are annealed with the same DNA strand of the template, no PCR products will be produced. The primers should be analyzed by an appropriate computer program before they are used for PCR. We routinely use the Oligo Version 4.0 or CPrimer f program to check the quality of primers, including GC contents, T_m value, potential formation of intraloops/dimers and interduplex of the primers, and potential annealing between the primers and the region flanked by the primers. In general, primers should be >17 bases with <60% GC content. The first base at the 5′ or 3′ end should be G or C followed by A or T for efficient annealing. No more than 4 bases of intraprimer or interprimer complementary are allowed.

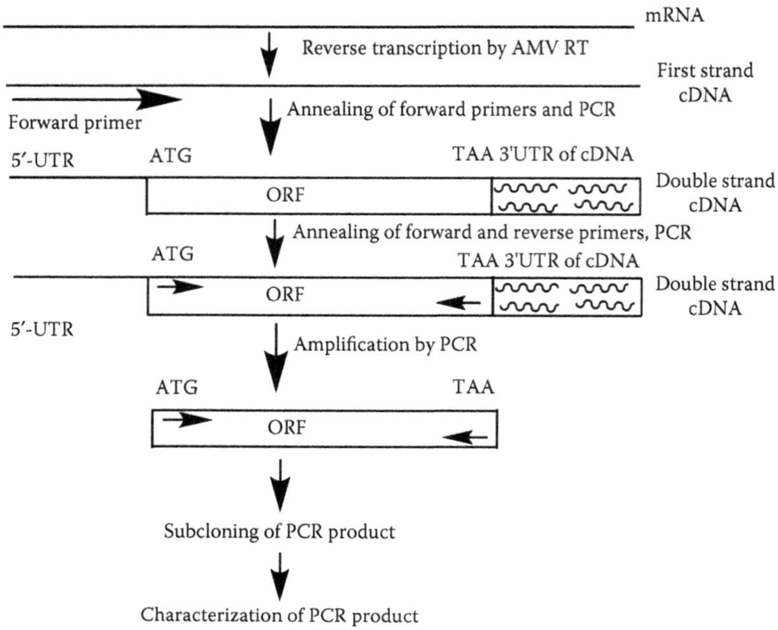

FIGURE 2.10 Diagram of isolation of full-length cDNA by RT-PCR using primers designed from known amino acid sequences.

The 2 different restriction enzyme sites are designed at the 5′ end of the forward primer and the 3′ end of the reverse primer. This is invaluable in facilitating efficient subcloning of PCR products for characterization, such as DNA sequencing and *in vitro* transcription. The enzyme-cutting site depends on the particular multiple cloning sites of an appropriate vector to be used for subcloning of PCR products.

Amplification of cDNA of Interest by RT-PCR

Target mRNA can be selectively transcribed to cDNA using a specific primer that anneals to the 3′ portion near the target region of the mRNA. It is recommended that oligo(dT) primers be used to transcribe all of the mRNA species into cDNAs. Oligo(dT) primers anneal to the 3′ poly (A) tails of mRNA molecules and facilitate the synthesis of the first strand cDNAs. Avian myeloblastosis virus (AVN) reverse transcriptase and Moloney murine leukemia virus (MoMuLV) reverse transcriptase are commonly used RT reactions. RT kits are commercially available. A standard reaction volume is 25 μL.

1. Perform synthesis of the first strand cDNA from mRNA.
   a. Anneal 10 μg of total RNA or 1 μg of mRNA template with 1 μg of oligo(dT) primers in a sterile RNase-free microcentrifuge tube. Add nucleasefree ddH_2O to a total volume of 15 μL. Heat the reaction at 70°C for 5 min and allow it to slowly cool to room temperature to finish annealing. Briefly spin down the mixture to the bottom of the microcentrifuge tube.

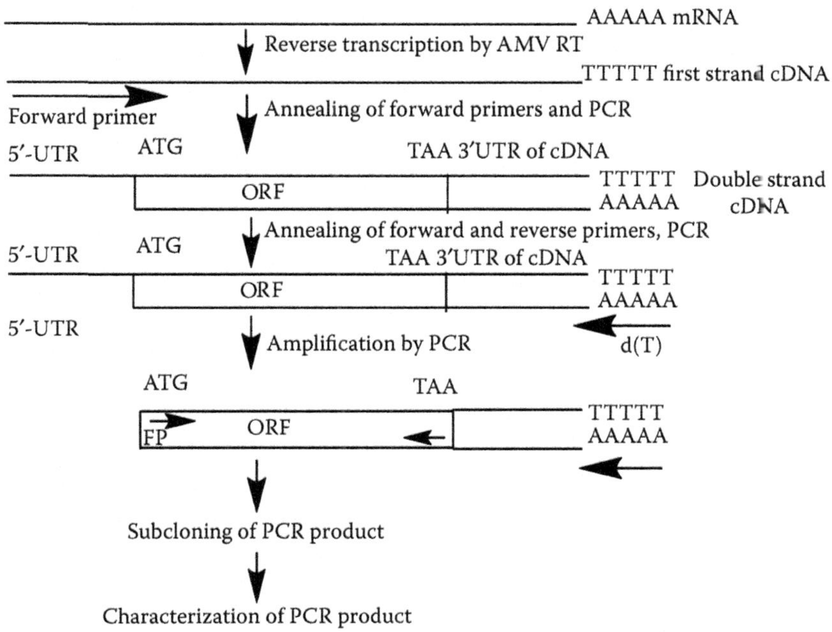

FIGURE 2.11 Diagram of isolation of full-length cDNA by RT-PCR using primers designed from known amino acid sequences and a poly (A) tail. FP, forward primer.

b. Add the following reagents to the annealed primer/template mixture in the order shown below:

First strand 5× buffer	5 μL
rRNase in ribonuclease inhibitor	50 units (25 units/μg mRNA)
40 mM sodium pyrophosphate	2.5 μL
AMV reverse transcriptase	30 units (15 units/μg mRNA)
Add nuclease-free ddH$_2$O to a total volume of = 25 μL	

c. Incubate the reaction at 42°C for 60 min. At this point, the synthesis of the first strand cDNA is complete.
d. Stop the reaction by adding 2 μL of 0.2 M EDTA to the mixture and place it on ice.
e. Precipitate the cDNA by adding 2.5 volumes of chilled (−20°C) 100% ethanol to the tube. Gently mix and allow precipitation to occur at −20°C for 2 h.
f. Centrifuge at top speed for 5 min, carefully remove the supernatant, briefly rinse the pellet with 1 mL of cold 70% ethanol, and gently drain away the ethanol.
g. Dry the cDNA pellet under vacuum for 10 min, and dissolve the cDNA pellet in 10 μL of TE buffer. Take 2 μL of the sample to measure the concentration of cDNA before PCR. Store the sample at −20°C until use.

Reagents needed

- 10× AMV or MoMuLV buffer
  0.1 M Tris—HCl (pH 8.3)
  0.5 M KCl
  25 mM $MgCl_2$
- Stock dNTPs
  10 mM dATP
  10 mM dTTP
  10 mM dATP
  10 mM dTTP
- Working dNTPs solution
  10 μL dATP (10 mM)
  10 μL dTTP (10 mM)
  10 μL dATP (10 mM)
  10 μL dTTP (10 mM)

2. Carry out amplification of specific cDNAs by PCR in a 0.5 mL microcentrifuge tube.
   a. Prepare a PCR cocktail as follows:

Forward primer (36–43 mer, 10–30 ng/μL)	5–8 pmol (30–40 ng) depending on the size of primer
Reverse primer (36–43 mer, 10–30 ng/μL)	5–8 pmol (30–40 ng) depending on the size of primer
cDNAs	1–2 μg
10× amplification buffer	4 μL
dNTPs (2.5 mM each)	4 μL
Taq or Tth DNA polymerase	5–10 units
Add ddH_2O to a final volume of 40 μL	

   b. Overlay the mixture with 30 μL of mineral oil to prevent evaporation of the samples during PCR amplification. Place the tubes in a thermal cycler and perform PCR.

		Cycling (30 Cycles)			
Profile	**Predenaturation**	**Denaturation**	**Annealing**	**Extension**	**Last**
Template (<4 kb) primer is <24 bases or with <40% G–C content	94°C, 3 min	94°C, 1 min	60°C, 1 min	70°C, 1.5 min	4°C, forever
Template is >4 kb primer is >24 bases with >50% G–C content	95°C, 3 min	95°C, 1 min	60°C, 1 min	72°C, 2 min	4°C, forever

Purification Products by High-Speed Centrifugation of Agarose Gel Slices of PCR

1. Load the amplified PCR mixture into a 1%–1.4% agarose gel, depending on the sizes of the PCR products, and carry out electrophoresis.
2. When electrophoresis is complete, quickly locate the DNA band of interest by illuminating the gel on a long-wavelength (>300) UV transilluminator. Quickly slice out the band of interest using a sharp, clean razor blade.
3. To enhance the yield of DNA, trim away extra agarose gel outside the band and cut the gel slice into tiny pieces with a razor blade.
4. Transfer the fine slices into a 1.5 mL microcentrifuge tube (Eppendorf).
   a. If one does not need to elute DNA out of the gel slices, the slices do not need to be further sliced into tiny pieces. They can be directly placed in a tube.
   b. At this point, there are 2 options for eluting DNA from the agarose gel pieces. The first option is to immediately carry out high-speed centrifugation. The second option is to elute the DNA as high yield as possible.
5. Centrifuge at 12,000–14,000 g or at the highest speed using an Eppendorf centrifuge for 15 min at room temperature.

Note: With high-speed centrifugation, the agarose matrix is compressed and/or even partially destroyed, by the strong force of centrifugation. The DNA molecules contained in the matrix are released into the supernatant fluid.

6. After centrifugation, carefully transfer the supernatant fluid containing the DNA into a clean microcentrifuge tube. The DNA can be used directly for ligation, cloning, labelling, and restriction enzyme digestion without ethanol precipitation. Store the DNA solution at 4°C or –20°C until use.

Note:

1. In order to confirm that the DNA is released from the gel pieces, the tube containing the fluid can be briefly illuminated with long-wavelength UV light after centrifugation. An orange-red fluid color indicates the presence of DNA in the fluid.
2. The supernatant fluid should be immediately transferred from the agarose pellet; within minutes, the temporarily compressed agarose pellet may swell, absorbing the supernatant fluid.

High Yield and Cleaner Elution of DNA

1. Add 100–300 μL of distilled deionized water (ddH_2O) or TE buffer (10 mM Tris-HCl, 1 mM EDTA, pH 8.0) to the gel pieces described step 4 above.

Note: ddH_2O or TE buffer is added to the gel pieces to help elute the DNA and to simultaneously wash EtBr from the DNA during high-speed centrifugation.

2. Centrifuge at 12,000–14,000 g at room temperature for 20 min; when centrifugation is complete, briefly visualize the eluted DNA in the fluid by illumination using a long-wavelength UV light.

3. Immediately transfer the supernatant the contains the eluted DNA into a fresh tube.
4. To increase the recovery of DNA, extract the agarose pellet by resuspending the pellet in 100 μL of ddH_2O or TE buffer followed by another cycle of centrifugation.
5. Pulll the DNA supernatant and precipitate the DNA with ethanol. Dissolve the DNA in 10 μL of ddH_2O or TE buffer. Store the DNA at 4°C or –20°C until use. Proceed to subcloning and characterization of the PCR products.

CONSTRUCTION OF GENETIC LINKAGE MAPS AND THE MAPPING OF QUANTITATIVE TRAIT LOCI (QTLs)

This protocol gives an example of how to handle molecular data for the construction of a genetic linkage map. It is also aimed at presenting a way to prepare phenotypic data for QTL analysis and subsequent QTL mapping. This protocol does not represent the only way to proceed, but it gives some general rules and advice when the following software types are used for analyses.

Mapmaker for the Construction of the Genetic Map, Plabstat for the Analysis of Phenotypic Data, PlaybQTL for the Interval Mapping QTL Analysis

Mapmarker, Plabstat, and PlabQTL, as well as Mapchart (used for the drawing of the map) are freely available and can be downloaded from the Web site given below.

The data used as the example for this protocol correspond to real data recorded on 240 F5 random inbred lines (RHs) of a population derived by single-seed descent of a cross between 2 Swiss winter wheat varieties. The genetic map of this population was realized using restriction fragment length polymorphism (RFLP) and simple sequence repeat (SSR) markers. The phenotypic data correspond to scorings of different traits on the same population in field trials with an incomplete block design and 2 replications.

The present protocol does not include the description of the commands and/or programs used for the different analyses. Electronic versions of the different manuals (Mapmarker, Plabstat, PlabQTL) are available.

1. Molecular data:
   a. Data recording

      Data must be recorded in an Excel file, in a table in which each column corresponds to a genotype and each line by a marker locus. The name of each locus must be preceded by a star that is essential for the Mapmarker data file.

      All the alleles coming from 1 of the parents are coded as "A." The alleles coming from the other parent are coded as "B." Heterozygous loci are coded as "H" and missing data as "-". Data have to be checked before use: the best way is to have them checked twice, once at least by an independent scorer.

In the data file, a chi-square analysis must be computed for each locus to test the existence of a distortion (deviation from the theoretical ration) or not. This information, which is not computed by Mapmaker, is essential for the construction of the map.

b. Preparation of the data for Mapmaker

   The Mapmaker data file comprises 2 lines dedicated to the description of the data. Followed by the data themselves. Each line of data corresponds to 1 locus, with a space separating each data point. This file can be easily obtained by saving the Excel file into a *csv file and replacing in winword the commas by spaces. The 2 lines of description of the data can be written at that point. To be used in Mapmarker, the file must then be saved as a *raw file.

c. Construction of the map with Mapmaker

   Mapmaker manuals can be downloaded from the Web site http://www.genome.wi.mit.edu/edu/genome_software/ and the software package is available at ftp://ftpgenome.wi.mit.edu/distribution/software/mapmaker3/.

d. Checking of the data: use of the graphical genotype method

   Even if data have been checked properly, it is not rare to find remaining errors in the data set. These errors lead to an inaccuracy of calculated distances and can even be responsible for marker inversions on the map. A way to check for these errors is to order the data in an Excel file in the order found for marker loci on the map. By using a different color for the "A" and "B" alleles, suspect data (close crossing over) can be identified. These data must then be checked and corrected if necessary.

e. Drawing of the map

   The Mapchart software (R.E. Voorrips, Plant Research International, Wageningen, The Netherlands) is of great help for the construction of the map, and its use is very simple. This software can be downloaded from the Web site http://www.joinmap.nl. The free use of the software is obtained by a simple license request to sales@biometris.nl.

2. Phenotypic data analysis

   Before being used for the QTL analysis, phenotypic data must be analyzed, and the adjusted means obtained through these analyses will be used for the QTL analyses. Plabstat is free software for data analyses, which can be downloaded from the Web site http://www.uni-hohenheim.de/-ipspwww/soft.html. The output data are nearly directly usable for QTL analyses with PlabQTL.

3. QTL analysis

   PlabQTL is free software that can be downloaded from the same internet site as Plabstat: http://www.uni-hohenheim.de-ipspwww.soft.html.

   a. Preparation of the data file *.qdt

      The data file comprises 3 lines of program giving indications of the data (population type, number of markers, genotypes, traits, etc., and eventually heritabilities) followed by:

      i. The molecular data themselves (same structure as I Mapmarker).
      ii. The phenotypic data.

iii. The description of each linkage group: name, list of markers, and distances (in centiMorgan—cM) between adjacent markers.

b. Description of controlling file *.qin
The *qin data file contains statements that control the analysis and allows for the choice between simple and composite interval mapping analyses. The different statements are presented in the manual and main statements can be used to perform a QTL analysis.

c. Interval mapping with PlabQTL
PlabQTL allows simple interval mapping (SIM) and composite interval mapping (CIM) for populations of individuals of selfing generations (F2 and later generations) or doubled haploids derived from an F1 cross between 2 homozygous lines.
Agenetic linkage map (for the 240 F5–RILs population) can be used to perform both SIM and CIM on phenotypic data.

d. Some words on single marker analysis by ANOVA
A first approach to detect QTLs is to compute an analysis of variance (ANOVA). This allows the effect of each single marker locus to be obtained for each considered trait. The advantage of this method is that it can be computed with any statistical software and allows the detection of effects even for unlinked markers. On the other hand, this method does not allow a precise estimation of the position and effect of the QTLs.

Note: A very useful link (http://linkage.rockefeller.edu/soft/list3.html#m) gives an alphabetical list of genetic analysis software with links to the related sites.

TRANSPOSONS AND THEIR MUTAGENESIS

Prokaryotic transposons are constituents of R plasmids and determine antibiotic resistance properties. Transposons are DNA sequences of varying sizes with inverted repeats at the ends and a gene encoding transposition. Properties of moving from one place to another in the genome are shown by all transposons, and this results in the induction of polar mutation. For this reason, transposons serve as a good source of mutagenic agents.

Transposon Tn5 is a commonly used mutagenic agent in a number of bacteria. Its various characteristics are shown in Figure 2.12. It is 5.6 kb in size and has a number of antibiotic resistance properties. It does not have any sites for EcoRI, but a number of sites are present for other important enzymes. It has been used both for general and site-specific mutagenesis in bacteria.

For using a transposon as a mutagenic agent, 2 things are required:

a. A replicon is needed that carries the transposon and can efficiently be introduced into the organism causing mutation.
b. A means is needed for selecting against stabilization of the transposon carrier replicon in the recipient strains.

Methods developed for the introduction of transposons into different bacterial genera are (i) by plasmid incompatibility, (ii) by the use of suicide vector, and (iii) using bacteriophage PI (generalized transducing phage).

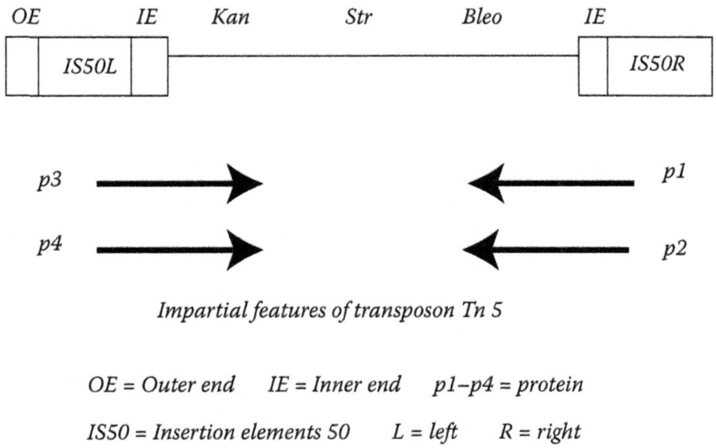

FIGURE 2.12 Structure of T5 transposon.

Site-Specific Transposon (Tn5) Mutagenesis

Strain

HBlOl: Tn5 hb 101 (pRK 290.7)
HB 101 (Prk 2073), C2110 nal

Method

a. Transformation of HB 101: Tn5 following the procedure given earlier
   1. Isolate plasmid DNA from HB 101 (pRK290.7) cells (follow the procedure given before).
   2. Find out the amount of DNA and store in refrigerator.
   3. Mix 1 μg plasmid DNA with 0.5 mL competent cells in Eppendorf tube.
   4. Put on ice for 30 min.
   5. Heat shock at 42–44°C for 3 min.
   6. Dilute cells in 10× volume of LB in test tube.
   7. Shake at 37°C for 60 min.
   8. Centrifuge at 10,000 × g for 10 min, resuspend in 1/10 volume of LB.
   9. Plate 10^{-1}, 10^{-2}, 10^{-3} dilutions on LB+ tet (12 μg/mL) and kan(50 μg/mL).
   10. 24 hours before streaking, set the culture of HB 101 (prk 2073) and C2110 NAL in 10 mL LB liquid medium in test tubes and shake culture at 37°C overnight.
   11. Pool together >500 transformations from step 9 by adding 1 mL of LB liquid to the plate; collect the cells in sterile test tube.
   12. Perform mating on LB plate by mixing together 0.1 mL cells of the following 3 strains from step 10 and 11: HB 101::Tn5 (pRK290.7)=pooled cellsX Hbl01 (pRK 290.7)X C2110nal.
   13. Incubate overnight at 37°C.

14. Scrap the growth area from the plate, suspend in 2 mL of LB liquid medium, vortex, and plate LB + Tet_{12} + Kan_{50} + Nal_{10} solid medium.
15. Incubate plates at 37°C overnight.
16. Collect 100 transformations individually and inoculate them in Sml LB + Tet12 liquid medium.
17. Maintain each transformant by streaking on LB + Tet 12 plates before proceeding for plasmid DNA isolation.
18. Isolate plasmid DNA through the procedure at b, given below. Restrict them with EcoRI and run in agarose gel following the usual procedure. Keep 2 lanes of control plasmid DNA (pRK 290.7) cut with EcoRI, 1 in the first late and 1 in the center.

b. Rapid plasmid DNA isolation

19. Transfer 5 mL cells from test tubes (from step 16) to polypropylene centrifuge tubes and centrifuge them at 10,000 × g for 5 min; decant the supernatant.
20. Thoroughly suspend in 100 μL TE and transfer to Eppendorf tubes.
21. Add 200 μL NaOH/SDS and shake well to mix, keep for 5 min at 0°C.
22. Add 150 μL 3 M sodium acetate and invert to mix, keep for 10 min at 0°C.
23. Spin 3 min, remove 400 μL supernatant into Eppendorf tubes.
24. Add 1 mL ice-cold ethanol and let stand for 30 min at –20°C.
25. Spin 3 min, aspirate off supernatant, dry with ethanol.
26. Dissolve pellet in 400 μL 0.1 M sodium acetate tris, vortex, add 2 volumes cold ethanol, keep for 10 min at –20°C.
27. Spin for 3 min. Remove ethanol by aspiration, then add 0.5 mL 95% cold ethanol to wash, spin 3 min, dry the pellet, and dissolve in 30 μL water.

Generalized Transposon (Tn5) Mutagenesis

The goal of this experiment is to carry out transposon (TnS) insertion mutagenesis in *Rhizobium cicer* strain (pR200S) to isolate mutants defective in dicarboxylic acid transport pathway and to characterize them.

Nitrogen fixation by the *Rhizobium*-legume symbiosis is a high-energy-demanding process. The infected host plant supplies bacteroids with photosynthate, which is converted to ATP, and reducing power for the support of nitrogen fixation and the maintenance of nodule function. The nitrogen-fixing capacity of the *Rhizobium*-legume symbiosis is thought to be limited by the amount of photosynthate available to the bacteroid. Therefore, knowledge of the carbon compounds supplied to the bacteroids is a prerequisite for formulating any strategy aimed at improving the nitrogen-fixation process.

The isolation of mutants defective in utilization of the TCA cycle intermediated in a number of *Rhizobium* strains has indicated the important role played by these carbon sources in symbiotic nitrogen fixation. But, to date, nothing is known about the C-4 dicarboxylate pathways in *Rhizobium ciceri* strains. To elucidate the role C-4 dicarboxylate in biological nitrogen fixation in *Rhizobium cicer,* experiments dealing with the isolation and characterization of *dct* negative mutant in this bacteria are undertaken.

Requirements

1. Bacterial culture
   a. *Mesorhizobium ciceri* strain (Pr2005) Rif
   b. *E. coli* strain (*WA803/p* gs9:Tn5) Cam;Cb, Met, Thi
2. Sterile water
3. Luria agar plate with neomycin 100 μg/mL and chloramphenicol 30 μg/mL
4. TY agar plate with rifampicin 50 μg/mL
5. RMM +TZ agar medium with neomycin 100 μg/mL
6. Glass spreader
7. Ethyl alcohol
8. Sterile 0.45 μm filter paper
9. Sterile Eppendorf tube
10. Ty agar plate

Medium Composition

Luria Agar		RMM+TZ Medium		TY Agar Medium	
Tryptone	10.0 g/L	K_2HPO_4	0.2 g/L	Tryptone	5.0 g/L
Yeast extract	5.0 g/L	KH_2PO_4	0.15 g/L	Yeast extract	3.0 g/L
NaCl	10.0 g/L	NaCl	0.15 g/L	$CaCl_2$	0.01%
pH	7.0 g/L	NH_4NO_3	0.5 g/L	Water	1000 mL
Water	1000 mL	$MgSO_4 \cdot 7H_2O$	0.5 g/L	Agar powder	20.0 g/L
Agar powder	20.0 g/L	$CaCl_2$	0.01 g/L	pH	6.8
		Tetrazolium salt	25 mg/L	Tryptone	5.0 g/L
		pH	6.8		
		Water	1000 mL		
		Agar powder	20.0 g		

The following ingredients will be filter sterilized and can be added aseptically to the above medium at the concentrations given below:

Biotin	1 mg/L	Succinate	2.7 g/L = 10 mM
Thiamine	1 mg/L	Glucose	1.8 g/L = 5 mM

Preparation of Antibiotic Stock Solution

1. Neomycin 10 mg/mL in sterile distilled water
2. Rifampicin 10 mg/mL in methanol
3. Chloramphenicol 10 mg/mL in ethanol

Procedure

1. Grow the recipient strain (pR2005) in TY broth at 300 C in a shaker overnight, and grow the donor strain WA803IPgs9::Tn5 in Luria broth at 37°C in a shaker for 12 h.

2. Mix 0.5 mL of each (donor and recipient) in an Eppendorf tube and spin down the cells by centrifugation. Wash the pellet 2 times with sterile water. Finally, the pellet will be dissolved in100 μL of sterile water.
3. Transfer the mixed culture (100 μL) on to the filter paper (0.45 μm) placed on the TY agar plate or directly on the TY agar plate.
4. Incubate the plate at 30°C for 36–40 h for conjugation.
5. Wash off the mated cultures off from the plate with 1 mL of sterile water or normal saline.
6. Plate 1/10 mL of the dilution on TY agar with rifampicin 60 μg/mL for recipient count, with neomycin 100 μg/mL and chloramphenicol 30 μg/mL on a Luria agar plate for the donor count and on TZ medium with neomycin 100 μg/mL for the selection of transconjugants.
7. After 4 days of incubation at 30°C, transconjugants appear as neomycin-resistant single colonies and can be observed by the pink and white colors.
8. Only the white colonies will be picked up onto master plates for the postmating selection.
9. Each master plate will be a replica plated onto the agar medium, such as
   a. RMM + TZ + succinate
   b. RMM + TZ + malate
   c. RMM + TZ + fumarate
   d. RMM + TZ + aspartate
   e. RMM + TZ + glucose
   f. RMM + TZs + succinate + glucose.

   Each of the above plates will contain neomycin 100 μg/mL.
10. Colonies defectives in dot pathway will not appear on the plates supplemented with only succinate, malate, fumarate, or aspartate. This is because cultures that grow on glucose as a sole carbon source produce acid and hence white colonies are formed in TZ (2,3,5-triphenyl tetrazolium salt) medium, while culture grown on medium containing succinate with or without glucose produce alkali and they turn red on TZ medium.

ON 2-DIMENSIONAL GEL ELECTROPHORESIS FOR ANALYSIS OF GENE EXPRESSION

Two-dimensional polyacrylamide gel electrophoresis using isoelectric focusing (IEF) in the first dimension separates proteins by charge and size. Although simple in concept, the technique has been underused for 3 main reason: (i) the difficulty in obtaining highly reproducible patterns of isoelectric focusing in the first (pH) dimension using tube gels and soluble carrier ampholytes, (ii) the lack of good analytical tools for analysis and assignment of differential patterns of gene expression, and (iii) the inability to obtain sequencing information from any other than the most abundant proteins. In the last few years, however, the practical and analytical tools have changed to such an extent that analysis of the proteome by 2-dimensional gels can (and should) now be considered an accessible technique. In 2-dimensional gel electrophoresis, different steps are involve, such as

1. Sample preparation
2. Rehydration of IPG strips
3. Isoelectric focusing
4. Equilibration of strips before transfer to SDS-PAGE
5. Casting the 2-dimensional gel
6. Drying of polyacrylamide gels

Sample Preparation

The quality of a 2-dimensional gel separation depends almost entirely on the quality of the sample. A good sample for 2-dimensional separation must: (i) retain its native characteristics as much as possible, (ii) contain no precipitate, and (iii) be as enriched as possible with the proteins being analyzed (Figure 2.13).

Equipment and Reagents

- Phosphate buffered saline (PBS): 150 mM NaCl, 5 mM KCl, 10 mM Na-phosphate buffer pH 7.4
- Lysis buffer: 9 M urea, 2% (w/v); CHAPS, 3-[(3-cholamidopropyl) dimethylammonio]-1-propane sulphonate, 0.5% (v/v); Triton X-100; 20 mM DTT; 1.2% pharmalytes pH range 3–10; 8 mM PMSF
- Liquid nitrogen-cooled mortar and pestle (for grinding tissues)
- Sonicator (for grinding tissues)
- Quick method for protein determination (such as Bradford reagent)

Method

1. Wash tissue quickly in ice-cold PBS, grind in cooled mortar with pestle, and suspend immediately in lysis buffer. Sonicate mixture for a further minute on ice.
2. For cultured cells, wash plates (if cells are adherent) thoroughly with ice-cold PBS, place dish on ice bucket at an angle to drain all excess PBS for 1–2 min, warm plate to room temperature and add lysis buffer, collect sample by scraping with a rubber policeman (can be the plunger of a 1–5 mL syringe). For cells in suspension, follow the same wash protocol, except that washes should be done by centrifugation.
3. Determine protein concentration; a concentration of 2–5 mg/mL is best for analytical detection by silver staining, although up to 10 mg/mL samples have been applied to IPG strips with good results.
4. Spin sample at 15,000 g for 10 min to remove all particulate material.
5. Save supernatant sample in aliquots of 100 μL at −80°C until use.

Rehydration of IPG Strips

Rehydration of IPG strips can be performed separately from and prior to sample application using a commercial set-up or an easily assembled homemade device. Alternatively, it is possible to swell the strip with the solution that already contains the

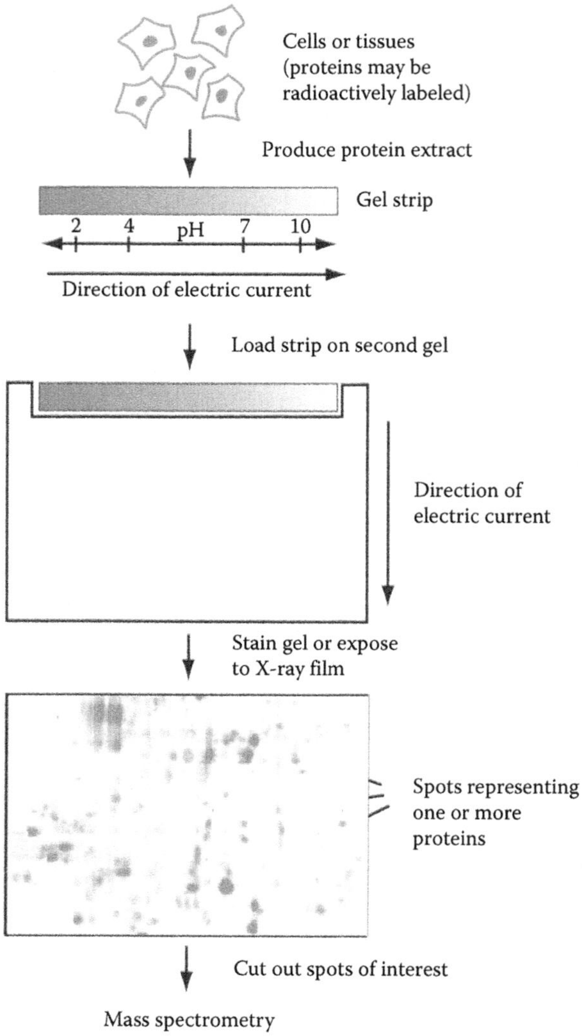

FIGURE 2.13 2-dimensional gel electrophoresis.

sample (rehydration loading). The latter is a neat way of ensuring smooth impregnation of the gel with the sample without getting deposits that can occur when the sample is applied in 1 spot (at the anode or cathode). We have used a sample volume of 500 μL for an 18 cm × 0.5 cm strip (slightly in excess of calculated volume of 450 μL), using an nuclear magnetic resonance (NMR) glass tube. This has served as our swelling cassette with excellent results. However, it is possible that some proteins in the sample will not enter the gel easily during passive swelling, so if your sample is scarce, it may be advisable to use the more conventional route of loading the sample onto a preswelled gel as loading is then performed under voltage and so increases the capacity of the gel for sample entry.

Equipment and Reagents

- IPG strips come in various pH ranges, gradient shapes, and sizes. They are cast on a strong plastic backing, and are covered with a thinner plastic protective sheet. Keep flat in –20°C until use. Fully swelled, they should be 0.5 mm thick.
- Rehydration solution contains 8 M Urea, 2% CHAPS (3-[3-cholamidopropyl] dimethylammonio)-1-propane sulphonate, 2% Pharmalytes pH 3–10, 10 mM DTT, and a few grains of bromophenol blue.
- Silanizing solution, which should be used in a fume hood.
- Rehydration cassette (2 homemade glass plates separated by 0 7 mm waterproof u-shaped PVC gasket with spring clips to hold assembled plates together can be used). Glass plates should be cleaned thoroughly before use (wiped so that no specs to dust or dirt adhere to the plate surfaces). Gasket should also be cleaned thoroughly to remove any protein contaminants.

Method

1. Rehydration in a cassette.
2. Put 1 of the 2 glass plates in a fume hood. Use a plastic dropper to place a few drops of sigmacote on the plate and use a tissue to spread it lightly over the glass surface. Allow the solution to evaporate. Repeat process 3 times. Wash with glass surface. Allow the solution to evaporate. Repeat process three times. Wash with 70% ethanol to remove residue and let dry. Solution should be reapplied every few months.
3. Take non-silanized glass plate, lay it flat, place u-shaped gasket, and moisten glass surface tightly with distilled water.
4. Remove protective (thin) plastic strip off and lay strips flat with plastic backing side down on moistened glass surface so that the moisture traps them securely in place.
5. Place second (silanized) glass plate on top, silanized face down, clip both plates together and stand cassette vertically.
6. Pour rehydration buffer into cassette ensuring that the entire surface of the strip is covered; cover with cling film to prevent evaporation and leave overnight to swell.
7. Rehydrate sample directly in the gel-rehydration cassette.
8. Calculate the precise volume of solution required for swelling an individual strip.
9. Put the sample (in lysis buffer) in a shallow purposely made well, or use an NMR tube and gently place strip with the free surface facing downward into the sample.
10. Cover the well to avoid evaporation and gently shake on a side by side rotator overnight.

Isoelectric Focusing

This method is useful because it is advisable to begin with a low voltage setting (100–300 V) to facilitate sample entry before switching to the final voltage. However, the main principle is to obtain voltages of 3.5–5 kV in order to accumulate between 30 and 100 kVh total voltage × time for good steady-state separations to be completed within 24 h. Any power supply that can give such output is sufficient. If the strips were swelled without the sample, the sample must then be applied separately and become absorbed into the gel. If the sample is a small volume, then it can be added directly on to the rehydration strip. We have added dilute, labeled samples of sympathetic neuron proteins in small volumes (18 μL) directly on to the rehydrated strips (at the anode end) with good results. Alternatively, the Multiphor system comes with sample cups. These can be tricky to operate because they must touch the gel surface so as to form a closed chamber without gouging into it, thus causing discontinuity, so care is required. The biggest problem with a sample added at 1 gel end in a small volume is that proteins can precipitate at the site of deposition. To offset this problem. On should try loading sample at the anode and at the cathode ends. It is also important to begin with a gentle application of 100–300 V for 1–3 h to facilitate sample entry. Finally, running temperature should be about 20°C. At lower temperatures, urea can crystalline out, causing water to follow, ultimately distorting the separation pattern and not allowing a completely steady-state condition.

Equipment and Reagents

- Multiphor II horizontal electrophoresis apparatus with Immobiline drystrip tray (which can be reused)
- Power supply that yields 3.5 kV
- 3 MM Whatman paper for making wicks
- Chilling water bath that can maintain a temperature of 15°C
- Sample cups
- Paraffin oil

Method

1. Pour off rehydration solution and open swelling cassette.
2. Blot each strip between 2 wetted filter papers that are just moist.
3. Align strips in the immobiline drystrip tray such that the pointed end is facing the anode (+) electrode and the blunt end is facing the cathode (−) end.
4. For wicks, cut 2 strips of 3 MM Whatman paper about 0.5–7 cm wide and moisten with deionized water. Place 1 strip across the (+) end and 1 strip across the (−) end of the gels.
5. Lay appropriate electrodes on wicks.
6. Place sample cups according to the manufacturer's instructions. Be sure to center the cups over the strips and press then down slowly so that they touch the surface of the swelled strips firmly but do not gouge out the gel. This is a bit tricky.
7. Optional; cover sample cups and strips with light paraffin oil.

Note: this step is useful because oil will leak thorough cups that are not adhering to the gel properly and so indicate a problem. If this happens, press the cups down again gently and make sure there are no leaks. Oil seepage doesn't affect the end result. However, oil is not absolutely necessary to get good separations.

8. Add samples to cups under the oil layer. It is best to have a dilute sample in 100 μL rather than a concentrated sample in a small volume because this avoids precipitation of the sample at the point of entry. Do not use less than 20 μL per cup.
9. Connect circulating water bath set at 15–20°C.
10. Connect power pack and begin sample run.
11. Optional: sample in <20 μL can be added directly to the strip (we added the samples at the anode end). In this case, begin the IEF and add paraffin oil only after the samples have entered the gel.
12. At end of the run, disconnect power supply, remove IPG strips, and either process them directly for second dimension or wrap each strip carefully, without damaging surface, in aluminium foil and store flat at –20°C or –80°C.

Equilibration of Isoelectric Focusing Strip before Transfer to SDS-PAGE

To obtain reproducible multiple 2-dimensional gels, the same set of proteins in each sample must be transferred from the first to the second dimension SDS-PAGE. Ideally, none of the sample should be left in the IEF strips. Various protocols have been describe that:

1. Facilitate SDS binding prior to loading strips on to the second-dimension gel.
2. Lower concentration of the urea, which can obstruct transfer, especially if it is crystalline.
3. Prevent oxidation of the thiol groups and formation of S–S bonds (second-dimension gels may retain some ammonium persulphate [APS], which will also oxidize proteins, so blocking SH groups with iodoacetamide will prevent this. If, however, proteins are to be sequenced, covalent modification may not be a good idea). It should be noted, however, that some proteins will invariably leach out into the equilibration buffer so that a balance must be struck between the time required for proper equilibration and the possible loss incurred.

Equipment and Reagents

- Large petri dish or Perspex block containing wells of 0.8 cm × 20 cm × 1 cm
- Equilibration buffer 1:50 mM Tris-HCl buffer pH 6.8, 6 M urea, 30% (w/v) glycerol, 1% (w/v) SDS, 2.5 mg mL^{-1} DTT
- Equilibration buffer 2:50 mM Tris HCl buffer pH 6.8, 6 M urea, 30% (w/v) glycerol, 1× (w/v) SDS, 45 mg mL^{-1} iodoacetamid, 30 μg mL^{-1} bromophenol blue

Method

1. Submerge strips in equilibration buffer 1 for 10 min at room temperature.
2. Wipe tightly on tissue to remove excess fluid.
3. Submerge strips in equilibration buffer 2 for 5 min at room temperature
4. Optional: cut off 1–1.5 cm from either end as usually these areas are covered by the wicks and no focusing would have occurred around these regions. For an 18 cm strip, this leaves a strip length of 15 cm, which will fit most medium-size PAGE systems.

Casting the 2-Dimensional Gel

Equipment and Reagents

- Any PAGE system (thickness 0.75 mm) that will support the length of the IPG strips. A square formate gel seems to work best.
- Resolving gel: acrylamide/bisacrylamide (a mix of 10% total acrylamide+bis acrylamide, 2.67% bisacrylamide relative to total will show an actins tot, 43 kDa, about halfway down the gel), 375 mM Tris-HCl pH 8.8, 0.1% SDS (w/v), 0.05% (v/v) N′N′N′N′ tetramethylethylenediamine.
- 1% (w/v) low melting point agarose in 150 mM Bis-Tris, 0.1 M HCl, 0.2% SDS (w/v).
- Small paper strips (0.5 × 1.5 cm) as identifying labels. Use an HB pencil to mark each strip.
- Fixing solution (I) 50% ethanol, 12.5% acetic acid (1 L for six gels, 20 cm × 20 cm) for drying down directly or as a prelude to liver staining.

Method

1. Place a label in a bottom corner of the glass cassette and cast the resolving gel (regular paper does not float but some filter papers may float depending on gel density). After polymerization, wash surface gently with distilled water.
2. Heat the agarose solution to melting, cool to about 60°C, and pour on to resolving gels.
3. Immediately immerse the IPG strips through the agarose until they just touch the surface of the gel.
4. Run sample as you would normally for a 1-dimensional gel. The faster the run, the more focused the spots. However, spot yawing may occur if the temperature rises.
5. At the end of the run, remove each gel carefully.
6. If gels are to be dried down without staining, or stained with silver nitrate, gel can be immersed in fixing solution 1 h to overnight at room temperature with gentle agitation.
7. If gels are to be blotted, transfer on to PVDF membrane without fixing.
8. Gels may be fixed and stained in Coomassie brilliant blue or colloidal solutions of Coomassie.

Drying of Polyacrylamide Gels

Drying large slab gels can cause large distortions that vary between gels because of uneven shrinkage. To avoid these problems as much as possible, it is best to avoid drying gels rapidly at high temperatures. We devised a slow-drying protocol that is outlined below.

Equipment and Reagents

- Wash solution: 30% (v/v) ethanol
- Glycerol impregnation solution: 30% ethanol containing 4% (v/v) glycerol (to provide gel with some flexibility)

- Silylated glass plates (22 × 22 cm)
- Squares of cellophane about 5 cm larger than the glass plate on which the gel will be dried
- Squares of Whatman 3 MM Paper (about 5 cm bigger than the glass plates)
- Photocopy film

Method

1. Wash gel in 30% ethanol for at least 1 h (to remove acetic acid fixative).
2. Impregnate gels with glycerol solution 2 × 30 min.
3. Place 1 gel onto the silylated surface of the glass plate (written surface of paper label facing glass plate).
4. Immerse cellophane in the glycerol impregnation solution until it is completely wet.
5. Place sheet of cellophane on top of the gel, remove all air bubbles and warp the excess cellophane around the plates such that the surface is smooth and taut.
6. Place the cellophane-wrapped glass plate (gel facing up) firmly on top of 2 Whatman 3 MM sheets and allow to dry overnight at room temperature or for a few hours at 35°.
7. Optional: spray photocopy film with Spraymount (3M) and glue to the cellophane side of the dried gel.
8. Remove glass plate, trim excess cellophane/photocopy sheet. Store dried gels in A4 plastic pocket.

METABOLIC LABELING FOR DETECTION OF DIFFERENTIAL GENE EXPRESSION

Often, it may be desired to obtain a differential fingerprint of proteins induced within a short time after a treatment (e.g., with a growth factor). In such case, *in vitro*, it is possible to capture the production of these proteins by metabolic labeling. Metabolic labeling with (^{35}S) methionine is compatible with matrix-assisted laser desorption/ionization methods or N-terminal sequencing of proteins from PVDF blots. Administration of carrier-free (^{35}S) methionine will give maximum specific activity of proteins and thus increase the sensitivity of detection but may not yield enough protein for analysis by microsequencing. High sensitivity may be combined with higher yields by diluting carrier-free labeling medium with a methionine-containing medium or spiking a cold protein extract with the "hot" extract. There are 2 main products for labelling with (^{35}S) amino acids. One in which methionine alone is used in a highly purified form, and one which contains a mixture of (^{35}S)–labelled methionine and cysteine (Translabel, ICN). The latter is derived from a dydrolysate of bacteria fed with ^{35}S sulphur and may contain some trichloroacetic acid (TCA)-precipitable material. This may add background when a TCA precipitate obtained from the labelled material is used to calculate total incorporation. Different steps are involve for the metabolic labelling.

Sample Preparation

Equipment and Reagents

- L-(^{35}S)methionine/cysteine >1000 ci mmol^{-1}
- Methionine-free medium
- Glass fiber filters
- BSA solution (2 mg mL^{-1} bovine serum albumin fraction V)
- 25% trichloroaceticc acid (w/v) in water (take care: solution is very corrosive)
- Scintillation fluid

Method

1. Incubate tissue/cells for 10 min in methionine-free medium to lower endogenous methionine levels.
2. Dilute 500 μCl (^{35}S) into 1 mL of methionine-free medium and add back 1/20 volume of full medium (to avoid reduction in rates of proteins synthesis due to limiting concentrations of amino acids).
3. Label cells for the desired period.
4. Wash out excess label using a normal medium containing methionine.
5. Wash cells thoroughly with ice-cold PBS, place dish on ice bucket at an angle to drain all excess PBS for 1–2 min, warm plate to room temperature, and add lysis buffer; collect sample by scraping with a rubber policeman (can be the plunger of a 1–5 mL syringe).
6. Estimate amount of (^{35}S) incorporated by adding a known volume (1%–5% of total) to 250 μL of BSA solution; then add 250 μL 25% TCA and leave at 4°C for at least 10 min.
7. Decant onto a glass fiber filter prewetted with 10% TCA solution and maintained under suction, wash filter 3 times with 10 mL ice-cold 10% TCA, dry filter, and count using an aqueous grade scintillation fluid.

Preparation of Calibration Strip for Quantitation of (^{35}S) Methionine-Labelled Spots

If a set of 2-dimensional gels are used to compare between several samples of (^{35}S) methionine-labelled proteins in which differences in intensities are used to indicate different levels of expression, it is imperative to prepare calibration strips with known amounts of radioactivity to offset any differences in exposure time, quality of film, or storage of phosphoimaging screens (Figure 2.14). Hence, the sum total intensity of all the spots can be scaled to that of the calibration strip before spot quantitation. Methionine has a half-life of 84 days, so [^{14}C]-labelled proteins (half-life ~5000 years) are used instead. To obtain [^{14}C]N-ethylmaleimide (proteins were a mix containing 1 mg/mL cytochrome c, RNase A, creatine kinase, actin, BSA, and β-galactosidase). However, the most inexpensive [^{14}C]-labelled protein product from any commercial supplier can be used instead as long as one can achieve 2–4 × 10^6 c.p.m.mL^{-1}.

Equipment and Reagents

- Minigel apparatus
- Capable of yielding 2–4 × 10^6 c.p.m.mL^{-1}

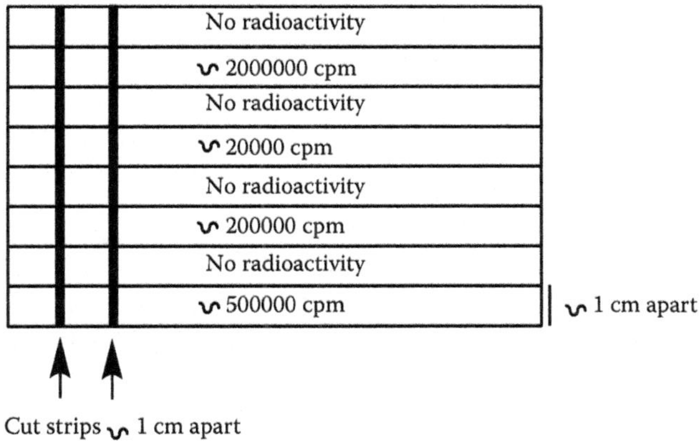

FIGURE 2.14 Preparation of calibration strips.

- 10% TEMED (N′N′N′N′ tetramethylethylenediamine)
- 8 mL of incomplete gel mixture (acrylamide/bisacrylamide 10% T/2 67% C in water, 0.0075% APS) but no TEMED

Method

1. Prepare 1 mL aliquots of the incomplete gel mixture.
2. To each of 4 aliquots, add [^{14}C]-labelled proteins to obtain a range of gel solution containing a spread of values from approximately ~20,000 to 2,000,000 c.p.m.mL^{-1}.
3. Collect a 10 μL sample from each solution to measure the precise radioactivity in the sample (in c.p.m.mL^{-1}).
4. Cast sequential layers as follows: add TEMED (0.03% final concentration) only to the layer being cast. Be sure to let each layer polymerize before adding the next layer.
5. Dry the gel.
6. Cut out strips at 90° to the direction of the layers.
7. Use 1 calibration strip for each gel being exposed.

Silver Staining for Mass Spectrometry

Most proteins are colorless and will not be visible in the gel without specific histochemical coloring procedures. For general protein staining, there are several more-or-less-sensitive methods. Many of the stains were originally developed by the textile industry (e.g., the widely used Coomassie brilliant blue). In recent years, more sensitive techniques (e.g., silver staining) have been developed. Commercial kits are available for silver staining, but the technique is also thoroughly described and can easily be adopted from Westermeier (1993).

Reagents

- Fixing solution 1:50% methanol, 5% acetic acid
- Fixing solution 2:50% methanol

- Sensitizing solution: 0.02% thiosulphate
- Staining solution: 0.1% silver nitrate in water (chill to 4°C)
- Deionized water
- Developer: 2% sodium carbonate containing 0.04% formaldehyde
- Stopping and storage solution: 5% and 1% acetic acid

Method

1. Fix gel in fixing solution 1 for 20 min.
2. Wash out acetic acid in fixing solution 2 for 10 min.
3. Wash 3 times with deionized water to remove all traces of acid.
4. Sensitize proteins with thiosulphate for about 1 min.
5. Rinse 2 × 5 min with water.
6. Submerge gel in silver nitrate solution at 4°C for 20 min.
7. Transfer the gel to a new clean dish that hasn't been in contact with silver nitrate. and rinse gel with water 2 times for 1 min each.
8. Add developer under constant supervision, making sure that developer is changed promptly if a yellowish color develops.
9. When spots are observed, stop development by replacing developer with 5% acetic acid.
10. Wash gel once in stopping solution, and store gel in 1% acetic acid.

3 Recombinant Techniques

DIGESTION OF PLASMID DNA WITH VARIOUS RESTRICTION ENDONULEASES AND SEPARATION OF PRODUCTS IN AGAROSE GELS

PLASMIDS

pBR322 is an *E. coli* cloning vector that carries genes conferring resistance to ampicillin and tetracycline and an orgin of replication. The vector has a length of 4.36 kb, and within the ampicillin and tetracycline genes, there are a number of recognition sites for restriction endonucleases. For example, PstI cuts within the ampicillin-resistance gene, whereas BamHI, HindIII, SalI, and ClaI all cut within the tetracycline-resistance gene. These genes are therefore suitable as selection markers in cloning experiments.

pBLUESCRIPT KS(+)

pBluescript KS(+), a 2.9 kb derivative of the pUC plasmid series, contains a multiple cloning site (polylinker) with recognition sites for 21 restriction endonucleases (Figure 3.1). The recognition sites are so arranged that those for enzymes creating 5′ overhangs are flanked on both sides by recognition site for endonucleases creating 3′ overhangs. Furthermore, the presence of T3 and T7 promoters on either side of the polylinker allows *in vitro* transcription with either T3 or T7 RNA polymerases. Yet another advantage of this plasmid is that it carries an intergenic sequence of M13, permitting the isolation of single-stranded DNA with the aid of a suitable helper phage.

KS(+) describes the orientation of the polylinker region (KpnI,SacI) with respect to the direction of transcription of the *lacZ* gene. The vector pBluescript KS(+) and the pUC series of vectors contain part of the *lacI* gene and the first 492 nucleotides of the *lacZ* gene (lacIOPZ), as well as the gene encoding β-lactamase (amp^r) and an origin of replication from *E. coli*. The polylinker is inserted into the *lacZ* gene so that the reading frame is not affected.

The enzyme that is encoded by the *lacZ* gene of *E. coli* plays an important role in molecular biology (blue/white selection). The inactivation of the *lacZ* gene by insertion of foreign β-glactosidase (the first 146 amino acids of β-galactosidase) is enzymatically inactive, but in an aggregate with another likewise-inactive fragment of the enzyme lacking amino acids 11–41, the result yields an active complex. This phenomenon is known as α-complementation and depends on the noncovalent association of peptide fragments which allows 2 inactive peptides to aggregate in such a way that an active β-galactosidase is created. This property of β-glactosidase is used as the

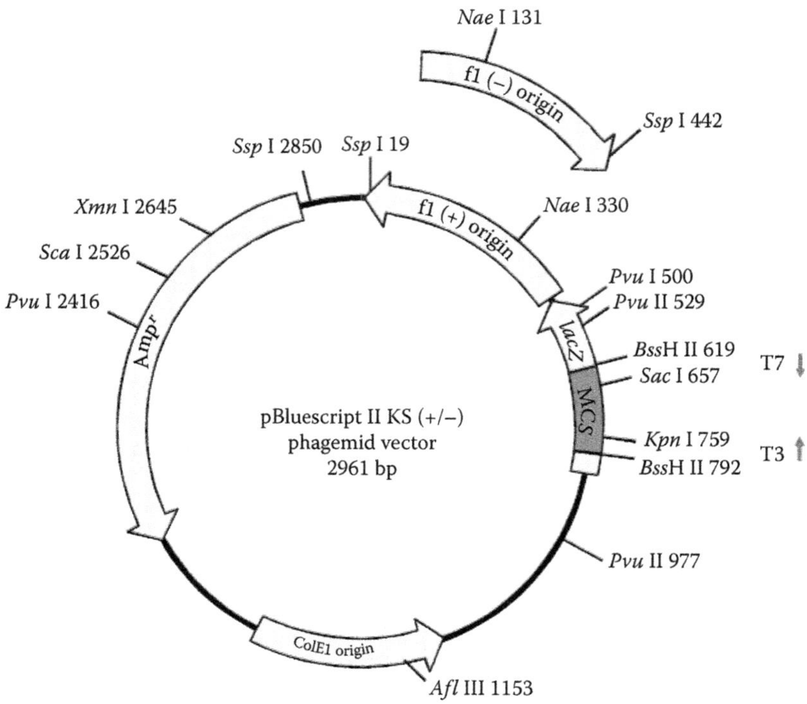

FIGURE 3.1 pBluescript KS(+).

selection system for pBS and pUC vectors. For selection purposes, the host contains a so-called F plasmid (fertility factor) on which the *lacZ* gene containing the M15 delection (spanning the codons for amino-acids 11–41) is located. In the presence of the chromomeric substrate for β-galactosidase X-Gal (5-chloro-4-chloro-3-indolyl-β-galactoside), bacterial colonies expressing β-galactosidase are blue in color. However, if the DNA region encoding the α-peptide is interrupted by the insertion of foreign DNA, α-complementation cannot occur, and as a consequence, no active β-galactosidase is present in the cell. These cells containing recombinant plasmids give rise to white colonies on X-Gal plates, hence the term blue/white selection.

Experimental Outline

1. Digestion of plasmid DNA with several restriction enzymes.
2. Perform agarose gel electrophoresis.
3. Carryout restriction mapping.

Materials and Pre-Lab Preparations

- Electrophoresis chamber (75 × 75 mm or 225 × 115 mm) and accessories
- Polaroid MP 4 land camera

- Power supply
- UV transilluminator
- Sterile reaction tubes
- Agarose (e.g., Ultrapure, Life Technologies)
- 5× TBE (Tris-base 0.45 M, boric acid 0.45 M, EDTA 0.125 M, pH 8.3; the pH of the buffer is automatically 8.3 when the substances have dissolved, but checking is advisable)
- 1× TBE (5× TBE diluted 1:5 with dH_2O)
- Loading buffer (glycerol 59%, bromophenol blue 0.03%, xylene cyanol 0.03%)
- DNA: bacteriophage λ DNA, vectors (pbr322, PUC series, pBluescript KS(+) from Stategene. Life technologies: e.g., EcoRI, Bam HI, PstI Hind III, PvuII, Sam I, etc.
- 10× TAM (O'Farrell buffer)
- Potassium acetate 0.66 M
- Tris-HCL 0.33 M
- Magnesium acetate 0.1 M
- Dithiothreitol 5 mM
- Bovine serum albumin 100 mg/mL (pH 7.9)
- Ethidium bromide
- Restriction enzyme digest

Pipette the following into a sterile reaction tube:

1 μL of DNA (1–2 μg)
2 μL of 10× TAM (the final concentration of the buffer is 1×)
1 μL of enzyme (2–5 U)
20 μL of H_2O

Do not forget a control tube (a possible control is a tube from which the restriction enzyme is omitted). When this sample has been run on the gel, several bands will be visible: the lowest 1 corresponds to the supercoiled form of the plasmid, which has migrated furthest into the gel; the band immediately above this the "open" circle form. Any bands higher up in the gel are due to concatameric forms of the plasmid. In some plasmid preparations, it is possible that "shearing" of the DNA has occurred, and in this case bands corresponding to linear plasmids appear between the supercoiled (closed circle) and "open" circle plasmid species.

Method

1. Digest plasmid DNA with several restriction enzymes (e.g., HindII + EcoRI; HindII + BamIII; HindIII + PstI, etc.).
2. λ DNA, as standardly cut with HindIII or HindIII/EcoRI. An aliquot of one or both of these standards are loaded onto the gel to allow the size of the

restriction fragments obtained with the plasmid DNAs to be determined. The length of the fragments (kb) of λ DNA obtained are:

a. HindIII-restricted λ DNA; 23.1, 9.4, 6.5, 4.3, 2.3, 2.0, 0.56, 0.125.
b. HindIII/EcoRI-restricted λ DNA: 21.2, 5.1, 4.9, 4.3, 2.1, 1.6, 1.3, 0.98, 0.83, 0.56, 0.125

3. Incubate the sample at 37°C for 1–2 h.
4. Stop the reaction by placing the samples at 65°C for 10 min.
5. Add 1/10 volume loading buffer to each sample, place again at 65°C for 10 min (this second incubation at the higher temperature is particularly important for the λ DNA digest to disrupt any concatemers that may have formed), and keep samples on ice until the gel is to be loaded.
6. Subject the samples to electrophoresis at 40–20 V, either in horizontal "minigels" (75 × 75 mm) with 14 wells or in larger gels measuring 225 × 155 mm with 14 or 32 wells. The buffer used for the electrophoresis is 0.8% in 1× TBE.
7. Dissolve the agarose by heating on a Bunsen burner or in a microwave oven. Allow the agarose to cool to approximately 55°C before pouring into the gel from and inserting the comb to form the wells. The gel should be about 0.5 cm in depth.
8. After the gel has set (the process can be sped up by placing the gel in the refrigerator or cold room), the comb is removed carefully and the gel is placed gently in the electrophoresis chamber. Pour in sufficient 1× TBE to cover the gel to a depth of 2–3 mm.
9. When the electrophoresis is complete, place the gel in a 0.1% solution of EtBr for 5 min to stain the DNA in the gel (safety note: wear disposable gloves). Visualize the DNA by exposing the gel to UV light. This is done by placing the gel on a transilluminator. Document by photographing the gel in UV light using a MP4 land camera and Polaroid 667 film. (Wear a protective face shield when using the transilluminator.)

Restriction Mapping (Figure 3.2)

The determination of the positions of restriction sites in the plasmid DNAs used in the restriction sites in the plasmid DNAs used in the restriction digests described above is termed restriction mapping and can be carried out as follows. The plasmid DNA is digested with various restriction enzymes, and the fragments obtained are separated by electrophoresis as described. With the help of appropriate double digests, the positions of the restriction sites can be determined. The lengths of the DNA fragments obtained are determined from the standard curve drawn by plotting the Log_{10} of the distance migrated for each of the λ DNA restriction fragments obtained with HindIII and/or HindIII/EcoRI against the length of the fragment.

Decontamination of EtBr-Containing Solutions

Ethidium bromide is a powerful mutagen, and therefore disposable gloves must always be worn when handling any solutions containing it. In addition, EtBr-containing solutions should never be disposed of by pouring down the sink.

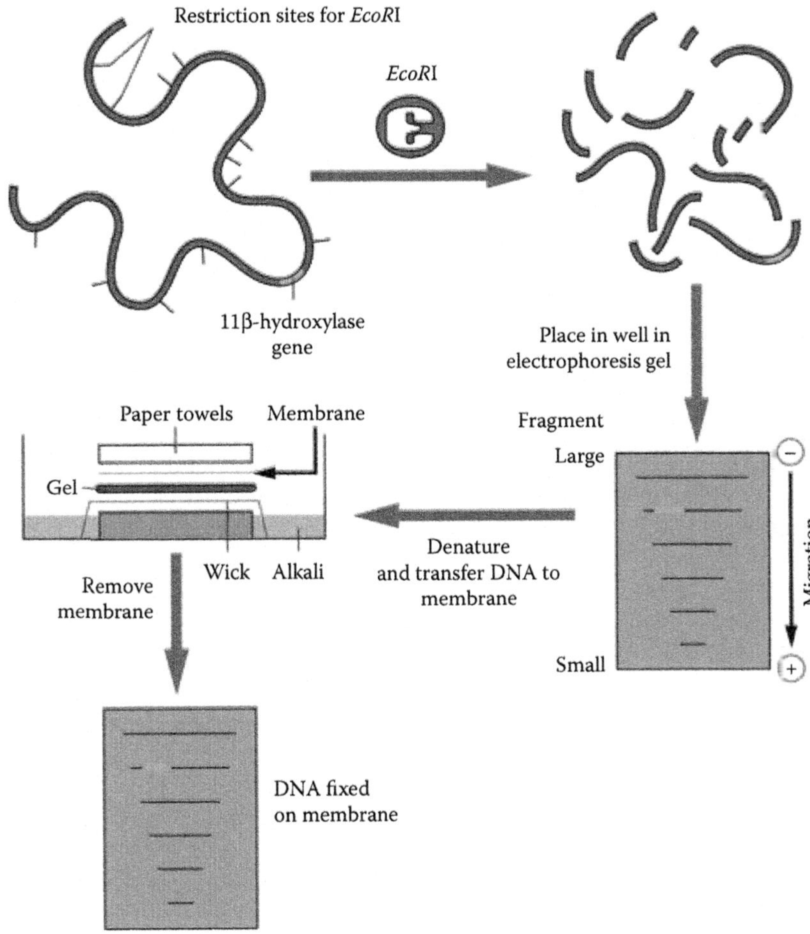

FIGURE 3.2 Restriction mapping.

Decontamination of used electrophoresis buffer containing around 0.5 μg/mL EtBr:

1. Add 1.0 g active charcoal per 1.0 mL of EtBr-containing buffer
2. Mix thoroughly for 1 h.
3. Filter the suspension through Whatman paper no. 1 and discard the filtrate. The active charcoal can be reused several times.
4. Label and dispose of the used active charcoal and filters as recommended for halogen-containing waste.

DIRECTIONAL CLONING INTO PLASMID VECTORS

Most plasmid vectors in common use contain multiple cloning sites that have recognition sequences for many different restriction enzymes. Given the large variety of multiple cloning sites currently available (as many as 46 unique sites are present in some polylinkers (e.g., pSE 280 from Invitrogen; and still longer polylinkers have been assembled), it is almost always possible to find a vector-carrying restriction site that is compatible with the termini of a particular fragment of foreign DNA.

Directional cloning usually requires incompatible termini at the opposite ends of both vector and target DNAs. However, in certain circumstances, directional cloning can be achieved when both the target and plasmid DNAs carry identical termini at both ends. For example, the restriction enzymes BamHI and BglII, which recognize different hexanucleotide sequence (GGATCC and AGATCT, respectively), generate restriction fragments with identical 3′-protruding termini. If a DNA fragment carrying BamHI and BglII termini is ligated into a vector that has been cleaved with the same 2 enzymes, then the foreign DNA can be inserted in either orientation. However, if 1 of the 2 restriction enzymes is included in the ligation mixture, or if the enzyme is used to digest the ligated DNA before transformation, then only those ligation events in which the BamHI end is joined to the BglII end and vice versa (which destroys the recognition sites of both enzymes) will give rise to recombinant products in *E. coli.* This strategy takes advantage of the observation that closed, circular DNAs transform bacterial cells with a much higher frequency than liner DNAs. Variation on this theme can also be used to improve the efficiency of cloning blunt-ended DNAs.

Occasionally, it is impossible to find a suitable combination of vectors, target DNA, and restriction enzymes that will allow directional cloning. There are several solutions to this problem:

1. Synthetic linkers or adaptors can be ligated to the termini of the liberalized plasmid and/or fragment of foreign DNA.
2. The fragment of foreign DNA can be amplified by PCR using oligonucleotide primers that add the desired restriction sites to 1 or both termini.
3. DNA fragments with recessed 3′ termini can be partially filled on controlled reactions using the Klenow fragment of *E. coli* DNA polymerase 1 (Figure 3.3).

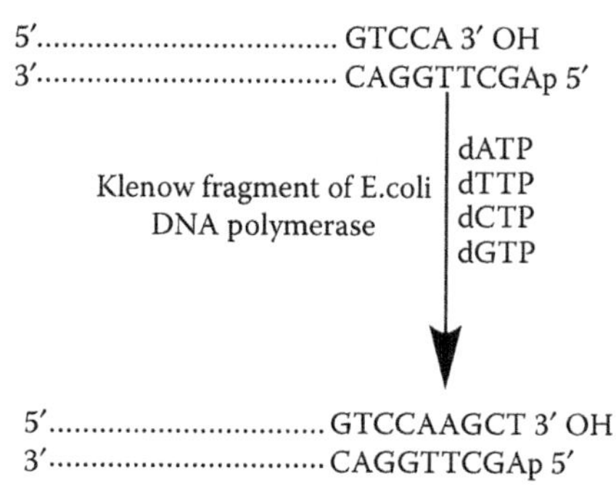

FIGURE 3.3 Filling of recessed 3′ termini by the Klenow fragment of *E. coli* DNA polymerase.

The procedure often generates complementary termini from restriction sites that are otherwise incompatible, thus facilitating ligation of the vector and foreign DNAs. Because partial filling eliminates the ability of termini on the same molecule to pair with one another, reactions are also reduced (Hung and Wensink 1984; Zabarovsky and Allikmets 1986). Keep in mind that micromolar concentrations of dATP can inhibit bacteriophage T4 DNA ligase. Thus if dATP is used as a substrate in partial end-filling reactions, the modified DNA product should be purified by spin-column chromatography or by 2 rounds of ethanol precipitation in the presence of ammonium acetate. This removes incorporated dATP from the DNA preparation. This protocol describes a standard procedure for cloning DNA fragments with protruding ends.

Experimental Outline

1. Digest vector and foreign DNA with restriction enzyme.
2. Purify the foreign DNA by extraction with phenol:chloroform and standard ethanol.
3. Purify the vector DNA separately.
4. Look for transformation of *E. coli*.

Materials and Pre-Lab Preparations

Buffers and Solutions

- Please see Appendix 1 for components of stock solution, buffers, and reagents.
- ATP (10 mM)
- Ethanol
- Phenol:chloroform (1:1:v/v)
- Sodium acetate (3 M, pH5.2)
- TE (pH 8.2)

Enzymes and Buffers

- Bacteriophage T4DNA ligase
- DNA ligases
- Restriction endonucleases

Gels

- Agarose gels
- Polyacrylamide gel

Nucleic Acid and Oligonucleotides

- Vector DNA (plasmid)
- Foreign or target DNA fragment
- Adaptors may be added to the target DNA

Equipment

- Equipment for spin-column chromatography
- Water bath set to 16°C

Method

1. Digest the vector (10 μg) and foreign DNA with the 2 appropriate restriction enzymes. Closed, circular plasmid vectors are prepared for directional cloning by digestion with 2 restriction enzymes that cleave at different sequences and generate different termini. Wherever possible, try to avoid using restriction enzymes that cleave within 12 bp of each other in the multiple cloning sites. After 1 of these sites has been cleaved, the second site will be located too close to the end of the linear DNA to allow efficient cleavage by the second enzyme. An excellent tabulation of the efficiency with which different restriction enzymes cleave sites located near the ends of DNA molecules is presented in the Appendix of the New England Biolabs catalog (www.neb. com/neb/frame_tech.html).
2. Read the manufacturer's instructions to determine if the 2 restriction enzymes work optimally in the same digestion buffer. If so, digestion of the vector DNA can be carried simultaneously with both enzymes. If the 2 restriction enzymes require different buffers, it is best to carry out the digestion sequentially. In this case, the enzymes that prefer the lower concentration of salt should be used first. At the end of the reaction, DNA molecule convert from circular to linear from and then analyze. Then adjust the salt concentration appropriately and add the second enzyme.
3. Purify the digest of foreign DNA by extraction with phenol:chloroform and standard ethanol precipitation.
4. This procedure eliminates from the vector preparation the small fragment of DNA generated by digesting the plasmid at 2 closely spaced restriction sites within the multiple cloning sites.
5. Reconstitute the precipitated DNA separately in TE (pH 8.0) at a concentration of 100 mg/mL. Calculate the concentration of the DNA (in pmol/mL), assuming that 1 bp has mass of 660 Da.
6. Confirm the approximate concentration of the 2 DNAs, analyzing aliquots by agarose gel electrophoresis.

Tube	DNA
A and D	Vector (30 fmol) [~100 ng]
B	Insert (foreign) (30 fmol [~10 ng])
C and E	Vector (30 fmol) plus insert (foreign) (30 fmol)
F	Superhelical vector (3 fmol [~10 ng])

Transfer the appropriate amount of the DNA to sterile 0.5 mL microfuge tubes as follows.

The molar ratio of plasmid vector to insert DNAfragment should be –1:1 in the ligation reaction. The final DNA concentration should be –10 ng/μL.

a. To tubes A, B, and C, add

10× ligation buffer	1.0 μL
Bacteriophase T4	0.1 weiss
DNA ligase	unit
10 mM ATP	1.0 μL
H_2O	to 10 μL

b. To tubes D and E, add

10× ligaion buffer	1.0 μL
10 mM ATP	1.0 μL
H_2O	to 10 μL
No DNA ligase	

Add the DNA fragments and the H_2O to the tubes and warm at 45°C for 5 min to melt any cohesive termini that have reannealed during fragment preparation. Chill the DNA solution to 0°C before the remainder of the ligation reagents are added. To achieve the maximum efficiency of ligation, set up the reaction in as small a volume as possible (5–10 μL). Adding ATP as a component of the 10× ligation buffer leaves more volume for vector or foreign DNA in the reaction mixture. Some commercial ligase buffers contain ATP. When using such buffers, the addition of ATP is no longer required. For a definition of Weiss units, please see the information on DNA ligases.

7. Incubate the reaction mixture overnight at 16°C or for 4 hours at 20°C.
8. Transform competent *E. coli* with dilution of each of the ligation reactions as described in Table 3.1. As controls, include known amounts of standard preparations of superhelical plasmid DNA to check the efficiency of transformation.

Transformants arising from ligation of vector DNA alone are due to failure of 1 or both restriction endonucleases to digest the DNA residual amounts of the small fragment of the multiple cloning sites.

Precautions

The following general cautions should always be observed:

1. Use proper disposal procedures for all chemical, biological, and radioactive waste.
2. For specific guidelines on appropriate gloves, consult the local safety office.

TABLE 3.1
Competent *E.coli* Dilution for Ligation Reaction

Tube	DNA	Ligase	Expected Number or Transformed Colonies
A	Vector	+	~0 (~0[4] fewer than tube F) 1
B	Insert	+	0
C	Vector and insert	+	~10 fold more than tube A or D
D	Vector	+	~0 (~10^4 fewer than tube F)
E	Vector and insert	–	Some, but fewer than tube C
F	Super helical vector	–	>2 × 10^5

3. Handle concentrated acids and bases with great care. Wear goggles and appropriate gloves, as well as a face shield if handling large quantities.
4. Never pipette solutions using mouth suction. This method is not sterile and can be dangerous. Always use a pipette aid or bulb.
5. Keep halogenated and nonhalogenated solvents separately (e.g., mixing chloroform and acetone can cause an unexpected reaction in the presence of bases). Halogenated solvents are organic solvents such as chloroform, dichloromethane, trichlorotrifluoroethane, and dichloroethane. Some nonhalogenated solvents are pentane, heptanes, ethanol, methanol, benzene, toluene, N,N-dimetylformamide (DMF), dimethyl-sulfoxide (DMSO), and acetonitrile.
6. Photographic fixatives and developers contain harmful chemicals. Handle them with care and follow manufacturer's directions.
7. Power supplies and electrophoresis equipment pose serious fire hazards and electrical shock hazards if not used properly.
8. Use extreme caution when handling cutting devices such as microtome blades, scalpels, razor blades, or needles. Microtome blades are extremely sharp. If unfamiliar with their use, have an experienced person demonstratedproper procedures. For proper disposal, use a sharps disposal container in the lab. Discard used needles unshielded, with the syringe still attached. This method prevents injuries (and possible infections), while manipulating used needles causes many accidents while trying to replace the needle shield. Injuries may also be caused by broken Pasteur pipettes and cover slips, or slides.

The Klenow fragment of *E. coli* DNA polymerase I catalyzes the template-directed addition of deoxynucleotide triphosphates to a recessed 3′ hydroxyl group. Synthesis occurs in a 5′→3′ direction until the recessed terminus is completely filled.

CLONING OF THE PHAGEMID VECTORS

Phagemids felicitously combine features of plasmids and filamentous bacteriophages. Stripped to their bare essentials, these vectors are conventional high-copy-number plasmids that carry a modified version of the major intergenic region of a filamentous bacteriophage (e.g., this region is 508 bp in its wild-type form) and encode no proteins but contain all of the cis-acting sequences that are indispensable for initiation and termination of viral DNA synthesis and for morphogenesis of bacteriophage particles.

Segments of foreign DNA can be cloned in phagemids and propagated as plasmids in the usual way. However, when a male strain of *E. coli* carrying a phagemid is infected with a suitable filamentous bacteriophage, the mode of replication of the phagemid changes in response to the gene products expressed by the incoming virus. The gene II protein encoded by the helper virus introduces a nick at a specific site in the intergenic region of the phagemid and hence initiates rolling circle replication, which generates copies of 1 strand of the plasmid DNA. These single-stranded copies of the plasmid DNA are packaged into progeny bacteriophage particles, which are then extruded into the medium. The secreted particles can easily be recovered by precipitation with polyethylene glycol and the single-stranded DNA purified by extraction with phenol.

Single-stranded DNAs produced by phagemids are used for the same purposes as single-stranded DNAs of bacteriophage M13 recombinants: for DNA sequencing, for synthesis or strand-specific radiolabelled probes, for subtractive hybridization, and for oligonucleotide-directed mutagenesis. In addition, phagemid vectors can be used in appropriate strains of *E. coli* (e.g., BW313) to produce single-stranded DNAs that contain uracil in place of thymine residues. These uracil-substituted DNAs are excellent substrates for certain types of oligonucleotide-directed mutagenesis.

Phagemids have several attractive features that overcome problems commonly encountered with cloning in bacteriophage M13, including

1. A positive selectable marker that can be used to select bacteria transformed by the phagemid.
2. Higher yields of double-stranded DNA
3. Elimination of the time-consuming process of subcloning DNA fragments from plasmids to filamentous bacteriophage vectors.
4. A significant reduction in the frequency and extent of deletions and rearrangements in single-stranded DNA
5. The ability to allow segments of DNA several kilobases in length to be isolated in single-stranded form.

In addition, it is possible for a phagemid to contain, for example, a strong promoter, the gene of cDNA of interest, and a transcription terminator. Expression phagemids of this type can be isolated in single-stranded-DNA form, subjected to site-directed mutagenesis, and then used to transform *E. coli* or yeast for phenotypic expression.

Improvements in Phagemids and Helper Viruses

The first generation of vectors, the pEMBL vectors, gave phagemids a bad name because, after super infection with a helper virus, the yield of single-stranded DNA was generally poor and was affected by many factors, including the density of the culture at the time of infection, the multiplicity of infection, and the length of time after infection. Even when super infection was carried out under optimal conditions, the yield of progeny virus all too often consisted predominantly of helper viruses rather than packaged single-stranded phagemid DNA. These problems were ameliorated by constructing helper viruses encoding a mutated version of gene II that preferentially activates the phagemid origin of replication. In addition, novel strains of *E. coli* have been engineered (e.g., DHIIs, TG2, and MVII84) that are efficiently transformed by plasmids, are easily infected by commonly used helper viruses, and yield preparations of single-stranded phagemid DNA that are free from contamination with bacterial DNA and helper phages.

The particular strand of the foreign DNA that is packaged into the bacteriophage particles depends on its orientation in the polylinker site of the phagemid vector and on the orientation of the bacteriophage origin of DNA replication carried in the vector. As a consequence, most commercially available phagemid vectors (e.g., pBluescript II from Strategene, pGEMZf from Promega) come in 4 chiralities in which the orientation of the polylinker sequence is opposite in 1 pair of vectors (e.g., pBluescript II SK and pBluescript II KS), and the orientation of the intergenic II SK (+) and pBluescript II SK (–). The (+) and (–) orientations of the intergenic region allow the rescue by helper bacteriophages of sense and antisense single-stranded DNAs. For first-time users of phagemids, the safest choice for helper virus, phagemid vector, and host is a well-tested and reliable helper virus such as M13K07 and a dependable phagemid (e.g., 1 or more of the Strategene SK or Promega pGem Z series) in an *E. coli* strains such as DH 115.

Experimental Outline

1. Culture *E. coli* strain DH 115 on YT culture medium. Infect *E. coli* strain DH 115 with M13K07 plasmid to culture the *E. coli* with the plasmid.
2. Examine for the foreign DNA insert with the help of agarose gels.

Materials

Buffers and Solutions

- Kanamcycin (10 mg/mL): Kanamycin is used in this experiment to ensure that all bacterial cells containing a phagemid genome are infected by the helper M13K07 bacteriophage. During propagation of a phagemid genome and infection by the helper M13K07 (e.g., steps 1–3), there is selection for bacteriophage genomes that have lost the p15 A origin and the Tn903 transposons. For this reason, it is essential to include kanamcycin in the medium used to prepare the stock of helper virus.
- SDS solution (2% w/v)
- Sucrose gel-loading buffer

Gel

Agarose gel (0.7%) cast in 0.5× TBE, containing 0.5 μg/mL ethidium bromide

Media

- Supplemented M9 minimal agar plates: When using *E. coli* strains that carry a deletion of the proline biosynthetic operon (Δ[lac-pro AB]) in the bacterial chromosome and the complementing proAB on the F′ plasmid, use supplemented M9 minimal medium.

- 2× YT medium
- 2× YT medium containing 60 μL/mL ampicillin
- 2× YT medium containing 25 μL/mL kanamycin
- The addition of Mg^{2+} (5 mM) to medium (Reddy and McKenney, 1996) is reported to improve the yield of bacteriophage M13 cultures infected at low multiplicity.

Special Equipment

Water bath preset to 65°C

Vectors and Bacterial Strains

Bacteriophage M13K07 (helper)—M13K07 may be obtained commercially (e.g., from Pharmacia or New England Biolabs) and propagated as described in steps 1–3 below. Store stock of helper virus at 4°C in growth medium or at −20°C in growth medium containing 50% (v/v) glycerol. M13K07 should not be used to superinfect cultures of *E. coli* JM 109 transformed with phagemid vectors. For reasons that are not understood, this strain undergoes significant lysis when infected with M13K07 (Lin et al. 1992).

E. coli F′ Strain

In principle, it should be possible to produce bacteriophage particles that contain single-stranded copies of phagemid DNA in any male strain of *E. coli.* Strains MV1184 (derived from Jm83) and MV1190 (derived from jMI 01) produce satisfactory yield, whereas MV1304 (derived from jM105) does not. The biological reasons for varying yield are poorly understood. Several strains (MV1184, DHII84, XL I-Blue, XL I Blue MRF) have been used by many laboratories and generally yield workable quantities of single-stranded phagemid DNAs (Vieira and Messing 1987). With all strains, maximum yields of phagemid DNA are obtained when infected cultures are well-aerated.

E. coli Strain DH 115

- DH 115 (Lin et al. 1992) is available from Life Technologies. MV1184, a less prepared option, can be obtained from J. Messing both strains should be plated on supplemented minimal agar plates.
- *E. coli* strain DH 115, transformed with bacteriophage M13 phagemid vector
- *E. coli* strain DH I115, transformed with bacteriophage M13 recombinant phagemid vector clone carrying foreign DNA. Transform *E. coli* with the recombinant phage mid-vector as described in the next experiment.

Method

Preparation of a High-Titer Stock of Helper Bacteriophage

The key to success in using phagemids is to prepare a stock of helper virus whose titer is accurately known.

1. In 20 mL of 2× YT medium, establish a culture of *E. coli* strain DH 115 from a single colony freshly picked from supplemented minimal agar plates. Incubate the culture at 37°C with moderate agitation until the OD_{600} reaches 0.8.
2. Prepare a series of 10-fold dilutions of bacteriophage M13K07 in 2× YT medium, and plate aliquots of the bacteriophage as described above to obtain well-isolated plaques on a lawn of DH 115 cells.
3. Pick well-separated, single plaques and place each plaque in 2–3 mL of 2× YT medium containing kanamycin (25 μg μL/mL) in a 15 mL culture tube. Incubate the infected cultures for 12–16 h at 37°C with moderate agitation (250 cycles/min).

Note: Use stock of M13K07 derived from single, freshly picked plaques in the following steps.

4. Transfer the infected cultures to 1.5 mL sterile microfuge tubes and centrifuge at 4°C.
5. Measure the titer of each to the bacteriophage stocks by plaque formation of 1 strain of *E. coli* F′ (TGI, DH 115, NM 522, or XLI-Blue) that supports the growth of bacteriophage M13.

The titer of infectious bacteriophage particles in the stock should be 10^{10} pfu/mL. Discard any stock with a lower titer.

Growth of Recombinant Phagemids with Helper Virus

6. Streak DH 115 cells transformed by (i) the recombinant phagemid and (ii) the empty (parent) phagemid vector onto 2 separate YT agar plates containing 60 μg/mL ampicillin. Incubate the plates for 16 h at 37°C.
7. Pick (i) several colonies transformed by the recombinant phagemid and (ii) 1–2 colonies transformed by the parent vector into sterile 15 mL culture tubes that contain 2–3 mL of 2× YT medium containing 60 μg/mL ampicillin. Because the variables that affect the yield of single-stranded DNA are poorly define, pick multiple isolates from each phagemid recombinant to increase the chances of success.
8. To each culture, add M13K07 helper bacteriophage to achieve a final concentration of 1.0–1.5 mL. Continue incubation for an additional 1 h at 37°C with strong agitation (300 cycles/min).

   The bacterial cultures should be only slightly turbid after this short incubation. If growth is too florid, dilute the cultures with prewarmed 2× YT medium until the turbidity is only just visible.
9. Add kanamycin to the cultures to a final concentration of 25 μg/mL. Continue incubation for a further 14–18 h at 37°C.

   Because the bacteriophage M13K07 contains a kanamycin resistance gene, only those cells that are infected will survive the addition of th antibiotic in this step. Other helper bacteriophages (e.g., R408) do not carry an antibiotic to the medium.

10. Transfer the cell suspensions to microfuge tube by centrifugation at maximum speed for 5 min at room temperature. Thereafter transfer the supernatants to fresh tubes and store them at 4°C.

Estimation of the Yield of Single- Stranded Phagemid DNA by Gel Electrophoresis

11. Combine 40 μL of each supernatant with 2 μL of 2% SDS in 0.5 mL microfuge tubes. Mix the contents of the tube by tapping and then incubate the tubes for 5 min at 65°C.

    Alternatively, estimate the yield of virus particles containing single-stranded copies of the phagemid by infecting *E. coli* DH 115 cells with dilutions of the supernatants and then plating the infected cells on YT agar containing ampicillin (60 μg/mL). The number of ampicillin-resistant colonies that arise after 24-h incubation at 37°C is a measure of the number of virus particle in the supernatant that contain single-stranded phagemid DNA. Supernatants that contain 2×10^{11} to 5×10^{11} cfu/mL will generate satisfactory yields of purified single-stranded phagemid DNA.

    Add 5 μL of sucrose gel-loading buffer to each sample of the phagemid DNA, mix the samples, and load them in separate wells of an 0.7% agarose gel.
12. Carry out electrophoreses for several h at 6 V/cm until the bromophenol blue has migrated approximately half the length of the gel. Examine and photograph the gel by UV light.
13. Yields vary depending on the size and nature of foreign DNA in the phagemid, but are generally 1 g/mL of culture volume.
14. Isolate single-stranded phagemid DNA from the supernatants containing the largest amount of single-stranded DNA. Follow the steps outlined above scaling up the volumes 2- to 3-fold.

    In phagemids, as in bacteriophage M13 vectors, the yield of single-stranded DNA can vary over a 5- to 10-fold range depending on the size and nature of the foreign DNA. In general, the larger the fragment, the poorer the yield. Furthermore, for reasons that are not understood, foreign DNAs of equivalent size can suppress the yield of single-stranded DNA to varying extents. For example, most segments of yeast DNAs of equivalent size may produce disappointing yields of single-stranded DNA. The orientation of both the foreign DNA insert and the bacteriophage origin or DNA replication in the phagemid vector can also dramatically affect yields. Thus, recloning a fragment in the opposite orientation or in a vector with the bacteriophage origin or replication in the opposite orientation will sometimes solve a problem of low yields.

Precautions

The following general cautions should always be observed.

1. Become completely familiar with the properties of all substances used before beginning the procedure.

2. The absence of a warning does not necessarily mean that the material is safe, because information may not always be complete or available.
3. If exposed to toxic substances, contact the local safety office immediately for instructions.
4. Use proper disposal procedures for all chemical, biological, and radioactive waste.
5. For specific guidelines on appropriate gloves, consult the local safety office.
6. Handle concentrated acids and bases with great care and wear goggles and appropriated gloves, as well as a face shield if handling large quantities. Do not mix strong bases with halogenated solvents as they may form reactive carbenes, which can lead to explosions. When preparing diluted solutions of acids from concentrated stocks, add acid to water.
7. Never pipette solutions using mouth suction. This method is not sterile and can be dangerous. Always use a pipette aid or bulb.
8. Keep halogenated and nonhalogenated solvents separately (e.g., mixing chloroform and acetone can cause unexpected reactions in the presence of bases). Halogenated solvents are organic solvents such as chloroform, dichloromethane, trichlorotrifluoroethane, and dichloromethane. Some nonhalogenated solvents are pentane, heptanes, ethanol, methanol, beneze, toluene, N, N-dimethylformamide (DMF), dimethylsulfoxide (DMSO), and acetonitrile.
9. Laser radiation, visible or invisible, can cause severe damage to the eyes and skin. Take proper precautions to prevent exposure to direct and reflected beams. Always follow manufactures safety guidelines and consult the local safety office.
10. Flash lamps, due to their light intensity, can be harmful to the eyes and may exploded on occasion. Wear appropriate eye protection and follow the manufacturer's guidelines.
11. Photographic fixatives and developers contain harmful chemicals. Handle them with care and follow manufacturer's directions.
12. Power supplies and electrophoresis equipment pose serious fire hazards and electrical-shock hazards if not used properly.
13. Microwave ovens and autoclaves in the lab require certain precautions. If the screw top on the bottle is not loose enough, and there is not enough space for the steam to vent, the bottle can explode when the containers are removed from the microwave or autoclave. Always loosen bottle caps before microwaving or autoclaving.
14. Use extreme caution when handling cutting devices such as microtome blades, scalpels, razor blades, or needles. Microtome blades are extremely sharp! If unfamiliar with their use, have an experienced person demonstrate proper procedures. For proper disposal, use a "sharps" disposal container in the lab. Discard used needles unshielded, with the syringe still attached. This method prevents injuries (and possible infections) while manipulating used needless causes many accidents while trying to replace the needle shield. Injuries may also be caused by broken Pasteur pipettes, coverslips, or slides.

GENE EXPRESSION IN *E. COLI* AND ANALYSIS OF GENE PRODUCTS

Plasmids carrying isopropyl thiogalactoside (IPTG)-inducible vectors are capable of expressing proteins at a level that exceeds 30% of total mass of bacterial protein. These plasmids are well-suited for small-scale laboratory experiments. But the high cost of IPTG prevents their use for large-scale production of foreign proteins. Variables that influence the efficiency of expression of foreign proteins from these and other expression plasmids are discussed in the panel on troubleshooting and optimization of protein expression from an inducible promoter at the end of this experiment.

The pGEM-3Z vector may be used for *in vitro* transcription and protein expression. The vector encodes the Lac Z a peptide preceded by lac operon sequences flanking the multiple cloning sites and the b-lactamase coding sequence that confers ampicillin resistance.

MATERIALS AND PRE-LAB PREPARATIONS

Buffers and Solutions

Buffer I (4×)	1.5 M Tris—HCl pH 8.8
Buffer II (4×)	0.5 M Tris—HCl pH 6.8
30% acrylamide	29.2% acrylamide
	0.8% bis-acrylamide
Sample buffer (2×)	0.125 M Tris—HCl pH 6.8
	20% glycerol
	4% SDS
	2% β-mercaptoethanol
	0.02% bromophenolblue
Running buffer (1×)	25 mM Tris
	192 mM glycine
	0.1% SDS
Coomassie brilliant blue stain or Silver stain IPTG (1 M)	

Prepare 80~1 of CAT reaction mixture 1 per 50~1 of cell lysate to be assayed. Assemble the reaction mixture immediately before use.

Acetyl-CoA is the S-acetylated form of coenzyme A. Two carbon units from fat and carbohydrate enter the citric acid cycle in this form. It is used as an acetylating cofactor in many biological reactions, as in the protocol, for acetylating of chloramphenicol by CAT.

Ethyl Acetate

Ethyl acetate ($CH_2COOCH_2CH_3$) is an organic solvent used in thin-layer chromatographic assays of CAT activity. Mix 1 Part ethyl acetate with 19 parts chloroform to make a solvent that separates acetylated and nonacetylated forms of chloramphenicol on silica gel plates. Ethyl acetate is an extremely flammable (flash point = −4°C) and volatile ester with a scent of ripe fruit.

Lysis Buffer

0.1 M Tris-Cr(pH 7.8) 0.5% v/v) Triton.

The same lysis buffer can also be used to prepare cell extracts for measurement of μ-galactrosidase activity.

Note: Use a highly purified preparation of Triton X-100 in the cell lysis buffer Lower grade detergent samples can inhibit CAT enzyme activity. The detergent Nonident P-40 at 0.125% (v/v) can be substituted for Triton X-100 in the lysis buffers.

Phosphate-buffered saline (PB5) without calcium and magnesium salts, thin-layer chromatography (TLC) solvent.

190 mL of chloroform
10 mL of methanol
Preaper 200 mL per TLC tank. Two TLC plates can be developed per tank.
Tris–C (1 M, pH 7.8)
Radioactive compounds
Radioactive ink

Nucleic Acids and Oligonucleotides

Gene or cDNA fragment of interest

Media

LB agar plates containing 50 μg/mL ampicillin

Equipment

- Centrifuge and rotors: Sorvall GSA rotor or equivalent
- Boiling water bath
- Shaking incubator

Vectors and Bacterial Strains

- *E. coli*: Strain suitable for transformation and carrying either the same allele. Some IPTG-inducible expression vectors carry the lac 19 allele on the expression plasmid (e.g., pMAL and Pgex). These can be used in any laboratory strain of E .coli (e.g., JM 101, dh51, and tgl). For details please see the panel on troubleshooting and optimization of protein expression from an inducible promoter.
- IPTG-inducible expression vector
- Another example includes pGEM-3Z (Figure 3.4)

Method

1. Resuspend the nuclear pellet in 1 mL ice-cold HSB buffer and 10 mg DNase. Pipette the nuclei up and down until they are resuspended and the viscosity of the solution is reduced (1–5 min). Add 2 mL of stop buffer.
2. DNase treatment serves to reduce the viscosity of the solution and to prevent the trapping of nascent radiolabelled RNA transcripts in genomic

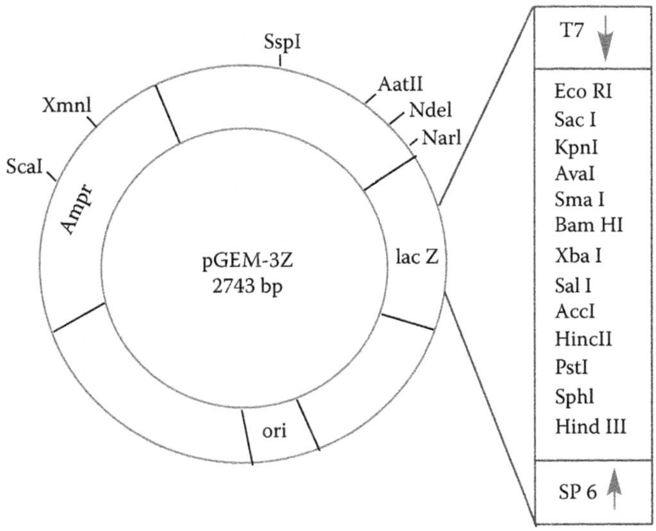

FIGURE 3.4 pGEM-3Z vector.

DNA. The viscosity never completely disappears and clumps of nuclei often remain.

3. To increase the yield of radiolabelled RNA, add proteinase K after the DNase step. Following the addition of 2 mL of stop buffer, add proteinase K to a final concentration of 100 mg/mL. Incubate the solution for 30 min at 42°C.
4. Add 3 mL of phenol and incubate the mixture for 15 min at 65°C with vortexing every 5 min. Add 3 mL of chloroform isoamyl alcohol, vortex, and separate the organic and aqueous phases by centrifugation at 1900 *g* (3000 rpm). Extract the aqueous layer again with 3 mL chloroform. Centrifuge as before and transfer the aqueous layer to a fresh tube.
5. Add 0.3 mL of 5 M LiCl and 2.5 volumes of ethanol, mix well, collect precipitated nucleic acid by centrifugation at 120,000 *g* (10,000 rpm) for 10 minutes.
6. Resuspend the pellet in 0.4 mL of H_2O and transfer it to a 1.5 mL microfuge tube. Add 40 mL of 5 M LiCl and 2.5 volumes of ethanol. Centrifuge the solution for 10 min at maximum speed in a microfuge.
7. Resuspend the pellet in 100 mg of H_2O and measure the cpm/mL in a liquid scintillation counter.

Hybridization of the RNA to Immobilized DNA

1. Linearize to pg of recombinant plasmid DNA containing the cDNA or gene of interest and 10 pg of empty plasmid vector using a restriction enzyme for which sites of cleavesge are present in the vector sequences.
2. Recover the cleaved DNAs using standard ethanol precipitation and resuspend each of the pellets separately in 20 ~d of DNA denaturation

solution. Boil the resuspended DNA for 2 min. Add 180 mg/mL of 6× SSC to each tube.

3. Cut a piece of nylon or nitrocellulose membrane to the appropriate size for use in a dot- or slot-blotting apparatus. Wet the membrane in H_2O and then soak it for 5–10 min in 6× SSC. Clamp the wet membrane in the blotting apparatus, attach a vacuum line, and apply suction to the device.
4. Filter the denatured DNA through separate slots and wash each filter.
5. Dismantle the device, dry the membrane by baking off or by exposure to UV light. Place the membrane in prehybridization solution and incubate for at least 16 h at an appropriate temperature (e.g., 42°C for solvents containing 50% formamide)
6. Add the radiolabelled probe (2×10^6 to 4×10^6 cpm/mL of 32 p-labelled RNA) directly to the prehybrization solution and incubate the filter for a further 72 h.
7. Wash the membrane at high stringency and expose it to X-ray film or a phosphorimager plate. Typical exposure times are 24–27 h for X-ray film and 4–24 h for phosphorimager plate.

Hybridization Controls and Troubleshooting

The inclusion of α-amanitin in the labelling step at 0.5 μg/mL can be used to demonstrate that an observed hybridization signal is due to transcription by RNA polymerase II (Marzluff and Huang 1984).

Several hybridization controls should be included in a nuclear run-on experiment. As a negative spurious hybridization, isolate and analyze unclear RNA from cells that do not express the gene of interest. Immobilize an empty plasmid vector on the nylon/nitrocellulose filter to act as a control for spurious hybridization. Prepare 2 positive hybridization controls including a gene that is expressed in the cells but is known not to respond to the regulatory input (e.g., actin, glyceraldehyde 3-phosphate dehydrogenase, and cyclophilin) and a gene that is known to respond to the stimulus.

The cDNA of the gene of interest in the immobilized DNA should not contain repetitive DNA sequences and should represent as much of the gene as possible. DNA fragments containing separate regions or exons of the gene can be used to assay for differences in the rate of elongation (Bentley and Groudine 1986) and/or to confirm a given hybridization result. It should also be kept in mind that, in rare instances, both strands of the DNA can be transcribed. For this reason, use a single-stranded DNA probe from 1 or more regions 01′ exons as control in at least 1 experiment.

The overall sensitivity of the hybridization assay can be enhanced by trimming the excess membrane from the immobilized DNA (i.e., leaving just the DNA-containing slot or dot) and carrying out the hybridization in a test tube in a small volume under mineral oil in a water bath set at the appropriate temperature.

In some cases, partial hydrolysis of nascent RNA transcripts will enhance hybridization. Resuspend the final RNA pellet from step 7 of Method in 100 mg of 10 mM Tris HCl (pH 7.4), 10 mM EDTA, and 200 mM NaCl, and add 25 mg of 1 N HCl and 500 mg of 1.2× Denhardt's solution and set up the hybridization reaction as described.

If substantial background is encountered in these protocols, treat the washed filter with RNase T1 (10 units/mL) and RNase A (20 mg/mL) 2× SSC for 1 h at 37°C to reduce non-specific binding of the probe to the membrane.

Precautions

1. Become completely familiar with the properties of all substances used before beginning the procedure.
2. The absence of warming does not necessarily mean that material is safe, because information may not always be complete or available.
3. If exposed to toxic substances, contact the local safety office immediately for instructions.
4. Use proper disposal procedures for all chemical, biological, and radioactive materials.
5. For specific guidelines on appropriate gloves, consult the local office.
6. Handle concentrated acids and bases with great care. Wear goggles and appropriate gloves, as well as a face shield if handling large quantities.
7. Do not mix strong acids with organic solvents as they may react. Sulfuric acid and nitric acid, especially, may react highly exothermically and cause fires and explosions.
8. Do not mix strong bases with halogenated solvents as they may form reactive carbenes, which can lead to explosions.
9. When preparing diluted solutions of acids from concentrated stocks, add acid to water.
10. Never pipette solutions using mouth suction. This method is not sterile and can be dangerous; always use pipette aid or bulb.
11. Keep halogenated and nonhalogenated solvents separately (e.g., mixing chloroform and acetone can cause unexpected reactions in the presence of a base). Halogenated solvents are organic solvents such as chloroform, dichloromethane, trichlorotrifluoroethane, and dichloromethane. Some nonhalogenated solvents are pentane, heptanes, ethanol, methanol, benzene, bendichloroethane, toluene, N,N-dimethyl formamide (DMF), dimethylsulfoxide (DMSO), and acetonitrile.
12. Laser radiation, visible or invisible, can cause severe damage to the eyes and skin. Take proper precautions to prevent exposure to direct and reflected beams. Always follow manufacturer's safety guidelines and consult the local safety office.
13. Flash lamps, due to their light intensity, can be harmful to eyes and may explode on occasion. Wear appropriate eye protection and follow the manufacturer's guidelines.
14. Photographic fixatives and developments contain harmful chemicals. Handle them with care and follow the manufacturer's directions.
15. Power supplies and electrophoresis equipment are serious fire hazard and electrical-shock hazards if not used properly.
16. Microwave ovens and autoclaves in the lab require certain precautions. If the screw top on the bottle is not loose enough, and there is not enough

space for the steam to vent, the bottle can explode when the containers are removed from the microwave or autoclave. Always loosen bottle caps before microwaving or autoclaving.

17. Use extreme caution when handling cutting devices such as microtome blades, scalpels, razor blades, or needles. Microtome blade are extremely sharp. If unfamiliar with their use, have an experienced person demonstrate proper procedures. For proper disposal, use "sharps" disposal container in thelab. Discard used needles unshielded, with the syringe still attached. This method prevents injuries (and possible infections) while manipulating used needles because many accidents occur while trying to replace the needle shield. Injuries may also occur because of broken Pasteur pipettes, cover slips, or slides.

ASSAY OF THE REPORTED CAT GENE

Reporter genes or markers provide a convenient means to identity and analyze the regulatory elements of genes. Standard recombinant methods are used to join the regulatory sequence of interest to a reporter gene or marker carried in an expression vector. The resulting recombinant is then introduced into an appropriate cell line, where its expression is detected by measurement of the reporter mRNA or the reporter protein (in the case of enzyme reporters, by assaying for the relevant catalytic activity). Whatever the method of detection, the reporters typically have an easily measured output that can be distinguished from the background proteins in the host cell. Most of the detection methods commonly used are sensitive, specific, quantitative, reproducible, and rapid to perform.

Reporter systems are used to measure transcriptional activity, in particular, to study promoters and enhancers and their interaction with cis-actin elements and transacting protein as well as their response to environmental changes.The fine structures of a regulator region can be analyzed by a series of promoter-bashing experiments in which selected mutations are introduced into the sequence under study, and its consequent activity (or ability to express the reporter) is measured and compared with that of the wild-type, or normal, sequence.

Reporter Genes Used in Analysis of Regulator Elements

After Gorman et al. (1982) first described the use of chloramphenicol acetyltransferase (CAT) as a reporter, the CAT gene became widely adopted as a reporter for many of the standard lines of cultured mammalian cells. CAT remains in common use and fulfills many of the criteria required of a reporter. Several different assays are available to measured CAT activity using procedures and equipment that are available in most laboratories. Despite its virtues, CAT is not an ideal reporter in all circumstances.

Naturally occurring resistance to chloramphenicol in bacteria is due to the activity of the enzyme chloramphenicol acetyltransferase (encoded by the *CAT* gene), which catalyzes the transfer of an acetyl group from acetyl coenzyme A (CoA) to the C3-hydroxyl group of the antibiotic. The product of the reaction, 3-acetyloxy chloramphenicol, neither binds to the peptidyl transferase center of the 70S ribosomes nor inhibits peptidyl transferase.

In field strains of Enterobacteriaceae and other Gram-negative bacteria, The *CAT* gene is constitutively expressed and is usually carried on plasmids that confer multiple drug resistance.

Several variants of the *CAT* gene product have been described, all of which form trimers consisting of an identical subunit of M_r 25,000. The type I variant, which is encoded by a 1102 bp segment of transposon, is in wide use as a reporter gene. However, most kinetic and structural analyses have been carried out with the type III variant, which yields crystals suitable for X-ray analysis. The active site, which is located at the subunit interface, contains a histidine residue which is postulated to act as a general base catalyst in the acetylation reaction. The 2 substrates (chloramphenicol and acetyl-CoA) approach the active site through tunnels located on opposite sides of the molecule.

Experimental Outline

1. Precipitate transfected cell pellets.
2. Remove the medium from monolayer and wash with PBS.
3. Remove PBS of the cell gently into the microfuge tube.
4. Perception of cell extract.

Materials and Pre-Lab Preparations

Buffers and Solutions

CAT Reaction Mixture 1	
1 M Tris HCl (pH 7.8)	50 μL
[^{14}C] chloramphenicol (60 mCi/mmol, diluted in H_2O to 0.1 mCi/mL)	10 μL
Acetyl Co-A (freshly prepared at a concentration of 3.5 mg/mL in H_2O)	20 μL

Prepare 80 mL of CAT reaction mixture 1 per 50 mL of cell lysates to be assayed. Assemble the reaction mixture immediately before use.

Acetyl-CoA is the S-acetylated form of coenzyme A. Two carbon units from fat and carbohydrate enter the citric acid cycle in this form. It is used as an acetylating cofactor in many biological reactions, as in this protocol, for acetylation of chloramphenicol by CAT.

Ethyl Acetate

Ethyl acetate ($CH_2COOCH_2CH_3$) is an organic solvent used in thin-layer chromatographic assays of CAT activity. Mix 1 part ethyl acetate with 19 parts chloroform to make a solvent that separates and non-acetylated forms of chloramphenicol on silica gel plates. Ethyl acetate is an extremely flammable (flash point = −4°C) and volatile ester with a scent of ripe fruit.

Lysis Buffer

- 0.1 M Tris-HCl (pH 7.8)
- 0.5% (v/v) Triton X-100
- Use if lysing the cells with detergent (please see step 4)
- The same lysis buffer can also be used to prepare cell extracts from measurements of galactosidase activity.

Note: Use a highly purified preparation of Triton X-100 in the cell lysis buffer. Lower grade detergent samples can inhibit CAT enzyme activity. The detergent Nonidet P-40 at 0.125% (v/v) can be substituted for Triton X-100 in the lysis buffer.

- Phosphate-buffered saline (PBS) without calcium and magnesium salts, thin-layer chromatography (TLC) solvent.
- 190 mL chloroform
- 10 mL methanol
- Prepare 200 mL per TLC tank. Two TLC plates can be developed per tank.
- Tris-Cl (IM, pH 7.8)
- Radioactive ink to mark adhesive dot labels as guides for orienting the autoradiogram with the TLC plate

Special Equipment

- Dry ice/ethanol bath
- Use these items if lysing the cells by freezing and thawing
- Marking pen with ethanol-insoluble ink
- Rotary vacuum evaporator
- Rubber policeman

Cells and Tissues

Cultured mammalian cells transfected with pCAT vectors carrying the DNA of interest. The cells should be transfected (using 1 of the transfection protocols) with a CAT reporter construct (e.g., the P cat3 series) and an expression plasmid containing the β -galactosidase gene (PCMV-sport-β-gal) or another reporter gene suitable for normalizing the results of the CAT assay.

1. Preparation of transfected cell pellets
   a. Use gentle aspiration to remove the medium from transfected monolayers of cells growing in 90-mm tissue culture dishes. Wash the monolayers 3 times with 5 mL of PBS without calcium and magnesium salts.
   b. Stand the dishes at an angle for 2–3 min to allow the last traces of PBS to drain to one side. Remove the last traces of PBS by aspiration. Add 1 mL of PBS to each plate, and use rubber policeman to scrape the cells into microfuge tubes. Store the tubes on ice until all of the plates have been processed.
   c. Recover the cells by centrifugation at maximum speed for 10 s at room temperature in a microfuge. Gently resuspend the cell pellets in 1 ml of

ice-cold PBS, and again recover the cells by centrifugation. Remove the last traces of PBS from the cell pellets and from the walls of the tubes. Store the cell pellets at –20°C for future analysis or prepare cell extracts by either of the methods in the next section.

d. The PBS can be conveniently removed with a disposable pipette tip attached to a vacuum line. Use gentle suction, and touch the tip to the surface of the liquid. Keep the tip as far away as possible from the cell pellet while the fluid is withdrawn from the tube. Keep the pipette tip as far away as possible from the cell pellet and use the tip to vacuum the walls of the tube to remove any adherent droplets of fluid.

2. Preparation of cell extracts
   Lyse the cells either by repeated cycles of freezing and thawing or by incubating the cells in detergent-containing buffers.

Lysis of Cells by Repeated Freezing and Thawing

1. Resuspend the cell pellet from one 90-mm dish in 100 μL of 0.25 M Tris-Cl (pH 7.8); vortex the suspension vigorously to break up clumps of cells.
2. Disrupt the cell by 3 cycles of freezing in a dry ice/ethanol bath, thawing at 37°C; make sure that the tubes have been marked with ethanol insoluble ink.
3. Centrifuge the suspension of disrupted cells at maximum speed for 5 min at 4°C in microfuge. Transfer the supernatant to a fresh microfuge tube. Set aside 50 mL of this supernatant for the CAT assay and store the reminder of the extract at –20°C.

Lysis of Cells Using Detergent-Containing Buffers

1. To lyse cells with detergent, resuspend the cell pellets from step 3, above, in 500 μL of lysis buffer and incubate the mixture for 15 min at 37°C.
2. Use 100 μL of this lysis buffer per cell pellet for extracts prepared from cells grown in 35 mm dishes.
3. Remove the cellular debris by centrifuging the tubes at maximum speed for 10 min in a microfuge. Recover the supernatant. Assay CAT activity using 1 of the methods described. Snap freeze the remainder of the cleared lysates in nitrogen and store them at –70°C.

Detection of CAT Activity Using Thin-Layer Chromatography

1. Incubate a 50 μL aliquot of the cell extract for 10 min a 65°C to inactivate endogenous deacetylases. If the extract is cloudy or opaque at this stage, remove the particulate material by centrifugation at maximum speed for 2 min at 4°C in microfuge.
2. Mix each of the samples to be used in the assayed with 80 μL of CAT reaction mixture I, and incubate the reaction at 37°C. The length of the incubation depends on the concentration of CAT in the cell extract, which

in turn depends on the strength of the promoter and the cell type and investigation. In most cases, incubation between 30 min and 2 h is sufficient.

3. The reaction can be incubated for longer periods of time (up to 16 h) when the expression of the CAT gene is low in the transfected cells. However, in this case, it is advisable to add an additional 10 μg aliquot of acetyl-CoA to each reaction after 2 h of incubation.
4. Add 1 mL of ethyl acetate to each sample, and mix the solution thoroughly by vortexing for 3 periods of 10 s each. Centrifuge the mixtures at maximum speed for 5 min at room temperature in a microfuge.
5. The acetylated forms of chloramphenicol partition into the organic (upper) phase; unacetylated chloramphenicol remains in the aqueous phase.
6. Use a pipette to transfer exactly 900 μL of the upper phase to a fresh tube, carefully avoiding the lower phase and the interface. Discard the tube containing the lower phase in the radioactive waste.
7. Evaporate the ethylacetate under vacuum by placing the tubes in a rotary evaporator (e.g., Savant SpeedVac) for 1 h.
8. Add 25 μL of ethyl acetate to each tube and dissolve the reaction products by gentle agitation.
9. Apply 10–15 μL of the dissolved reaction products to the origin of a 26-mm silica gel plate.
10. The origin on the plate can be marked with a soft-lead pencil. Apply 5 μL at a time, and evaporate the sample to dryness with a hair dryer after each application.
11. Prepare a TLC tank containing 200 μL of TLC solvent. Place the TLC plate in the tank, close the chamber, and allow the solvent front to move—75% of the distance to the top of the plate. A number of different TLC plates and solvents can be used to separate the acetylated form of chloramphenicol. We routinely use chloroform methanol (95.5) and Sybron SIL. C/UV254 (Brinkmann) TLC plates.
12. Remove the TLC plate from the tank and allow it to dry at room temperature. Place adhesive dot labels marked with radioactive ink on the TLC plate to align the plate with film, and then expose the plate to X-ray film. Alternatively, enclose the plate in a phosphorimaging cassette. Store the cassette at room temperature for an appropriate period of time, then expose the plate to X-ray film.
13. Do not cover the TLC plate with Saran Wrap, as this coverage will block the relatively weak radiation emitted by the C isotope.
14. Develop the X-ray film and align it with the plate. Alternatively, expose the chromatogram, the imager plate of a phosphorimager device or subject the plate to scanning.
15. Typically 3 radioactive spots are detected. The spot that has migrated the least distance (from the origin) consists of nonacetylated chloramphenicol migrating that partitioned into ethyl acetate. The 2 faster-migrating spots are modified forms are of chloramphenicol that have been acetylated at 1 of 2 potential sites. Diacetylated chloramphenicol may be detected as a third,

even faster migrating spot only) when high concentrations of CAT or lengthy incubations with copious amounts of extract protein are used.

16. To quantitate CAT activity, cut the radioactive spots from the TLC plate and measure the amount of radioactivity they contain in a liquid scintillation counter. Use another aliquot of the cell extract (from step 3 above) to determine the concentration of protein in the extract, using a rapid colorimetric assay, such as the Bradford assay. Reduce the concentration of Triton X-100 from 0.2% to 0.1% by dilution before determining the concentration of protein to prevent interference with the assay. Express the CAT activity as pmol of acetylated product forms per unit time per mg of cell extract protein.
17. Thin-layer chromatography on silica gels, followed by scraping and scintillation counting of the modified chloramphenicol products, can become quite tedious when large numbers of samples are prepared. An alternate method for quantitating the amount of product in these cases is to use a phosphorimager to scan the TLC plates.

ASSAY FOR β-GALACTOSIDASE IN EXTRACTS OF MAMMALIAN CELLS

The *E. coli* β-galactosidase gene is often used as an internal reference in transfection studies. In a typical experiment, a test plasmid containing a eukaryotic gene promoter linked to CAT or luciferase is contransfected into a mammalian cell line with a small amount of plasmid that contains the β-galactosidase activity. A normalized expression value can be obtained (please see below). Constructs are available that carry the β-galactosidase gene downstream from promoters that express strongly in a wide variety of eukaryotic cell types (e.g., the SV 40 early promoter, the Rous sarcoma virus long terminal repeat promoter, or the immediate early region promoter of cytomegalovirus). Extracts of most types of cultured mammalian cells contain relatively low levels of endogenous β-galactosidase activity and an increase in enzyme activity of up to 100-fold can usually be detected during the course of a transfection. With effort, β-galactosidase can also be used as an internal control in certain specialized cells (e.g., gut epithelial cells and human embryonic kidney 293 cells) that express high endogenous levels of β- galactosidase activity. Because endogenous β-galactosidase is usually more heat-labile than the bacterial enzymes, a heating step can be used to eliminate the endogenous β-galactosidase activity while preserving the bacterial activity expressed from the control plasmid. In addition, mammalian β-galactosidase are usually associated with the lysosome and therefore have an acidic pH optimum. The *E. coli* enzyme has a neutral to slightly alkaline pH optimum. The contribution of the mammalian enzyme can therefore be reduced by carrying out the β-galactosidase assay at pH 7.5.

Several different approaches are used to normalize CAT or other reporter enzyme activities to β-galactosidase activity. In one approach, the amount of protein in individual extracts prepared from a series of transfected cells is first measured, and CAT and β-galactosidase are then assayed separately using a standard amount of protein in each assay. Finally, the CAT activity is normalized to the β-galactosidase activity. In another method, the β-galactosidase activity in a constant volume of extract is first measured, and CAT assays are then carried out using amounts of extract containing the same amount of β-galactosidase activity. Alternatively, both enzymatic assays are carried out in a constant volume of extract, and the results are then normalized to a defined level of β-galactosidase activity (i.e., the amount of CAT activity is divided by the amount of β-galactosidase activity). With some marker genes such as luciferase, the amounts of β-galactosidase and luciferase that are present in the same aliquot of cell lysate can be determined by using different luminescent substrates simultaneously.

This protocol describes the detection of β-galactosidase expressed from reporter vectors (Figure 3.3) transfected into mammalian cells. The assay is both simple and rapid and can be carried out using a visible light spectrophotometer. Please note that several manufacturers sell kits that can be used to assay β-galactosidase in eukaryotic cell lysates; that average cost of a single reaction is typically $1 US. Many of these kits contain the buffers listed here and adhere closely to the protocol. However, some kits gain a distinct advantage by enzyme activities to be assayed, including β-galactosidase and CAT.

Summarized here are the features of p β-gal reporter vectors that carry the coding sequence for β-galactosidase. All vectors in the series contain (i) the origin of replication derived from filamentous phage (f1 ori), (ii) an origin of plasmid replication in *E. coli* (ori), (iii) the origin of replication for SV 40 that allows replication in mammalian cells, (iv) an ampicillin-resistance marker (Amp^R) for selection in prokaryotic cells, and (v) a poly(A) addition signal located 5′ of lacZ to reduce the background of read-through transcription from upstream sequences. The sequence of interest is cloned into the multiple cloning sites (MCS) at the 5′ end of te lacZ gene.

Vectors include (A) p β-gal-Basic lacks eukaryotic promoter and enhancer sequences. This vector may be used as a negative control or as a vehicle to characterize cloned promoters; (B) p β-gal-Enhancet lacks the SV-40 promoter, but contains the SV 40 enhancer. This vector can be used to study cloned promoter sequences; (C) p β-gal-promoter lacks the SV 40 enhancer, but contains the SV 40 promoter. This vector can be used to study cloned enhancer sequences; and (D) p β-gal-control contains the SV40 early promoter and enhancer. This vector can be used as a positive control or as a reference when comparing the activities of different promoters and enhancers. (Figure 3.5).

Experimental Outline

1. Prepare cell extracts as described in the earlier experiment.
2. Prepare transfected cell pellets as described in the earlier experiment.
3. For the transfected cell lysate, mix ONPG extract and 0.1 sodium phosphate 100× Mg^{2+} solution
4. Incubate for 30 min–4 h.
5. Stop the reaction with 1 M Na_2CO_3 and read OD at 420 nm.
6. ONPG (o-nitrophenyl β-D galactophyranoside) is the most widely used substrate in assays for β-galactosidase in bacterial and eukaryotic cell

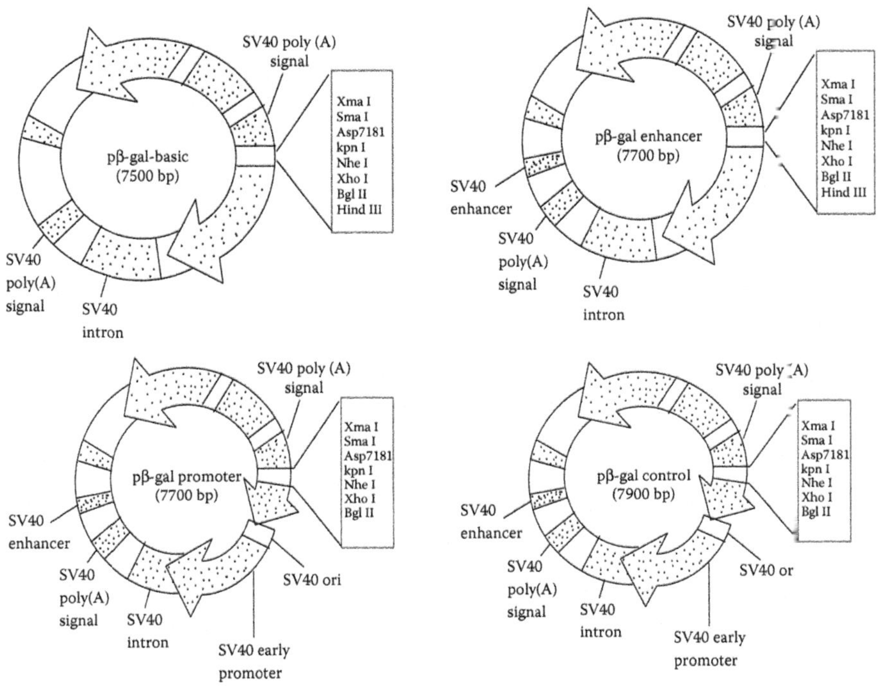

FIGURE 3.5 p-β gal vectors.

lysates. ONPG is colorless, but in hydrolysis, it yields o-nitrophenol, which is yellow in alkaline solution (max 420 nm at pH 10.2). When ONPG is in excess, the OD 420 of the assay solution increases linearly as a function of time and enzyme concentration. The reaction can be stopped by adding a concentrated solution of $NaCO_3$, which shifts the pH to 11, inactivates β-galactosidase, and maximizes the absorbance of o-nitrophenol at 420 nm.

Materials and Pre-Lab Preparations

Buffers and Solutions

Please see Appendix 1 for components of stock solutions. Dilute stock solutions to the appropriate concentrations.

- Lysis Buffer
  9.1 mL 1 M K_2HPO_4
  920 μL 1 M KH_2PO_4
  2 mL 10% Triton X-100
  88 mL DD water
  100× Mg^{2+} solution
  0.1 M $MgCl_2$
  4.5 M β-mercaptoethanol

Just before use, add an appropriate amount of β-mercaptoethanol from a 14.7 M stock.

- Na_2CO_3(1M)
  Dissolve 10.6 g of the anhydrous solid in 100 mL of H_2O.
  1× ONPG
  Dissolve ONPG at a concentration of 4 mg/mL in 0.1 M sodium phosphate (pH 7.5).
- Sodium phosphate (0.1 M, pH 7.5)
  Mix 41 mL of 0.2 M Na_2HPO_4. $2H_2O$ (M_r = 178.05; 35.61 g/L), 9 mL of 0.2 M Na_2HPO_4 (M_r = 156; 31.21 g/L), Tris-Cl (1 M, pH 7.8)
- *E. coli* β-galactosidase; the enzyme is commercially available (e.g., Sigma).
- Culture mammalian cells transfected with the DNA of interest

Use a transfection protocol to transfect the cells with a plasmid containing a galactosidase reporter gene (e.g., the p β-gal reporter series from Clontech; please see Figure 3.5).

β-Galactosidase Substrates

The assay relies on the ability of the *E. coli* β-galactosidase enzyme to hydrolyze ONPG to free o-nitrophenol and β-D-galacto. The substrate o-nitrophenol is yellow in aqueous solution and absorbs light of 420 nm wavelength. Since the invention of this assay by Joshua Lederberg, more sensitive substrates for the detection of β-galactosidase have been developed. These include 4-methylumbelliferyl β-D-galactoside, for which the hydrolysis product, 4-mehylumbellidrone, is detected by fluorometry (excitation wavelength = 364 nm; emission wavelength = 440 nm) and is blue in buffers with a basic pH and chlorophenol red β-D-galactoside (Boehringer Mannheim), for which the hydrolysis product is detected in a visible-light spectrophotometer at 570 nm weavelength. An assay using 4-methylumbelliferyl-β-D-galactoside as a substrate is about 10-fold more sensitive than one using ONPG, whereas an assay using chlorophenol red β-D-galactopyranoside is 20 times more sensitive than one using ONPG. Finally, chemiluminescent substrates for β-D-galactosidase for which the hydrolysis product can be detected in a luminometer or scintillation counter are sold by several companies. Assays using chemiluminescent substrates are 20–1000-fold more sensitive than ONPG-based assays.

METHOD

1. Prepare cell extracts from the transfected cells as described in above protocols, steps 1–4. Set aside 3 μL of the extract for the β-D-galactosidase assay. The exact amount of the extract required will depend on the strength of the promoter driving the expression of the β-D-galactosidase gene, the efficiency of transfection, and the incubation time of the assay. If a heat treatment is to be used to inactivate endogenous β-D-galactosidase, incubate the cell lysates for 45–60 min at 50°C before the assay. Luciferase activity

is also inactivated by preheating; assay luciferase and β-D-galactosidase activities in separate aliquots of cell lysate if a preheating step has been used.
2. For each Sample of transfected cell lysate to be assayed, mix

100× Mg^{2+} solution	3 μL
1× ONPG	66 μL
Cell extract	30 μL
0.1 M sodium phosphate (pH 7.5)	201 μL

It is essential to include positive and negative controls. These assays check for the presence of endogenous inhibitors and β-galactosidase, respectively. All of the controls should contain 30 μL of cell extract from mock-transfected cells. In addition, the positive controls should include 1 μL of a commercial preparation of *E. coli* β-galactosidase (50 units/mL). The commercial enzyme preparation should be dissolved at a concentration of 3000 units/mL in 0.1 M sodium phosphate (pH 7.5). Just before use, transfer 1 μL of the stock solution of β-galactosidase into 60 μL of 0.1 M sodium phosphate (pH 7.5) to make a working stock of the enzyme containing 50 units/mL. One unit of E.coli β-galactosidase is defined as the amount of enzyme that will hydrolyze 1 μmol of ONPG substrate in 1 min at 37°C.
3. Incubate the reaction for 30 minutes at 37°C or until a faint yellow color has developed. In most cell types, the background of endogenous β-galactosidase activity is very low, allowing incubation times as long as 4–6 h to be used.
4. Stop the reaction by adding 500 μL of 1 M Na_2CO_3 to each tube. Read the optical density of the solution at a wavelength of 420 nm in a spectrophotometer.

The linear range of the assay is 0.2–0.8 OD 420. If the assay falls outside of this range, repeat the experiment using less protein extract. The extract can be diluted in 0.25 M Tris-Cl (pH 7.8) to decrease the protein concentration.

Biochemists calculated the specific activity of the β-galactosidase enzyme and express this value as units of enzyme activity per milligram of cell protein, where 1 unit of *E. coli* β-galactosidase is defined as the amount of enzyme that will hydrolyze 1 μmol of ONPG substrate in normal expression of the marker gene, for which the specific activity has also been determined. Most moleclular biologists simply divide the amount of CAT or luciferase activity by the amount of β-galactosidase activity present in a given volume to normalize for transfection efficiency, thus ingnoring the calculation of specific activity.

Precautions

1. Never pipette solutions using mouth suction. This method is not sterile and can be dangerous. Always use a pipette or bulb.
2. Photographic fixatives and developers contain harmful chemicals. Handle them with care and follow manufacturer directions.
3. Use proper disposal procedures for all chemical, biological, and radioactive waste.

4. Power supplies and electrophoresis equipment pose serious fire hazards and electrical-stock hazards if not used properly.

DNA SEQUENCING

In fact, it is the sequence of nucleotides in DNA that direct the synthesis of RNA and proteins and also regulation of gene expression. Thus, the knowledge about DNA sequence is basic for unraveling the structure, function, and regulation of gene expression at the molecular level. The overall strategy for nucleic acid sequencing is analogous to that of sequencing of amino acids in proteins. It involves (i) degradation and fractionation number of the nucleic acid molecule to smaller fragments that can then be conveniently and fully sequenced, and (ii) ordering of the fragments by using a different degradation procedure that yields a set of nucleic acid fragments that overlap the cleavage points in the preceding degradation step.

The 2 most reliable techniques for DNA sequence determination are (i) the chemical degradation and modification method of Maxam and Gilbert (1977) and (ii) the enzymatic dideoxy chain termination method of Sanger et al. (1977). Sanger's method is widely used because the technique is simpler and amenable to automation. Both methods are based on generation of specific sets of radioactively labelled fragments having 1 end in common, but the methods differ with respect to a specific base at the other end. The fragments are separated by high-resolution polyacrylamide gel electrophoresis that allows separation of fragments differing from one another by only a single nucleotide, and the sequence is then directly read off the autoradiogram of the sequencing gel (Figure 3.6).

1. *Chemical cleave and modification method*—This method can be used for sequencing both single-stranded and double-stranded DNA. The first step in the chemical cleavage method is to radioactively label 1 end, usually the 5′ end. Radioactive fragments with one end that is labeled with ^{32}P is common, but the end terminates with a specific base along the polynucleotide chain. Perform 4 sets of reactions on separate aliquots of the nucleic acid samples as described as follows.
   a. G Cleavage—DNA is treated with dimethyl sulfate (DMS), which methylates the G residues at N7 of the base, thereby making the glycosidic bond of the methylated residue susceptible to hydrolysis. Subsequent treatment with piperidine cleaves the polynucleotide chain at a site just before the G residue.
   b. A + G Cleavage—In the second set, DNA is methylated at A and G bases using dimethyl sulfate. Under acidic conditions, both A and G are released resulting in a depurinated product. Piperidine treatment then breaks the strand at a point just before both A and G residues. The A residues are then identified by comparing the positions of the G and A + G cleavages.
   c. C + T Cleavage—In the third sample aliquot, DNA is made to react with hydrazine followed by piperidine treatment, and this cleaves the DNA molecule before both its C and T residues.

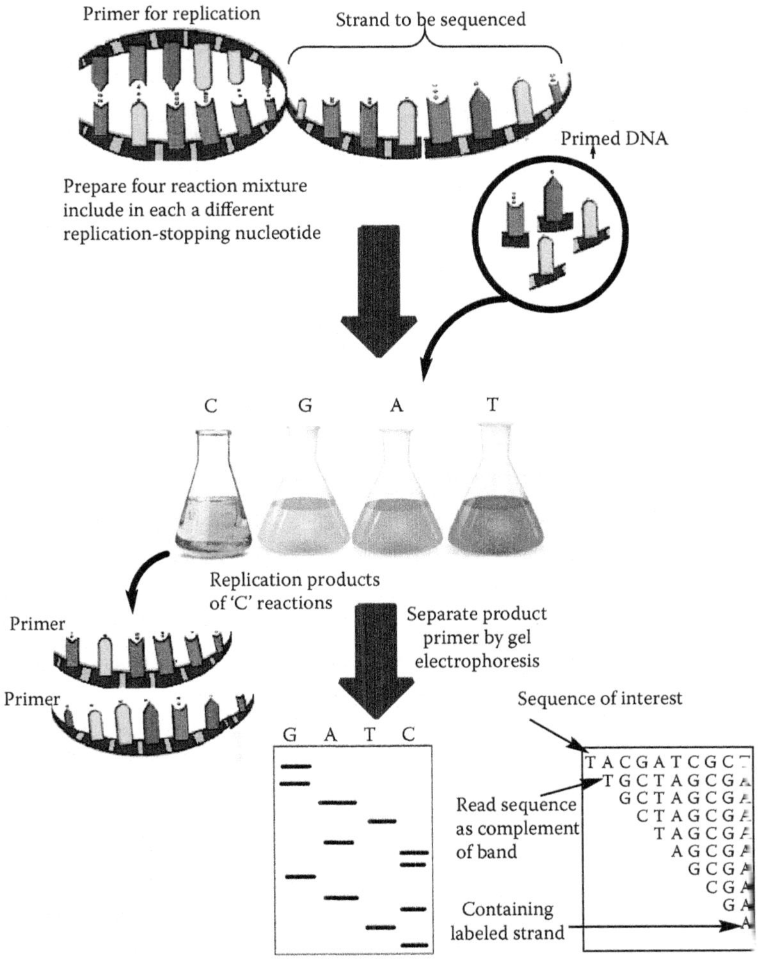

FIGURE 3.6 DNA sequencing method.

d. C Cleavage—If DNA is treated with hydrazine in the presence of 1.5 M NaCl, then only its C residues react appreciably, and thus the piperidine cleavage of C + T cleavage position facilitates identification of the T residues.

In all 4 reactions, the conditions are varied so that the strands are cleaved at an average of 1 randomly located position. These 4sets of reaction mixtures are simultaneously electrophoresed in adjacent lanes on a long, thin denaturating polyacrylamide slab gel (0.1 mm thick × 40 cm long). The DNA fragments separate according to their size. After electrophoresis is over, the gel is subjected to autoradiography and the sequence of the DNA can then directly be read from the autoradiogram in the $5' \rightarrow 3'$ direction (as bands from bottom to top).

2. *Dideoxy chain-termination method*—The method involves a primed synthesis of a complementary radioactive copy of a single-stranded DNA

template by a DNA polymerase. The base-specific ends of the oligonucleotide thus synthesized are produced by incorporating 2′, 3′-dideoxycleoside triphosphates (2′,3′ ddNTPs) into the newly synthesized DNA. 2′, 3′ ddNTPs differ from the normal dNTPs in that they lack a hydroxyl residue at the 3′ position of deoxyribose. They can be incorporated by DNA polymerase (Klenow enzyme) into a growing DNA chain through their 5′-triphosphate groups. However, the absence of a 3′ OH residue prevents formation of a phosphodiester bond with the succeeding dNTP. Further extension of the growing chain is performed in 4 different sets, each containing a different dideoxynucleotide, and this results in production of nested sets of fragments. The 4 reaction mixtures each terminate with a dideoxynucleotide being used. The 4 reaction mixtures are electrophoresed on a gel. The fragments get separated according to their autoradiogram. This indicates the sequence of complementary strands and from this, the sequence of the original (template) strand can be computed.

Since 1977, several advancements have occurred is Sanger's method, such as development of sequencing vectors, use of nonradioactive fluorescent dideoxy analogues, and PCR technology, which have led to its automation.

DETERMINATION OF NUCLEOTIDE SEQUENCE OF DNA BY DIDEOXY CHAIN TERMINATION METHOD

The method involves a primed synthesis of a complementary radioactive copy of a single-stranded DNA polymerase (Klenow enzyme) in such a way that nested sets of radioactively labelled fragments having 1 common end but differ at their other end by the specific base produced. The base-specific ends are produced by incorporating 2′3′-dideoxynucleoside triphosphates (2′3′ddNTPs) into the growing DNA chain; the ddNTPs lack hydroxyl groups at the 3′ end of the deoxyribose, and thus, when incorporated, further elongation of the chain is stopped. The fragments thus synthesized are separated by high resolution polyacrylamide gel electrophoresis that allows fragments differing by a single nucleotide to be resolved, and the sequence can be directly read as dark bands on the autoradiogram of the gel.

Experimental Outline

1. Carry out sequencing reaction.
2. Perform polyacrylamide gel electrophoresis.
3. Identify nucleotides on X-ray film.

Materials and Pre-Lab Preparations

- *DNA sequence apparatus*, which consists of the following:
    - 2 glass plates (1 notched and 1 unnotched) 40 × 20 cm each
    - 3 spacers: Two 50 cm long and One 30 cm long, each 1 cm wide and 0.4 mm thick

- Sharkstooth comb
- Electrophoretic chamber
- Power pack 3000 V capacity
- Gel loading pipette or syringe
- Clamps

- Gel dryer
- X-ray film
- Whatman filter paper
- *Termplate DNA*: M 13 mp 11SS-DNA containing plant DNA fragment
- *TE Buffer*
- *Universal primer*: Procure commercially available M 13 sequencing primer. Dissolve in TE solution at concentration of 2 mg/mL.
- 10× DNA polymerase reaction buffer. Prepare 70 mM Tris-HCl buffer (pH 7.5) containing 70 mM $MgCl_2$ and 500 mM NaCl.
- ^{35}S dATP (8 mCi/mL 650 Ci/mmol) or ^{32}dATP
- 0.1 M dithioreitol (DTT)
- Klenow enzyme (2 units/mL)
- *dNTP stock solution*: Prepare 0.5 mM solution of each of dATP, dCTP, dTTP, and dGTP in TE buffer (pH 8.0); store at –20°C.
- *dNTP working solution*: Prepare working solution of dNTPs mixing the 4 0.5 mM stock solutions of dNTP s in the ratios as shown.

dNTP (X^0) Working Solution	0.5 mM Stock Solution of dNTP s (μL)			
	dTTP	dCTP	dGTP	dATP
A°	20	20	20	–
C°	20	1	20	–
G°	20	20	1	–
T°	1	20	20	–

Note: Working solution of a dNTP contains lower concentration of a particular dNTP and higher concentration of other 3 dNTPs. Working solution containing high concentration of dATP is not required as dATP is supplied in the reaction mixture as (^{35}S) dATP. Solution may be stored at –20°C up to 2 wk.

- *ddNTP stock solution*: Prepare 10 mM solution of each of ddATP, ddCTP, ddTTP, and ddGTP in TE buffer (pH 8.0); store at –20°C.
- *ddNTPs (ddX) working solution*: Prepare by diluting 10 mM stock solution of ddNTPs with distilled water as follows.

ddNTP (ddX) Working Solution	10 mM ddNTP (μL)	H_2O (mL)
ddA (1 mM)	1	9
ddC (0.35 mM)	1	27.6
ddG (0.7 mM)	1	14.5
ddT (2 mM)	1	5

These solution can be stored at –20° for up to 2 wk.

- *Formamide-dye mix (gel-loading solution)*

Formamide	10 mL
Xylene cyanol (xylene cyanol FF)	10 mg
Bromophenol blue	10 mg
0.5 M EDTA (pH 8.0)	200 μL
Store at room temperature.	

- *Methanol-acetic acid solution*: Prepare by mixing 100 mL methanol with 100 mL acetic acid and make volume to $1 l^{-1}$ with distilled water.
- X-ray developer
- X-ray fixer
- *Acrylamide-bisacrylamide solution*: Prepare solution containing 35 g of solution through nitrocellulose filter (0.45 μpore size) and store in dark bottle at room temperature.
- *TBE buffer (10×)*, which consists of

Tris	108 g
Boric acid	55 g
EDTA	9.3 g

Make volume to 1 L with water, and adjust pH to 8.3.

- Urea
- TEMED as supplied
- *10% ammonium persulfate:* Dissolve 1 g ammonium persulphate (APS) in 10 mL water.
- Siliconizing fluid
- Sealing tape
- Paraffin (liquid)

Method

Sequencing Reaction

1. Add the following to an Eppendorf tube to make a template primer mixture:

Template DNA (700 ng)	7.0 μL
Primer DNA (5 ng)	2.0 μL
10× polymerase reaction buffer, pH 7.5	1.0 μL
Paraffin (liquid)	15.0 μL

   Keep in boiling water bath for 5 min; switch off the water bath and allow the contents to cool slowly to room temperature.
2. Add the following components to another Eppendorf tube:

[^{35}S] dATP	1.0 μL
0.1 MDTT	1.0 μL
Klenow enzyme	2–3 units
Sterile distilled water	1.0 μL
Template primer mix	10.0 μL
(Prepared in step1)	
Mix the contents and label it as enzymes-template-primer mix.	

3. Carry out the sequencing reaction as follows in 4 different Eppendorf tubes labeled as G, A, T, and C.
4. Incubate the tubes at room temperature for 20 min.
5. Add 1 mL 0.5 μM dATP (chase) to each tube and allow the tubes to stand for another 20 min at room temperature.
6. Add 10 μL formamide-dye mix (gel-loading solution) to all the tubes; keep at −15°C until use.

Electrophoresis

7. Wash the 2 plates (notched and untouched or smooth) with deionized water. Dry the surface of the plates with tissue paper. Rinse with deionised water and again wipe dry the surface with tissue paper.
8. Put 1 mL of saline (5% solution of dichlorodimethyl silane in chloroform) on a smooth plate, spread it evenly and keep at room temperature for 30 min. Rinse with ethanol and dry thoroughly with tissue paper (caution: ensure complete removal of all silane).
9. Place the spacers on the notched plate and align properly the smooth plate with the notched one, and fix tape at the bottom and on the sides (3–4 ins) of both the plates together; clamp the plates.
10. Make 8% of denaturating polyacrylamide-bisacrylamide solution with 5 mL 0× TBE buffer and 25 g urea. Keep in boiling water bath to dissolve urea, and make the volume to 50 mL. Filter the solution through nitrocellulose paper (0.45 μ) and deaerate.
11. Cool the solution and add 50 μL TEMED and 200 μL 10% ammonium persulfate.
12. Immediately transfer the solution between the 2 plates slowly and continuously using a large needle syringe, keeping the gel mold slightly slanted at an angel of 30°.
13. Insert the comb. Refill the acrylamide solution if the gel retracts.
14. Allow the gel to polymerize for about 1 h.
15. Remove the tape; clamp the plates to the electrophoretic unit.
16. Run the gel at 20–30 mA, ~2000 V for 30 min.
17. Heat samples at 90°C for 3 min and keep on ice; load 1.5 μL of each sample on the gel in different lanes.
18. Run the gel at 20–30 mA, ~2000 V, until bromophenol blue runs out (90 min).

Autoradiography

19. Take out the plates from the electrophoretic chamber and keep them on the table top. Remove both spacers, insert a spatula at 1 corner, and separate the plates carefully. Make sure that the gel is sticking evenly to the bottom plate.
20. Keep the gel in methanol-acetic acid solution for 10 min (this step is used when ^{35}S dATP is used as an isotope).
21. Transfer the gel onto a Whatman filter paper no. 3 and dry it at 80°C for 30 min.
22. Expose the dried gel to an X-ray film for 48–72 h at room temperature.
23. Develop the film for 3 min in an X-ray developer in the dark. Wash with water and keep for 2 min in fixer. Air dry and label the autoradiogram with date and name of the template. Mark each set of sequencing reactions clearly. Read the sequence from bottom to top of the autoradiogram as dark bands in different lanes. This will indicate the sequence of the complementary strand in the 5′→3′ direction. From this sequence, deduce the sequence of the template strand.

HYBRIDIZATION ANALYSIS OF YAC LIBRARY FILTERS

Hybridization-based screening of YAC libraries is carried out using membranes carrying yeast colonies. To help scoring of positively hybridizing YAC colonies, hybridization and washing conditions should be established so that a low level of nonspecific background hybridization remains on all colonies. Alternatively, ^{35}S-labelled YAC vector can be included in the hybridization buffer. This will hybridize to all YAC clones. Colonies hybridizing to a ^{32}P- labelled specific DNA fragment can be clearly scored against the uniform background of the ^{35}S-labelled colonies.

Equipment and Reagents

- 0.5 M phosphate buffer (pH 7.2): 0.5 M Na_2HPO_4, add 4 mL 85% (v/v) H_3PO_4 per liter
- Hybridization solution: 0.25 M NaCl, 0.25 M phosphate buffer, 10% (w/v) PEG 6000, 7% (w/v) SDS, 1 mM EDTA
- Hybridization boxes (8 × 12 cm)
- 20× SSC: 3 M NaCl, 0.3 M trisodium citrate
- Solution A: 3× SSC, 0.1% (w/v) SDS
- Solution A: 0.1× SSC, 0.1% (w/v) SDS

Method

1. Label 50 ng of gel-purified probe (PCR product, cDNA clone/EST, genomic DNA clone, or YAC DNA) using a random primers labelling kit. We usually use α-[^{32}P]dCTP, but other labels could be used. There is no requirement

to remove unincorporated ^{32}P-labelled nucleotides prior to hybridization; in fact, these help to provide background, which aids scoring of the filters.

2. Prehybridize membranes at 62°C for at least 1 h with 10 mL of hybridization solution in a small hybridization box.
3. Remove filter from the hybridization solution and add boiled denatured probe. Mix well and place filter back into the hybridization solution.
4. Hybridize overnight at 62°C with gentle shaking.
5. Transfer the filter into a large box and rinse briefly with cold solution A.
6. Add 250 mL of solution A and wash for 20–30 min at 62°C with gentle shaking.
7. Finally, rinse the filter in 3× SSC, blot dry to remove excess liquid, seal in a plastic bag, and expose to film. Adjust the length of exposure depending on the strength of the signal. An ideal autoradiograph should show a just-visible background of most of the colonies on the filter, with the positive signals clearly stronger.

PREPARATION AND GEL PURIFICATION OF YAC CHROMOSOMAL DNA

For many of the uses of individual YAC clones, such as for hybridization probes or for subcloning, the ability to prepare high molecular weight DNA from YAC-containing yeast strains is required. The method we routinely use, together with a method for the gel purification of YAC clones to be used as hybridization probes, is given here (Figure 3.7).

Equipment and Reagents

- YEPD medium: 1% (w/v) yeast extract, 2% (w/v) peptone, 2% (w/v) glucose; autoclave
- 1 M sorbitol
- InCert agarose
- 100 µL block formers
- SCEM: 1 M sorbitol, 100 mM sodium citrate (pH 5.8), 10 mM EDTA, 30 mM 2-mercaptoethanol
- 100 mg/ml novozyme 234 in SCE: 1 M sorbitol, 100 mM sodium citrate (pH 5.8), 10 mM EDTA; filter sterilize
- ESP (1.0): 100 mM EDTA (pH 8.0), 1% (w/v) sarkosyl, 1 mg/ml proteinase K
- ES: 100 mM EDTA (pH 8.0), 1% (w/v) sarkosyl
- ESP (0.5): 100 mM EDTA (pH 8.0), 1% (w/v) sarkosyl, 0.5 mg/mL proteinase K
- 1 mM phenylmethylsulphonyl fluoride (PMSF) in 1 mM EDTA (pH 8.0), freshly made up using a stock solution of 100 mM PMSF in isopropanol
- TE: 10 mM Tris-HCl, 1 mM EDTA (pH 8.0)
- 2× CA buffer: 40 mM Tris pH 8.0, 14 mM $MgCl_2$, 200 mM KCl, 400 µg/mL BSA

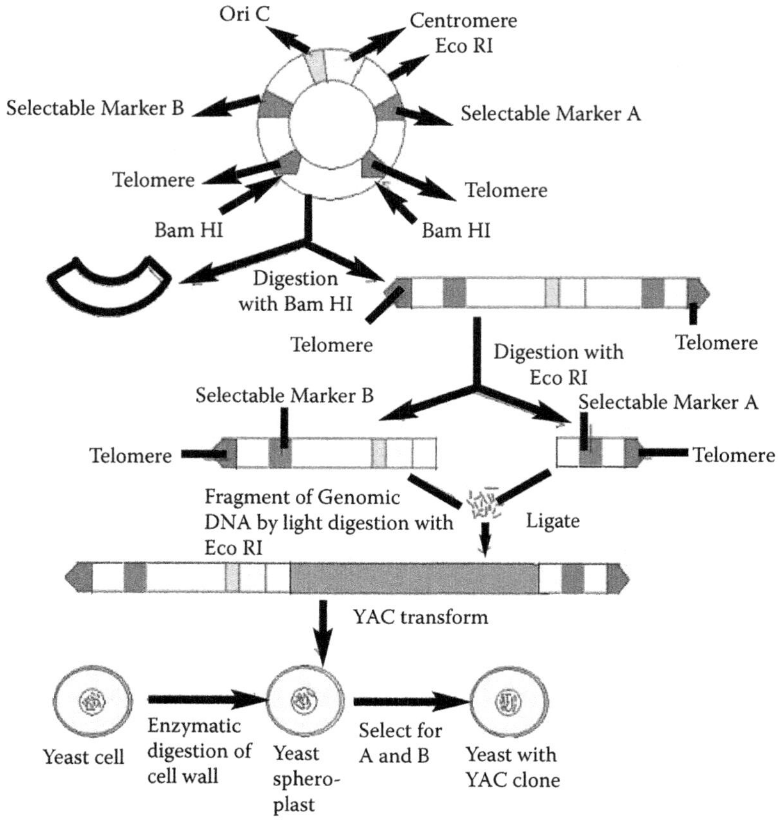

FIGURE 3.7 Formation of yeast clone.

- β-agarose I
- 10 mg/mL yeast RNA
- Seakem GTG agarose
- Phenol:chloroform:isoamyl alcohol (24:25:1)

Method

Preparation of Intact Yeast Chromosomes

1. Transfer a single yeast colony into a plastic centrifuge tube containing 50 mL YEPD medium and grow to stationary phase (2 d) in an orbital shaker at 30°C.
2. Pellets cells by centrifugation at approximately 2600 g for 10 min at room temperature.
3. Resuspend cells in 10 mL 1 M sorbitol and transfer to a 15 mL screw-capped centrifuge tube.
4. Centrifuge at approximately 2600 g for 10 min at room temperature.
5. Resuspend cells in 1 M sorbitol to a final volume of 0.6 mL.

6. Make up 1% (w/v) InCert agarose in 1 M sorbitol, cool to 50°C, and add 0.6 mL to the reuspended cells.
7. Mix thoroughly and quickly dispense into 100 μL block formers. Set on ice for 10 min.
8. Press blocks out into 13 mL SCEM in a 15 ml screw-capped centrifuge tube and add 120 μL 100 mg/mL novozyme 234 solution. Screw on cap and place on cell mixer at room temperature for 4 h.
9. Transfer block to a 15 mL screw-capped centrifuge tube containing 7.5 mL ESP, fill tube with ES, and incubate at 50°C overnight.
10. Drain blocks and add fresh ESP (0.5) and incubate at 50°C for an additional 24 h.
11. Transfer blocks to a 15 mL screw-capped centrifuge tube and wash twice for 1 h with 1 mM PMSF in 1 mM EDTA (pH 8.0), then 6 times for 20 min each with 1 mM EDTA (pH 8.0), and finally for 20 mi with TE. Store the blocks (containing intact YAC chromosomes) in TE at 4°C.

Gel Purification of YAC Clones to be Used as Hybridization Probes

1. Prepare a 1% (w/v) Seakem GTG agarose gel using 0.5× TBE buffer. Cast gel without wells and reserve some molten agarose.
2. Cut a trough near an end of the gel, and insert the blocks containing intact YAC chromosomes. Insert the blocks side by side and place in contact with the front face of the trough. Place lambda ladder markers in the trough alongside the plant DNA blocks. Fill in the remainder of the trough with the reserved molten agarose to seal in the blocks, and let the agarose set.
3. Resolve intact yeast chromosomes by pulsed field gel electrophoresis (PFGE) using pulse parameters to achieve maximum resolution of the YAC, visualize under long-wave UV, and excise the minimum-sized gel slice containing the YAC.
4. Wash gel slice in at least 10 volumes of water in an appropriately sized tube on a cell mixture.
5. Transfer gel slice to a 1.5 mL microfuge tube and place tube in a 65°C water bath for 10 min.
6. Add molten gel slice to an approximately equal volume of 2× CA buffer (prewarmed to 65°C) and mix.
7. Add 10 U Taq I and incubate at 65°C for 30 min.
8. Add 0.05 volume of EDTA, transfer tube to a 37°C water bath, and equilibrate for 10 min.
9. Add 1 U-β-agarase I per 150 μL original 1% gel slice and incubate at 37°C overnight.
10. Centrifuge tube briefly and heat to 65°C for 20 min.
11. Equilibrate tube at 37°C, add 0.5 U 1 U-β-agarase I per 150 μL original 1% gel slice, and incubate at 37°C for 4 h.
12. Heat to 65°C and transfer to a new microcentrifuge tube containing 0.1 vols for 3 M sodium acetate, 2 μL yeast RNA carrier, and an equal volume of phenol. Invert to mix and centrifuge at 12,000 g for 10 min at room temperature.

13. Transfer aqueous phase to a new tube and extract with an equal volume of phenol:chloroform:isoamyl alcohol. Centrifuge at 12,000 g for 2 min.
14. Transfer aqueous phase to a new microcentrifuge tube and add 2.5 vols of 100% (v/v) ethanol to precipitate DNA. Precipitate at –20°C overnight.

CONSTRUCTION OF BAC LIBRARIES

In many circumstances, BAC libraries of nonstandard *Arabidopsis* ecotypes or other plant species would be useful. For example, in the analysis of plant/pathogen interactions the cloning of complex disease-resistance loci from the different ecotypes may require large insert clones.

Two vectors have been developed for large insert library construction (pBelo BAC11 and pECBAC1) and 1 for large insert library construction and plant transformation (BIBAC2). These vectors are present at 1–2 copies per cell, resulting in only a low production off beta galactosidase and, therefore, only pale blue colonies and the yield of insert DNA is limited. Also, there are limited restriction sites available for cloning in these vectors, BIBAC2 only has a single cloning site. Recently, it has been reported that the cosmid vector pCLD04541 (designed as a binary vector for *A. tumefaciens*-mediated plant transformation) has been used to clone large DNA fragments (40–310 kb), and is well suited for large insert library construction. pCLD04541 is present at a higher copy number than BAC vector, 4–5 copies per cell, and the clones are highly stable. This vector also has a range of cloning sites available. pCLD04541 has been used for the construction of large insert DNA libraries in a range of plants, including sorghum, *Arabidopsis*, and maize.

Equipment and Reagents

- Nucleobond A × 10,000 kit
- 1× CA buffer: 20 m Tris pH 8.0, 7 mM $MgCl_2$, 100 mM KCl, 200 μg/mL BSA
- Calf intestinal alkaline phosphate
- Phenol:chloroform:isoamyl alcohol (24:25:1)
- ESP(0.5): 100 mM Tris (pH 3.0), 1% (w/v) sarkosyl, 0.5 mg/mL proteinase K
- 1 mM phenylmethylsulphonyl fluoride (PMSF) in 1 mM EDTA (pH 8.0), freshly made up using a stock solution of 100 mM PMSF in isopropanol
- 2× CA buffer: 40 mM Tris (pH 3.0), 14 mM $MgCl_2$, 200 mM KCl, 400 μg/mL BSA
- Seakem GTG agarose
- Gene Capsule
- 0.025 μm filters
- T4 DNA ligase, 2,000 U/μL
- Electromax DH10B cells
- Cell porator with voltage booster
- Isopropyl-1-thiogalactopyranoside
- 5-bromo-4-chloro-3indolylgalactopyranoside

- LB: 1% (w/v) tryptone, 0.5% (w/v) yeast extract, 1% tryptone, 0.5% yeast extract, 36 mM K_2HPO_4, 1.5 mM trisodium citrate, 0.75 mM $MgSO_4$, 7 mM $(NH_4)_2SO_4$, 13 mM KH_2PO_4, 0.48 M glycerol, pH 7.4, autoclaved.

Methods

Preparation of the Vector

1. Inoculate 4 L of LB medium with the bacterial culture harboring the BAC vector plasmid, and grow overnight at 37°C with shaking.
2. Isolate plasmid DNA using a Nucleobond A × 10,000 kit according to the manufacturer's instructions, except elute the DNA from the column with 4 25 mL aliquots of buffer N5, rather than with 1 100 mL aliquot, and discard the first 10 mL of buffer off the column.
3. Linearize 30 μg of plasmid DNA with 100 U of appropriate restriction endonuclease (e.g., HindIII for pBeloBAC11, BamHI for BIBAC2) in a total volume of 400 μL of 1× CA buffer at 37°C for 1 h.
4. Check linearization of the plasmid by analyzing a 3 μL aliquot of the digest by agarose gel electrophoresis.
5. Dephosphorylate by adding 30 U CIAP to the linearized plasmid and incubate at 37°C for 1 hr.
6. Inactivate the CIAP by incubation at 65°C for 20 min.
7. Extract once with an equal volume of phenol:chloroform:isoamyl alcohol. Add 40 μL of 3 M sodium acetate and 1 mL 100% (v/v) ethanol to the aqueous phase and precipitate for 30 min on ice.
8. Pellet plasmid DNA by centrifugation for 10 min at 12,000 g and resuspend vector DNA in 30 μL 0.1× TE.

Preparation of the Insert DNA

1. Prepare plant-cell nuclei and embed in InCert agarose blocks.
2. Transfer the blocks to a 15 mL screw-capped centrifuge tube containing 13 mL ESP (0.5), and incubate at 50°C overnight.
3. Replace buffer with fresh ESP (0.5) and incubate at 50°C for another 24 h.
4. Transfer the blocks to a new 15 mL screw-capped centrifuge tube and wash twice for 1 h each with 1 mM PMSF in TE, and then 6 times for 20 min with TE. Store blocks at 4°C in TE.

Calibration of Restriction Endonuclease for Partial Digestion of the DNA Preparation

1. Transfer 4 blocks to a 15 mL screw-capped centrifuge tube containing sterile water and place the tube on a cell mixer for 30 min.
2. Cut the blocks in half and place each into a microfuge tube.
3. To each tube, add 50 μL 2× CA buffer, and incubate at 37°C for at least 2 h.
4. Exchange the buffer with fresh 1× CA buffer and incubate on ice for 20 min.
5. Add 5 μL of the appropriate restriction endonuclease diluted in 1× CA buffer (e.g., HindIII for pBeloBAC11, BamHI for BIBAC2), and incubate

on ice for 30 min. Use a 2-fold dilution series to add 0, 0.125, 0.5, 1.0, 2.0, 4.0, and 8.0 units to each half block.

6. Transfer the tubes to a 37°C water bath for exactly 1 h.
7. Transfer the tubes to ice and immediately replace the buffer with 50 μL of ice-cold 50 mM EDTA (pH 9.0). Incubate on ice for 1 h.
8. Resolve the samples by PFGE, using lambda ladder markers and appropriate pulse parameters, in order to identify the partial digest that produced the maximum amount of DNA in the desired 130–200 kb size range.

Size Fractionation and Purification of the Insert DNA

1. Carry out a large scale digest on 12–16 full blocks, using the procedure described above, but with 1 whole block per tube, double volumes, and with the optimum quantity of restriction endonuclease added (remembering to take account of the doubling of solution volumes and the quantity of DNA in each tube).
2. Prepare a 1% (w/v) Seakem GTG agarose gel using 0.5× TBE buffer. Cast gel without wells and reserve some molten agarose.
3. Cut a trough near one end of the gel and insert the blocks of the partially digested plant DNA. Insert the blocks in 2 rows of 6 or 8 blocks, placed in contact with each other and the front face of the trough. Place lambda ladder markers in the trough alongside the plant DNA blocks. Fill in the remainder of the trough with the reserved molten agarose to seal in the block, and let set.
4. Run PFGE for 18 h using pulse parameters selected so as to resolve to approximately 250 kb.
5. Cut off the gel outer edges containing the lambda markers and the edges o the plant DNA smear. Stain the gel edges with ethidium bromide and visualize on a UV transilluminator.
6. Measure the exact migration distance of the DNA in the size range 130–200 kb.
7. Guided by the measurement taken, excise the slice of gel containing 130–200 kb DNA from the unstained portion of the gel.
8. Rotate the gel slice through 180°C and load onto a fresh 1% (w/v) agarose gel, such that the largest DNA is now furthest along the gel.
9. Run the second pulse field gel, use exactly the same parameters as used in step 4.
10. Analyze gel as describe in step 5.
11. The majority of the DNA should be concentrated into a sharp band. Measure the exact migration distance of the band.
12. Based on the migration distance, recover the DNA from the unstained portion of the gel by electroelution. We use Gene Capsules, following the manufacturer's instructions.
13. Pulll eluted DNA and analyze a 10 μL sample by PFGE to check integrity and estimate yield.
14. Float a 0.025 μm VS filter on 50 mL 0.1× TE in a 50 mL screw-capped centrifuge tube. Gently pipette the eluted DNA solution onto the membrane and place at 4°C overnight to dialyze DNA at 4°C.

Ligation, Electroporation, and Arraying the Library

1. Prepare ligation reactions using 200 ng of insert DNA, a 10-fold molar excess of vector DNA, 0.1 volume 10× ligation buffer, 2,000 U (1 μL) T_4 DNA ligase in a final volume of 100 μL. Incubate at 4°C overnight. Multiple 100 μL ligations should be set up in parallel. As a control, set up one ligation lacking insert DNA.
2. Float dialyze the ligations as described in step 14 above. Store at 4°C.
3. Electroporate 1 μL Electromax DH10B cells, using an electroporator with voltage booster and the protocol supplied by the manufacturer. Electroporation parameters: fast charge, low ohms, 330 μF, voltage booster set to 4000 ohms. Plate cells on appropriate selective medium (LB agar plus 0.5 mM IPTG, 40 mg/mL X-Gal, and 12.5 μg/mL chloramphenicol for pBelo BAC 11; LB agar plus 0.5 mM IPTG, 40 mg/mL X-Gal, 40 μg/mL kanamycin, and 5% sucrose for BIBAC2).
4. Check that the transformation efficiency is adequate for library construction and that the ligations including insert DNA result in at least a 50-fold increase in transformants over the control ligation without insert DNA. If the background of nonrecombinant is too high (indicated by a lower ratio), a new preparation of vector must be made.
5. Plates of transformed cells can be stored for several days at 4°C prior to picking. Constructs in pBelo BAC11 must be stored at 4°C for several days to allow the blue color of nonrecombinant clones to fully develop.
6. Use toothpicks to pick colonies and inoculate 384-well microtiter plates containing 50 μL per well of freezing broth and appropriate selection. Grow at 37°C for 2 days, stir and freeze on dry ice. Store at −80°C.

PREPARATION OF COLONY FILTERS FOR HYBRIDIZATION ANALYSIS

Hybridization-based screening is performed using membrane-carrying bacterial colonies. To maximize the number of colonies on each membrane, a multiprong replicator is used. This procedure can be carried out manually or by a robot. Economical handling of large numbers of colonies can be achieved by slightly offsetting each colony print, and up to 3 × 3 × 384 colonies can be transferred to a single 11 × 7.5 cm membrane. To make the identification of positively hybridizing colonies even more reliable, the colonies can be duplicated. The preparation of BAC colony filters by hand is describe in this protocol.

Equipment and Reagents

1. LB:1% (w/v) tryptone, 0.5% (w/v) yeast extract, 1% (w/v) NaCl, pH7.2 (add 1.5% bacto-agar for LB agar)
2. 8.6 × 12.8 cm Omni Tray
3. Hybond N^+ membrane
4. Whatman 3 MM paper

5. Denaturation solution: 0:5 M NaOH, 1.5 M NaCl
6. Neutralization buffer: 1 M Tris (pH 7.5), 1.5 M Nacl
7. Scrape solution: 5× SSC, 0.5% (w/v) SDS, 1 mM EDTA (pH 8.0)

Method

1. Pour LB agar plate containing appropriate selection using 8.6 × 12.8 cm Omni Trays, and allow to dry.
2. Cut 11.7 × 7.6 cm rectangles of Hybond N^+ membrane. Label using a pencil or fine, permanent marker pen (in position A1) and place on the surface of an LB agar plate.
3. Using a 384-prong replicator, spot the colonies of 1 384-well microtiter plate onto each filter (up to 9 plates may be spotted if an offsetting strategy is used). Prepare several replicates for simultaneous hybridization analysis.
4. Invert the plates, and incubate at 37°C overnight.
5. Carefully lift the filters from the agar plates, place on a sheet of 3 MM paper soaked in denaturation solution, and leave for 3 min.
6. Transfer the filters to a fresh sheet of 3 MM paper soaked in denaturation solution, and leave for another 3 min.
7. Transfer filters to a sheet of 3 MM paper soaked in neutralization solution, and leave for 6 min.
8. Rinse the filters in 2× SSC and air dry. Place filters ins sleeves of 3 MM paper when they begin to curl.
9. Bake at 80°C for 2 h.
10. Soak the filters for at least 2 h at 42°C in scrape solution.
11. Using a paper towel, scrape off the bacterial debris. Rinse filters in 2× SSC. The wet filter can be used immediately for hybridization or dried for use later.

HYBRIDIZATION ANALYSIS OF COLONY FILTERS

BAC libraries can be used for whole genome mapping or for the assembly of contigs in specific regions of the genome. For physical mapping in *Arabidopsis*, the extensive YAC-based physical maps provide a valuable resource to help assemble BAC contigs in defined genomic regions. The YAC maps are well integrated with the genetic maps of *Arabidopsis*. Chosen YACs can be used to identify clones in the BAC library by preparing gel purified YAC chromosomal DNA and hybridizing the isolated DNA to BAC colony filters. The screening of BAC filters by hybridization is described in this protocol.

Equipment and Reagents

1. Random prime labelling kit
2. 20× SSPE: 3 M NaCl, 0.2 M NaH_2PO_4, 25 mM EDTA (pH 7.4)
3. 50× Denhardt's: 1% (w/v) ficoll, 1% (w/v) polyvinylpyrolide, 1% (w/v) BSA
4. Hybridization boxes (8×12 cm, e.g., Bunzl 6064)
5. Hybridization buffer: 5× SSPE, 5× Denhardt's, 0.5% (w/v) SDS, 5 μg/mL salmon-sperm DNA

6. Hybridization wash buffer: 0.1% SSC, 1% (w/v) SDS
7. Membrane neutralization buffer: 0.1× SSC, 0.1% (w/v) SDS, 0.2 M Tris (pH 7.5)

Method

This protocol is designed for use with sets of 8 membranes of size 11.7 × 7.6 cm and should be scaled up or down as appropriate.

1. Label 50 ng of gel-purified probe (PCR product, cDNA clone/EST, genomic DNA clone, or YAC DNA) using a random primer labelling kit. We usually use α-[^{32}P]dCTP, but other labels could be used. There is no requirement to remove unincorporated ^{32}P-labelled nucleotides prior to hybridization; in fact, these help to provide background, which aids scoring of the filters.
2. Rinse the filters in 2× SSC and then place in a hybridization box containing 30 mL of hybridization buffer. Add filters 1 by 1, making sure each is fully submerged before adding the next.
3. Prehybridize at 65°C for at least 4 h with gentle shaking.
4. Add denatured probe to a fresh hybridization box containing 10 mL hybridization buffer, and transfer the filters 1 by 1 as previously.
5. Hybridize overnight at 65° with gentle shaking.
6. Transfer the filters, 1 by 1 to a larger box containing 250 mL hybridization wash buffer. Incubate at 65°C for 20 min with gentle shaking.
7. Discard the wash buffer and replace with fresh solution. Incubate as in step 6.
8. Repeat step 7 to give a third wash.
9. Blot dry and seal filters in a plastic bag and expose to film, with the film to the DNA side of the filters, adjust the length of exposure depending on the strength of the signal.
10. It may be necessary to rewash the filters to reduce excessive background. An ideal autoradiograph should show most of the colonies on each filter as a just-visible background, with the positive signals clearly stronger.
11. Filters can be stripped by incubation in 0.4 M NaOH twice for 20 min each at 45°C with gentle shaking, followed by a 15-min wash in membrane-neutralization buffer.

CONSTRUCTION OF A BINARY COSMID LIBRARY FROM *ARABIDOPSIS* GENOMIC DNA

Cosmid libraries can be easily constructed from a range of source DNA, such as genomic DNA, YAC insert, or BAC inserts. We have routinely used the binary cosmic vector pCLD04541 to construct cosmid libraries and a reliable method for construction involves the partial digestion of source DNA with *Sau* 3A, size selection to 15–25 kb. Treat the insert with phosphatase and ligate, at high concentration, into a *Bam* HI-cut vector. Package the ligated DNA into lambda phage using commercial packaging extracts. These can then be used to infect *E. coli*, in which the DNA is circularized and propagated as a plasmid. This protocol describes a detailed method for the construction of a library using *Arabidopsis* genomic DNA. This method can also be used for subcloning BAC DNA, and can be adapted for subcloning a YAC, as the source DNA in this case is limiting.

Equipment and Reagents

1. Cosmid vector pCLD04541
2. 100–150 μg CsCl-pure *Arabidopsis* genomic DNA
3. 10 μg CsCl-pure vector pCLD04541
4. LB: 1% (w/v) tryptone, 0.5% (w/v) yeast extract, 1% (w/v) NaCl (pH 7.2)
5. Phenol:chloroform:isoamyl alcohol (24:25:1)
6. CIAP (calf intestinal alkaline phosphate)
7. T4 DNA ligase and ligation buffer
8. Gigapack II XL or Gold lambda packaging extracts
9. *E. coli* sure Tet strain
10. IPTG (isopropyl-1-thiogalactopyranoside)
11. Freezing broth: 1% tryptone, 0.5% yeast extract, 36 mM K_2HPO_4, 1.5 mM trisodium citrate, 0.75 mM $MgSO_4$, 7 mM $(NH_4)_2$ SO_4, 13 mM KH_2PO_4, 0.48 M glycerol (pH 7.4); autoclaved.

Method

Calibration of Restriction Endonuclease for Partial Digestion of Plant Genomic DNA

1. To 10 μg genomic DNA, add restriction buffer (1×), spermidine (4 mM), BSA (1×), and water to a final volume of 150 μL; mix well.
2. Label 9 microfuge tubes 1–9, add 30 μL of the above mix to tube 1, 15 μL to tubes 2–8, and the remainder of the mix to tube 9.
3. Add 1 U Sau 3A to tube 1 and mix well.
4. Transfer 15 μL from tube 1 to tube 2 and mix well. Continue the 2-fold dilution through tube 8. Do not add enzyme to tube 9.
5. Incubate at 37°C for 1 h.
6. Stop the reaction by placing tubes on ice and add 1 μL of 0.5 M EDTA (pH 8.0).
7. Resolve the digests by 0.4% (w/v) agarose gel electrophoresis.

Note: Make a 1%–2% (w/v) agarose base and then put 0.4% (w/v) agarose on top, the 1%–2% gel supports the low-percentage gel during transfer to the UV transilluminator. Use DNA-sized markers that have fragments 9–30 kb in size.

8. Stain the agarose gel with ethidium bromide solution and determine which concentration of enzyme appears to best produce DNA of the required size, 15–25 kb. Half this amount of enzyme will give the maximum amount of molecules of the required size.

Large-Scale Partial Digestion of Genomic DNA

1. Using the same concentration of DNA (1 μg/15 μL) and components as in the calibration experiment, conduct a series of 10 μg partial digests on the remaining genomic DNA using half the concentration of enzyme that appeared to produce the most amount of DNA in the desired 15–25 kb range.
2. Incubate at 37°C for 1 h.

3. Remove 1 μg aliquots from the digestion mixes and add 1 μL 0.5 M EDTA (pH 8.0).
4. Stop the rest of the reaction by incubation at 65°C for 20 min.
5. Add 0.1 volumes of 3 M sodium acetate and an equal volume of phenol:chloroform:isoamyl alcohol and mix gently. Centrifuge at 12,000 g for 5 min.
6. Transfer the aqueous phase to a new microfuge tube, and precipitate the partially digested DNA with 2 volumes of 100% (v/v) ethanol.
7. Pellet the DNA in a microcentrifuge at 12,000 g for 10 min.
8. Wash the pellets with 70% (v/v) ethanol, and centrifuge briefly at 12,000 g.
9. Resuspend the pellets in 270 μL water to produce a pool.
10. Resolve 1 μg aliquots by agarose gel electrophoresis (0.4% gel on a 1%–2% base, as in step 7 of the previous procedure) to confirm that the digested fragments are of the correct size range.

Treatment of Partially Digested Genomic DNA with CIAP

1. To the 270 μL of partially digested DNA, add 30 μL 10× CIAP buffer and 2.5 U CIAP.
2. Incubate at 37°C for 20 min and then at 68°C for 15 min.
3. Add 0.1 volumes of 3 M sodium acetate and an equal volume of phenol:chloroform:isoamyl alcohol and mix gently. Centrifuge at 12,000 g for 5 min.
4. Transfer the aqueous phase to a new microfuge tube and add 2 volumes of 100% (v/v) ethanol to precipitate the partially digested DNA.
5. Pellet the DNA in a microcentrifuge at 12,000 g for 10 min.
6. Wash the pellet with 70% (v/v) ethanol and centrifuge briefly at 12,000 g; air dry.
7. Resuspend the DNA in 500 μL TE.

Size Fractionation and Purification of Genomic DNA

1. Prepare a sucrose gradient with 10%–40% (w/v) sucrose solutions (diluted in: 1 M NaCl, 20 mM Tris [pH 8.0], 5 mM EDTA).
2. Load the 500 μL of DNA obtained from the previous procedure onto the top of the gradient, and centrifuge at approximately 50,000 g (e.g., 20,000 rpm in an SW41 rotor) for 18 h.
3. Following centrifugation, collect 20 fractions of 300–500 μL from the bottom of the tube.
4. Analyse 10 μL of each fraction by agarose gel electrophoresis (0.4% gel on a 1%–2% base, as in step 7 of the first procedure in this Method section), and identify fractions containing genomic DNA fragments 15–20 kb in length

Note: it is necessary to add NaCl to the size markers so that the NaCl concentration is the same as in the fractions.

5. Dilute the fractions that contain DNA fragments 15–25 kb 1:4 with water.
6. Precipitate the DNA by adding 1/30 volume of 3 M sodium acetate (pH 5.5), 1 μL tRNA, and 0.7 volumes os isopropanol; incubate on ice for 60 min.

7. Pellet the DNA in a microcentifuge for 20 min at 12,000 g.
8. Wash and dry the pellet as described in step 6 of the previous procedure.
9. Resuspend each fraction in 5 μL of water.
10. Check the size of the purified fractions by resolving 0.5 μL from each on a 0.4% (w/v) agarose gel (0.4% gel on a 1%–2% base, as in step 7 of the first procedure in this Method section).
11. Pool fractions of the required size into a single tube.

Preparation of Vector

1. Digest 10 μg of pCLD04541 vector to completion with *Bam* HI. Use at least 5 U *Bam* HI/μg vector DNA, and incubate at 37°C for 2–3 h.
2. Remove a 100 ng aliquot, and add an equal volume of phenol:chloroform: isoamyl alcohol. Centrifuge at 12,000 g for 5 min.
3. Precipitate the vector DNA with ethanol, and wash with 70% (v/v) ethanol as in steps 4, 5, and 6 of the previous procedure.
4. Resolve the 100 ng aliquot by agarose gel electrophoresis to check that the vector is completely linearized.

Ligation of Vector and Insert DNA

1. For the ligation, combine 1 μg dephosphorylated, purified genomic DNA fragments and 0.25 μg vector DNA in a 7.5 μL volume.
2. Remove a 0.5 μL aliquot and store at –20°C.
3. Incubate the mix at 65°C for 5 min, at 42°C for 20 min, and finally at room temperature for 2 h.
4. To the tube, add 2 μL of ligation buffer and 1 μL (2000 units) of T4 DNA ligase.
5. Incubate at 12°C for 18 h.
6. To check the ligation, analyze 0.5 μL of the ligation mix against the unligated control, from step 2, by agarose gel electrophoresis.

Packaging and Plating the Library (Figure 3.8)

1. Perform a test packaging with 2 μ of ligation using Gigapack II SL or GOLD lambda packaging extracts. Perform according to the manufacturer's protocol.
2. Combine 10 μL of phage stock, 100 μL phage buffer, and 200 μL plating cells (*E. coli* strain).
3. Incubate at 37°C for 20 min.
4. Add 1 mL LB medium, and incubate at 37°C for 45 min.
5. Spread 0.1 and 0.5 mL aliquots on selective plates (LB agar + 10 μg/mL tetracycline) including 0.5 mM IPTG and 40 mg/mL X-Gal

Note: spin down cells and resuspend in fresh LB medium before plating to avoid adding $MgCl_2$ to tetracycline plates, as this inhibits tetracycline selection.

6. Incubate at 37°C overnight.
7. Count the colonies and calculate the titer of the library.
8. Plate out the library, as described above, at a density such that single colonies can easily be picked.
9. Pick colonies into 384-well microtiter plates containing 50 μL freezing broth plus 10 μg/mL tetracycline per well. Grow at 37°C for 2 d, and freeze at –80°C.

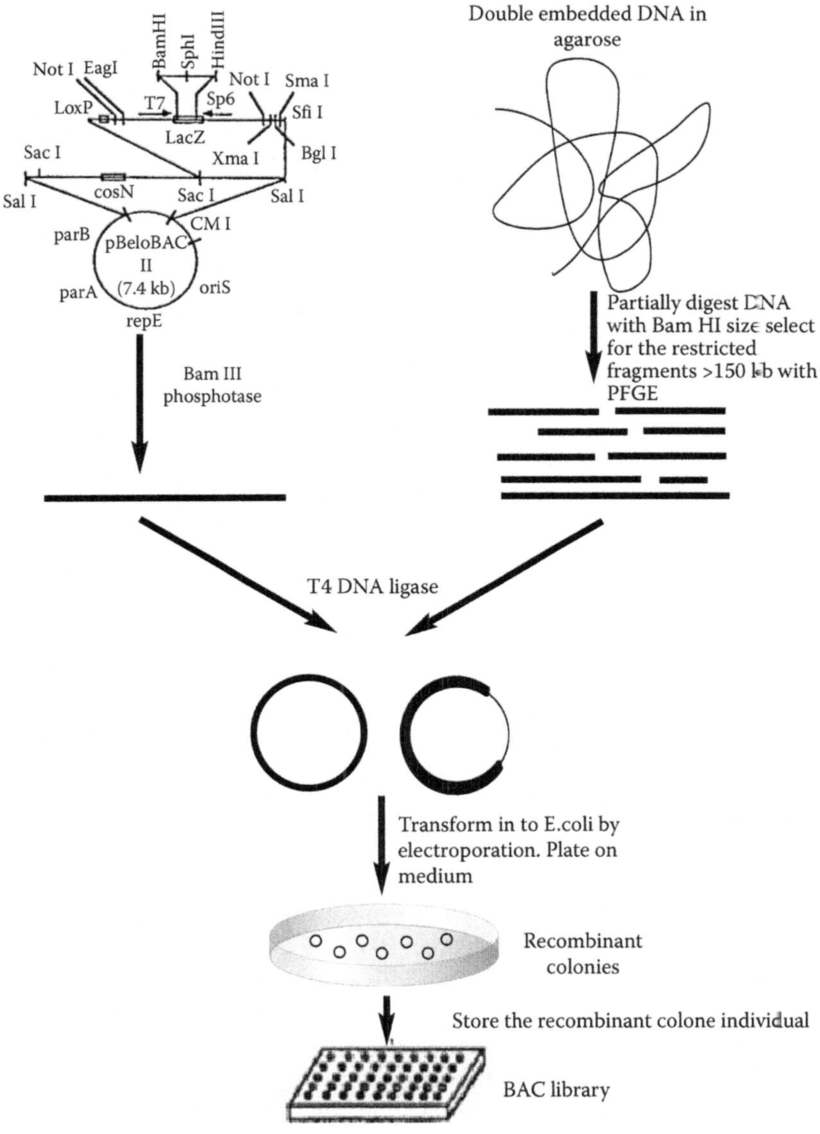

FIGURE 3.8 Construction of BAC library. *Note*: OriS is the origin for unidirectional replication and CM is chloramphenicol.

AGROBACTERIUM: MEDIATE TRANSFORMATION

In nature, the gram-negative soil bacterium *Agrobacterium tumefaciens* causes crown gall formation in dicots by a multistep transformation process. This naturally occurring mechanism of gene transfer has been extensively used in plant biotechnology. *A. tumefaciens* has long been used as a vector to transfer foreign genes into dicots, the preferred hosts of *Agrobacterium*. Although monocotyledonous plants are generally not considered within the host range of crown gall, it has been shown recently that *A. tumefaciens* can also be used as a vector for the successful transformation of monocotyledonous species, particularly.

Virulent strains of *A. tumefaciens* possess an extrachromosomal plasmid that is involved in crown gall formation. Because of its role in tumor induction, this plasmid is called the tumor-inducing plasmid. During infection, a small segment of the Ti plasmid, the T-DNA, surrounded by a 24 bp border repeat, is transferred to the plant cell and is integrated into a host chromosome in the nucleus. The T-DNA contains *onc* genes that encode phytohormone biosynthetic enzymes for which expression leads to tumor formation. Deletion of the *onc* genes of the Ti plasmid results in a nononcogenic strain. Wild-type *A. tumefaciens* can be classified as octopine, nopaline, succinamopine, or LL-succinamopine types according the opine-synthesizing capabilities encoded by their T-DNAs. Opines are tumor-specific products that are synthesized by plants infected with *A. tumefaciens* and used by the bacteria as carbon and nitrogen sources. Opine-biosynthetic genes have eukaryotic regulatory sequences for expression. Besides the *onc* genes, a large number of other genes involved in tumorigenicity are present on the Ti plasmid in a segment called the virulence (*vir*) region. The T-DNA transfer system is determined by the *vir* genes, while the flanking 24-bp direct repeats are essential as recognitions signals for the transfer apparatus.

Explants, Equipment, and Reagents

Callus Induction from Immature Embryos

- *Plant material*: Plant material is *Oryza sativa*: indica-type rice cultivar, IR64, IR72, Basmati 122, Tulsi, Vaidehi, and japonica-type rice cultivars Taipei 309; immature embryos (panicles collected 7–10 days after anthesis).
- *Soil and fertilizers*: 2.5 g $(NH_4)_2SO_4$, 1.25 g P_2O_5, and 0.75 g K_2O mixed per 2 kg of soil.
- *General equipment*: Laminar flow cabinet, stereomicroscope, incubator, autoclave, plastic petridishes (50 mm; 90 mm diameter), pipettes, greenhouse facilities.
- *Sterilizing solution*: Ethanol (70%) sodium hypochlorite solution (e.g., 100 mL 1.8% sodium hypochlorite with 2 drops Tween 20% or 50% chlorox containing 1–2 drops of Tween 20).
- *Media*: Callus induction and proliferation medium, for example, modified MS medium (Table 3.2).

Agrobacterium Culture for Transformation (Figure 3.9)

- *Agrobacterium strains*: A number of different strains of *A. tumefaciens* containing different binary vectors need to be considerd. *A. tumefaciens* strain LBA4404 (pTOK233) from Japan Tobacco Inc. contains pTOK 233, a superbinary vector, for which the T-DNA carries (a) the β—glucuronidse (gus) reporter gene with introns in the N-terminal regions, (b) a hygromycin-resistance gene, and (c) a kanamycin-resistance gene. The gus gene expresses Gus activity in plant cells, but not in *A. tumefaciens* cells. The binary vector contains vir B, vir C, and vir G gene derived from supervirulent Ti plasmid pTiB0542.

TABLE 3.2
MS Medium

Component	Quantity (mg/L)
NH_4NO_3	1,650
KNO_3	1,900
$CaCl_2.2H_2O$	440
$MgSO_4.7H_2O$	370
KH_2PO_4	170
KI	0.83
H_3BO_3	6.3
$Mn.SO_4.H_2O$	22.3
$Zn.SO_4.7H_2O$	8.6
$Na_2. MoO_4.2H_2O$	0.25
$CuSO_4.5H_2O$	0.025
$CoCl_2.6H_2O$	0.025
$FeSO_4.7H_2O$	27.8
Na_2 EDTA	37.3
Nicotinic acid	0.5
Pyridoxine HCl	0.5
Thiamine HCl	1.0
Glycine	2.0
Casein hydrolysate	300
Myo-inositol	100
2,4-D	2
Sucrose/maltose	30 g/L
Agar	8 g/L

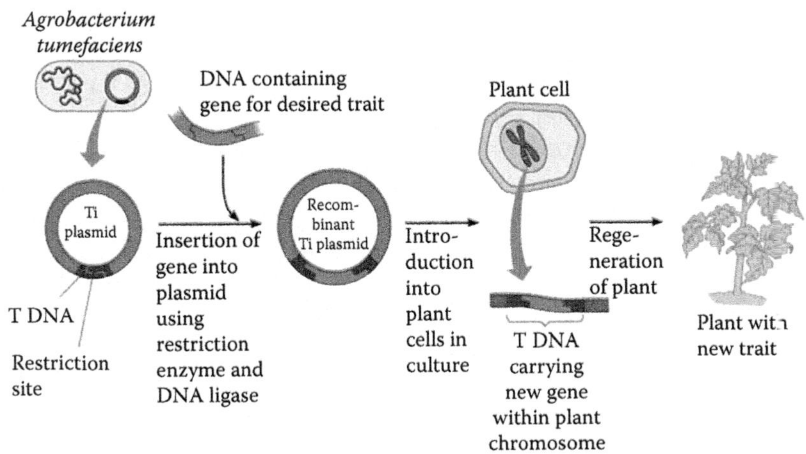

FIGURE 3.9 *Agrobacterium*-mediated transformation.

- *Laminar flow cabinet*
- Rotary shaker at 28°C
- 20 mm high petri dishes
- For growth of *Agrobacterium*, use liquid AAM medium (Table 3.3) or Luria broth medium [(1% Bacto-peptone), 0.5% Bacto yeast extract, 1% NaCl] together with suitable antibiotics for plasmid selection.
- For maintenance of *Agrobacterium*, use solid AB (Table 3.4) or LB medium with agar (15 g/L).

TABLE 3.3
AAM Medium

Component	Amount (mg/L)
Macronutrients	
$CaCl_2.2H_2O$	150.0
$MgSO_4.7H_2O$	250.0
$NaH_2PO_4.H_2O$	150.0
KCl	2,950.0
Micronutrients	
KI	0.75
H_3BO_3	3.0
$Mn.SO_4.H_2O$	10.0
$Zn.SO_4.7H_2O$	2.0
$Na_2.MoO_4.2H_2O$	0.25
$CuSO_4.5H_2O$	0.025
$CoCl_2.6H_2O$	0.025
Iron Composition	
Na_2 EDTA	37.3
$FeSO_4.7H_2O$	27.80
Vitamins	
Nicotinic acid	0.5
Pyridoxine HCl	0.5
Thiamine HCl	1.0
Glycine	2.0
Inositol	100.0
Others	
L- glutamine	876.0
Aspartic Acid	266.0
Arginine	174.0
Casamino acid	500.0
Sucrose	68.5 g/L
Glucose	36 g/L
Acetosyringone	200 μM
pH	5.2

TABLE 3.4
AB Medium

Component	
Stock I	**g/500 mL**
K_2HPO_4	30.0
NaH_2PO_4	10.0
Stock II	**g/500 mL**
NH_4Cl	10.0
$CaCl_2$	0.1
$MgSO_4.7H_2O$	3.0
$FeSO_4.7H_2O$	0.25
KCl	1.5
Stock III	**g/900 mL**
Glucose	5.0
Agar	15.0

Transformation and Regeneration

General equipment:

- Centrifuge (speed and temperature controlled)
- Spectrophotometer for measuring optical density of bacterial solution
- Vacuum infiltration apparatus
- Plastic centrifuge tubes for 50 mL capacity
- 10 mL pipettes
- Automatic pipettes
- Petri dishes (90 mm)

Solutions and Media

- 10 mM $MgSO_4$ solution
- N6-AS medium with acetosyringone for infiltration and cocultivation. For infiltration N6-AS, liquid medium (Table 3.5) should be used (without agar).
- Cefotaxime: 250 mg/L, filter sterilized and stored at –20°C
- GUS staining solution
- MS-CH medium (Table 3.5) for selection
- Regeneration medium (Table 3.5)
- Rooting medium-MS basal medium (Table 3.2) without hormones
- Culture solution (Table 3.6)

METHOD

Growth of Donor Plants

1. Break the dormancy of clean seeds of the desired pure-rice line by incubating them at 50°C for 3–5 d.
2. Sow the seeds in seed boxes in damp soil.

TABLE 3.5
Composition of Media for Growth of Transformants

Component	N6-AS (mg/L)	MS-CH (mg/L)	Regeneration Medium (mg/L)
$(NH_4)_2\ SO_4$	463.0	–	–
KH_2PO_4	400.0	170.0	170.0
KNO_3	2830.0	1900.0	1900.0
NH_4NO_3	–	1650.0	1650.0
$CaCl_2.2H_2O$	166.0	440.0	440.0
$MgSO_4.7H_2O$	185.0	370.0	370.0
Na_2 EDTA	37.3	37.3	37.3
$FeSO_4.7H_2O$	27.8	27.8	27.8
$Mn.SO_4.H_2O$	4.4	22.3	22.3
H_3BO_3	1.6	6.3	6.3
$Zn.SO_4.7H_2O$	1.5	8.6	8.6
KI	0.8	0.83	0.83
$CoCl_2.6H_2O$	–	0.025	0.025
$CuSO_4.5H_2O$	–	0.025	0.025
$Na_2MoO_4.2H_2O$	–	0.25	0.25
Thiamine-HCl	1.0	1.0	1.0
Nicotinic acid	0.5	0.5	0.5
Pyridoxine—HCl	0.5	0.5	0.5
Glycine	2.0	2.0	2.0
Myo-inositol	100	100	100
Casamino acids	1000	1000	1000
Kinetin	–	–	2.0
NAA	–	–	1.0
2,4-D	2.0	2.0	–
Cefotaxime	–	250	–
Hygromycine	–	50	–
Glucose	10 (g/L)	–	–
Maltose	–	30 (g/L)	–
Sucrose	30 (g/L)	–	30 (g/L)
Agar	8 (g/L)	7 (g/L)	–
Agarose	–	–	6 (g/L)
Acetosyringone	200 μM	–	–
pH	5.2	5.8	5.8

3. 21 d after sowing, transplant the seedlings to pots (1 seedling per pot) containing soil and fertilizer.
4. Grow the plants in a greenhouse and keep them well-watered. The greenhouse should have full sunshine and sufficient ventilation to maintain a daytime temperature of 27°C–29°C with a 90% humidity level.

TABLE 3.6
Culture Solution

Component	Quantity (mg/L)
NH_4NO_3	114.25
$NaH_2PO_4.H_2O$	50.37
K_2SO_4	89.25
$CaCl_2.2H_2O$	110.75
$MgSO_4.7H_2O$	405.0
$MnCl_2.4H_2O$	1.87
$(NH_2)_6Mo_7O_{24}4H_2O$	0.09
H_3BO_3	1.16
$Zn.SO_4.7H_2O$	0.04
$CuSO_4.5H_2O$	0.04
$FeCl_3.6H_2O$	9.62
Citric Acid (monohydrate)	14.87

Establishment of Embryogenic Calli from Immature Embryos

1. Sterilize dehulled immature (9–14 d after anthesis) seeds in 70% ethanol for 1 min and then in 1.8% sodium hypochlorite for 30 min. Wash 3 times with sterile double-distilled water.
2. Isolate the embryos from the sterile seeds and place them on petri dishes containing 20 mL modified MS medium. The distance between each plated embryo should be at least 5 mm. The scutellar tissue should be upward, because callus growth occurs only with this orientation.
3. Incubate the embryos in the dark at 25°C.
4. After 3–4 days, cut off the emerging shoots and roots, and subculture the remaining tissues on fresh medium.
5. Make at least 2 subcultures at 2-wk intervals until yellowish-white soft embryogenic calli develop in the surface of the sculella.
6. Select and collect the embryogenic calli from the 4–6 wk old culture only on the day of transformation.

Preparation of *Agrobacterium* Culture

1. Maintain the *Agrobacterium* culture in AB or LB solid medium and subculture once a month.
2. Preculture: 4 days prior to transformation, grow a single clone of the bacterium on 10 mL AAM medium (Table 3.3) or LB medium with the appropriate selective agent.
3. Incubate with shaking (250 rpm) at 28°C overnight.
4. Transfer 500 μL of the culture to 20 mL AAM or LB medium with appropriate antibiotics and acetosyringone.
5. From the preculture medium, streak the culture again onto AB or LB solid medium.

6. Incubate both solid and liquid cultures at 28°C for 2 d. Liquid cultures should be shaken at 250 rpm.

Transformation

1. Centrifuge the bacterial culture in liquid medium at 3800 g for 30 min at 10°C.
2. Discard the medium and harvest the bacteria in 20 mL 10 mM $MgSO_4$ in a 50 mL centrifuge tube. In the case of bacteria growing on solid AB medium, harvest directly in 10 mM $MgSO_4$ solution.
3. Prepare a homogeneous bacterial suspension in 10 mM $MgSO_4$ by pipetting up and down several times.
4. Pipette 2.7 mL 10 mM $MgSO_4$ solution into another tube and add 300 μL of the bacterial suspension in 10 mM $MgSO_4$ (total volume 3 ml).
5. Take optical density (OD) 600 nm measurement of the mixture from step 4.
6. Centrifuge the original harvested bacterial suspension in 20 mL 10 mM $MgSO_4$ at 3800 g for 30 min.
7. Discard the supernatant.
8. Resuspend the bacterial pellet by pipetting up and down in a small volume (5 mL) of liquid N6-AS infiltration medium (Table 3.5).
9. Adjust the final optical density (600 nm) to approximately 2 by adding N6-AS medium.
10. Add the embryogenic calli to the bacterial suspension in a petri dish.
11. For vacuum infiltration, place the petri dish in a vacuum dessicator for 10 min.
12. Let the embryogenic calli stand in the bacterial suspension for 20 min after removal from the vacuum dessicator.
13. Remove the bacterial suspension from the petri dish with a pipette.
14. Blot the calli on filter paper to remove excess innoculum and transfer the calli to cocultivation medium (N6-AS solid medium).
15. Keep the cultures in the dark at 25°C for 3 d.

Washing and Selection

1. Wash the embryogenic calli twice with sterile water containing 250 mg/L cefotaxime to remove the bacteria.
2. After washing, remove the excess water adhering to the calli with sterile blotting paper.
3. Place some of the calli in GUS staining solution to check GUS expression if the T-DNA carries the β-glucuronidase reporter. Keep overnight at 37°C.
4. Place the remaining calli into MS-CH selection medium containing cefotaxime (the bacteriostatic antibiotic) and hygromycin (the selective antibiotic). Keep for 15 d in the dark at 27°C.
5. Check the stained calli from step 3 for the presence of blue spots.

Regeneration

1. Following the first selection, put the growing calli through 3 more selection cycles in the same medium, without cefotaxime, under the same conditions.

2. Transfer the selection embryogenic colonies with developing somatic embryos to regeneration medium. Incubate in darkness at 25°C for 2 wk.
3. Transfer the developing embryos onto fresh plates of the same medium and culture in the light (24 μmol/m^2/s), with a 16 h photoperiod at 25°C to obtain plantlets.
4. Transfer the plantlets to the same medium without hormones to obtain well developed root systems.
5. Transfers the plants to a culture solution in a plastic tray, place in the phytotron or greenhouse (with 29°C light/21°C dark period) in a 16 h photoperiod and 70%–95% relative humidity. The plants should set seed 3–4 months following transfers to soil.

BIOLISTIC TRANSFORMATION

In a biolistic transformation system, metallic particles are coated with DNA and transferred to target plant cells by shooting with a particle bombardment machine. Immature embryos and embryogenic calli from immature embryos or mature seeds have been widely used as explants for biolistic transformation. Immature embryo explants, especially from monocotyledonous plants, have been frequently used for studies of gene delivery by biolistic bombardment. Stable transformants using this method have been reported for papaya, barley, wheat, and maize, using immature embryos as a recipient for gene delivery. In this protocol, we describe the genetic transformation of rice by particle bombardment using the PDS-1000/He system.

Explants, Equipment, and Reagents

- Immature embryos/immature embryo-derived calli, 4–6 weeks old
- Preparation and delivery of DNA coated gold particles and bombardment
  - Gold particles (approximately 1 μm diameter)
  - $CaCl_2$ (2.5 M, filter sterilized) in 1 mL aliquots, stored at −20°C
  - Spermidine-free base (0.1 M, filter sterilized) in 1 mL aliquots, stored at −20°C
  - Ethanol (100%)
  - DNA in TE (10 mM Tris-HCl, 1 mM EDTA, pH 8.0)
  - Helium gas cylinder
  - Particle bombardment machine

Methods

Preparation of Immature Embryos/Embryogenic Calli

1. Collect spikelets at 9–14 d after anthesis and sterilize.
2. Isolate immature embryos and arrange 100–120 of them, with scutellum facing upward, in a 2 cm diameter circle at the center of a petri dish containing MS medium supplemented with 100 mg/L each of benlate (antifungal), ceftaxime, and carbenicillin (antibacterial) to avoid contamination. Embryogenic calli can be generated following the method described in the

second protocol of the previous Method section. Antibiotics should not be used with embryogenic calli.

3. Incubate the samples overnight in the dark (Figure 3.10).

Sterilization of Consumables and Preparation of Micro-Carrier Particles

1. Soak macro-carrier holders, macro carriers, stopping screen, and rupture disks in 70% ethanol for at least 15 min and air dry in a laminar flow hood.
2. Suspend 30 mg gold particles in 0.5 mL 100% ethanol and vortex for 1–2 min.
3. Centrifuge at 3000 g for 1 min and discard the supernatant.
4. Wash the gold particles 3 times by vortexing, centrifuge in a microfuge at 10,000 g for 1 min. and finally resuspend in 0.5 mL sterile distilled water.
5. Aliquot 50 μL of the final suspension into 1.5 mL microtubes.
6. Store at 4°C or room temperature.

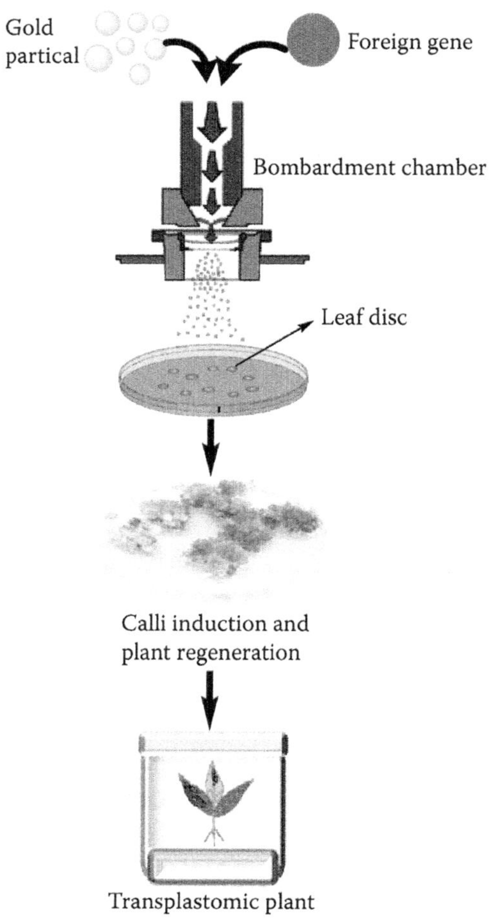

FIGURE 3.10 Biolistic transformation method.

DNA Precipitation

1. Under continuous vortexing, add to a 50 μL aliquot of gold particles the following components in the order given: 8 μL DNA (DNA concentration = 1 μg/μL), 50 μL 2.5 M $CaCl_2$, and 20 μL 0.1 M spermidine.
2. Vortex for 3 min, spin at 3000 g in a microfuge for 5–10 s.
3. After removing the supernatant, add 250 μL cold (–20°C) 100% ethanol
4. Vortex or 1–2 min and centrifuge in a microfuge at 4500 g for 1 min
5. Remove the supernatant and resuspend the micro carrier in 60 μL 100% ethanol.
6. Install macro carriers in the marcro-carrier holder.
7. Pipette 10 μL of the DNA coated micro carrier (gold particles) onto the center of each macro carrier. Each preparation is sufficient for 6 bombardments.
8. Dry for about 1 min in the laminar flow hood.

Operation of Particle Delivery System

1. Set the delivery pressure from the helium cylinder between 1300 and 1500 psi (200 psi above the desired rupture disc value).
2. Turn on the vacuum pump.
3. Place the rupture disc in the holder and screw the holder tightly.
4. Transfer the coated macro-carriers disk to the holder (particle side facing down), and place the macro-carrier assembly unit in the chamber at level 2 from the top.
5. Position the sample for bombardment at the desired level.
6. Close the chamber door.
7. Follow the instruction manual for the bombardment protocol.
8. Leave the plate overnight in the dark after bombardment.

Selection and Regeneration

1. Transfer the bombarded explants to a petri dish containing a selection medium for callus proliferation, as in the last protocol of the previous Method section.
2. After 4–5 selections, further steps for regeneration are the same as in the last protocol of the previous Method section.

PROTOPLAST TRANSFORMATION USING PEG

Direct gene transfer to protoplast exploits the efficient uptake of DNA into the protoplast. It can be enhanced by chemical treatment, for example, with polyethylene glycol (PEG) or by electrical pulses (electroporation). This method does not require any kind of biological vector. For stable transformation of protoplasts, the 2 most important prerequisites are the ability to isolate and culture protoplasts in large numbers and an efficient system for routine plant regeneration from protoplasts. For dicotyledonous species, a large number of protoplasts can be obtained from leaf mesophyll, from which plant regeneration can be routinely achieved; however, there is still no convincing evidence of sustained divisions in protoplasts isolated from the leaves or shoots of any cereal. Nonetheless, protoplasts isolated from embryogenic

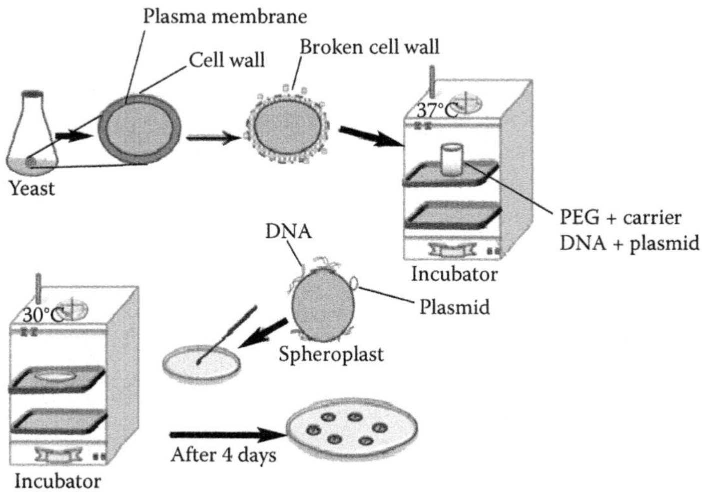

FIGURE 3.11 Protoplast transformation using PEG.

suspension culture can be induced to divide in culture. Microspore cultures and immature or mature embryos may be used to obtaining embryogenic calli that can be used to establish embryogenic cell suspensions.

PEG-mediated gene transfer to rice protoplasts appears to be an efficient, reliable, and inexpensive method. In this system, suspension culture from either microspores or immature embryo-derived embryogenic calli is used to derive a large population of protoplasts. Selection pressure is usually applied at an early stage of callus development from the protoplasts, and successive regeneration of transgenic plants is performed as in other methods. However, the tissue culture response may vary, depending on plant genotype, handling, and the condition of the suspension cells (Figure 3.11).

Explant, Equipment, and Reagents

Mature/Immature Embryo-Derived Calli

- Plant material—Plant material is *Oryza sativa*, 4–6 week old embryogenic calli developed by the method described in previous protocols.

Establishment of Suspension Culture

- Media for culture development and maintenance: AA medium and R2 medium

Protoplast Isolation, Transformation, and Regeneration

- General equipment:
  - Centrifuge
  - Temperature-controlled shaker
  - 12 mL round-bottom screw-cap centrifuge tubes

- Nylon sieves, 50 and 25 μm
- Hemocytometer for counting the protoplasts
- Solutions and media:
  - Enzyme solution: 4% w/v cellulose Onozuka RS, 1% w/v maccrozyme R10, 0.02% pectolyase Y23, 0.4 M mannitol, 6.8 mM $CaCl_2$, pH 5.6, filter sterilized
  - Wash solution: 0.4 M mannitol, 0.16 M $CaCl_2$, autoclaved
  - MaMg buffer solution: 0.4 M mannitol 15 mM $MgCl_2$, 1% (w/v) MES, pH 5.8, autoclaved
  - PEG lolution: 40% polyethylene glycol solution in 0.1 M Ca $(NO_3)_2.4 H_2O$, and 0.4 M mannitol
  - Protoplast culture medium, for example, P1 medium
  - Agarose protoplast medium: 600 mg sea plaque agarose, melted in 30 mL P1 medium
  - P2 medium
  - DNA for transformation (gene of interest, preferably linearized, together with calf-thymus DNA)
  - Nurse cells for rice transformation, OC cell line derived from seedling of *O. sativa* L.

Method

Establishment of Suspension Culture

1. Transfer the embryogenic yellowish-white, soft calli developed by the method described in the second protocol in the Method for *Agrobacterium* transformation to 6 mL of liquid AA medium (Table 3.7) in 50 mm petri dishes and incubate on a gyratory shaker at low seed (80 rpm) in diffuse light (3 μ mol/m²/sec) at 25°C.
2. Subculture the callus every 7 d with continued manual selection of small and densely cytoplasmic cells, which are transferred to R2 medium.

Protoplast Isolation

1. Add 20 mL of enzyme solution to approximately 3–5 g of suspension cells (3–4 d after subculture) and incubate without shaking at 30°C in the dark for 3–4 h.
2. Add an equal volume of wash solution to the protoplast enzyme mixture.
3. Pass the mixture through 50 and 25 μm sieves.
4. Transfer the filtrate to 12 mL round-bottom screw-cap centrifuge tubes and centrifuge for 10 min at 70 g to separate the enzyme solution.
5. Wash the pellet containing protoplasts 2 times by centrifugation and determine the density of protoplasts using a hemocytometer.

Transformation Using PEG

1. After the final washing, resuspend the protoplast pellet in MaMg buffer solution. Adjust the protoplast density to 1.5–2.0 × 10^6/0.4 mL.

TABLE 3.7
Media for Suspension Culture Development

Component	AA Medium (mg/L)	R_2 Medium (mg/L)
	Macronutrients	
$CaCl_2.2H_2O$	150.0	147.0
$MgSO_4.7H_2O$	250.0	250.0
$NaH_2PO_4.H_2O$	150.0	240.0
KCl	2950.0	–
KNO_3	–	4044.0
$(NH_4)_2SO_4$	–	330.0
	Micronutrients	
KI	0.75	–
H_3BO_3	3.0	0.50
$Mn.SO_4.H_2O$	10.0	0.50
$Zn.SO_4.7H_2O$	2.0	0.50
$Na_2.MoO_4.2H_2O$	0.25	0.05
$CuSO_4.5H_2O$	0.025	0.05
$CoCl_2.6H_2O$	0.025	–
	Iron Composition	
Na_2 EDTA	37.3	37.3
$FeSO_4.7H_2O$	27.80	27.80
	Vitamins	
Nicotinic acid	1.0	0.5
Pyridoxine HCl	1.0	0.5
Thiamine HCl	10.0	1.0
Glycine	7.5	2.0
Inositol	100.0	100.0
	Others	
L-glutamine	876.0	–
Aspartic acid	266.0	–
Arginine	174.0	–
Sucrose	20 g/L	20 g/L
2–4 D	1.0	–
Kinetin	0.2	–
GA_3	0.1	–
pH	5.6	5.6

2. Add 6–10 µg of sterile plasmid DNA and 20 µg of calf-thymus DNA (transformation can be done without calf-thymus DNA, if not available) to each 0.4 mL aliquot of the protoplast suspension.
3. Add 0.5 mL of the PEG solution, drop-wise, and mix gently. Incubate at room temperature for 10 min.

4. Slowly add 10 mL of wash solution and centrifuge at 70 g for _0 min. Discard the supernatant.
5. Resuspend the protoplast pellet in 0.4 mL P1 medium (Table 3.8).
6. Mix 0.6 mL of agarose protoplast medium (after melting, cool to 40°C) with 0.4 mL of P1 medium containing protoplasts and transfer to a 3.5 cm petri dish. Incubate at 20°C for 1 h.
7. Transfer the solidified agarose gel to a 5 cm Petridish containing P1 medium (bead type culture) with or without adding nurse culture. Incubate culture in the dark at 28°C with slow shaking (40 r.p.m)
8. After 10 days remove all the nurse cells and add 5 mL fresh P1 medium (Table 3.8). Keep in dark at 28°C with slow shaking.

Selection

1. First selection: 14 d after transformation, add the selective agent to the medium (concentration of selective agent depends on the selectable marker gene and the cultivars used). Incubate for 2 wk without shaking.

TABLE 3.8
Composition of Protoplast Culture Media

Component	P1 (mg/L)	P2 (mg/L)
$(NH_4)_2SO_4$	330.0	463.0
KH_2PO_4	–	400.0
KNO_3	4044.0	2830.0
$CaCl_2.2H_2O$	147.0	166.0
$MgSO_4.7H_2O$	247.0	185.0
Na_2 EDTA	37.3	37.3
$FeSO_4.7H_2O$	27.8	27.8
$NaH_2PO_4.2H_2O$	240.0	–
$Mn.SO_4.H_2O$	0.5	4.4
H_3BO_3	0.5	1.6
$Zn.SO_4.7H_2O$	0.5	1.5
KI	–	0.8
$CuSO_4$. $5H_2O$	0.05	–
Na_2MoO_4. $2H_2O$	0.05	–
Thiamine-HCl	1.0	1.0
Nicotinic acid	0.5	0.5
Pyridoxine-HCl	0.5	0.5
Glycine	2.0	2.0
Myo-inositol	100	100
2,4-D	2.0	2.0
Sucrose	136.92 (g/L)	60.0 (g/L)
Maltose	144.12 (g/L)	60.0 (g/L)
Agarose	–	3–6 (g/L)
pH	5.6	5.8
	Filter sterilized	Autoclaved

2. Second selection: for second selection pressure replace 4 mL of medium with fresh P1 medium with the same concentration of selective agent. Incubate for another 2 wk in the same conditions.
3. Third selection: transfer the visible calli onto P2 soft-agarose medium (0.3% agarose) with same concentration of selective agent and incubate in dark at 25°C for 2 wk.

Regeneration

1. Transfer the visible colonies to P2 medium (Table 3.8) containing 0.6% agarose without the selective agent. Keep in same condition for 2 wk.
2. Transfer the embryogenic colonies with developing somatic embryos to regeneration medium and follow the same procedure as described in the last protocol of the *Agrobacterium* transformation Method to regenerate transgenic plants.

AGROBACTERIUM: MEDIATED TRANSFORMATION OF ARABIDOPSIS ROOT EXPLANTS

In *Arabidopsis*, as in all other species, the use of tissue culture transformation techniques was first elaborated. These techniques require an initial amiplification of *Agrobacterium*-transformed cells. This can be conveniently achieved by selection for transformed microcalli that can then be regenerated into shoots and plants. Therefore, the choice of explants for cocultivation with *Agrobacterium* largely depends on the response of diverse plant tissues to those hormonal treatments that are employed to induce cell proliferation and subsequent shoot formation. In general, cell proliferation is induced by culturing the explants in media, providing a high-auxin to low-cytokinin ratio. Explants from various *Arabidopsis* ecotypes and mutants respond differently to auxin and cytokinin and therefore to the induction of cell division. Throughout the last decade, several methods for transformation were published. This experiment is useful to generate a large number of transgenic Arabidopsis plants and to search for recessive gene mutations using the T-DNA-aided gene-fusion technology.

Equipment and Reagents

i. Macro-salts: 20 g NH_4NO_3, 40 g KNO_3, 7.4 g $Mg_2SO_4.7H_2O$, 3.4 g KH_2PO_4, and 2 g $Ca(H_2PO_4)_2$ for 1 L
ii. Micro-salts: 6.2 g H_3BO_3, 16.9 g $MnSO_4.H_2O$, 8.6 g $ZnSO_4.7H_2O$, 0.25 g $Na_2MoO_4.2H_2O$, 0.025 g $CuSO_4.5H_2O$, 0.025 g $CoCl_2.6H_2O$ forl L
iii. Fe-EDTA: dissolve separately 5.56 g $FeSO_4.7H_2O$ and 7.46 g $Na_2EdTA.2H_2O$, mix the 2 solutions, and adjust the volume tol L
iv. $CaCl_2$: 7.5 g $CaCl.2H_2O$ per 1 L
v. KI: 375 mg KI per 1 L
vi. Vitamin: 50 g myo-inositol, 2.5 g thiamine.HCl, 0.5 g nicotinic acid, and 0.2 g pyridoxine.HCl for 1 L
vii. MSAR-medium: to prepare liquid medium, add for 1 L: 50 mL macro salts, 1 mL micro salts, 5 mL Fe-EDTA, 5.8 mL $CaCl_2$, 2.2 mL KI, 2 mL

vitamins, and 30 g sucrose. Adjust the pH to 5.8 with 0.5 M KOH. To prepare solid MSAR media, add for 1 L, either 2.2 g phytagel (essential for high-frequency *Arabidopsis* shoot regeneration) or 8 g high-quality agar for plant tissue culture. Autoclave the media for 15 min.

viii. Seed germination (SG) medium: regular MSAR medium, but contains half concentration of macro salts and 0.5% sucrose.

ix. Callus medium (MSAR I): MSAR medium supplemented with 2.0 mg/L indole-3-acetic acid (IAA), 0.5 mg/L 2,4-dichlorophenoxyacetic acid (2,4-D), 0.2 mg/L kinetin (6-furfurylaminopurine), and 0.2 mg/L N^6-[2-isopentenyl]-adenosine (IPAR)

x. Shoot medium (MSAR II): MSAR medium supplemented with 2.0 mg/L 6-benzylaminopurine (BAP), 0.1 mg/L indole-3-butyric acid (IBA), and 0.05 mg/L NAA.

xi. Seed sterilization solution: 5% $Ca(OH)_2$ containing 0.05% Triton X-100. If $Ca(OH)_2$ is not available, use 10% sodium hypochlorite solution containing 0.1% Triton X-100.

xii. *Agrobacterium* YEB medum: 5 g beef extract, 1 g yeat extract, 5 g peptone, and 5 g sucrose per 1 L water.

METHOD

Preparation of Plant Material

1. Sterilize 10 mg (about 500) seeds in a microfuge tube after adding 1 mL seed-sterilization solution by continuous shaking for 15 min.
2. Pellet the seeds by brief centrifugation (i.e., 1 min in a microfuge at room temperature), remove the sterilization solution, wash the seeds with 4 × 1 mL sterile water, and dry the tubes in a flow hood overnight.
3. Germinate the seeds on solid SG-medium in petri dish for 7–10 days using an 8 h light and 16 h dark period at 25°C.
4. Place 20 seedlings into a 250 mL Erlenmeyer flask containing 35 mL liquid MSAR medium with 3.0% sucrose and fix the flasks onto a rotary shaker adjusted to 120 rpm at 25°C under an 8 h light and 16 h dark period. Within 14–20 d, the flasks will be filled with roots. This material can be used either for direct infection with *Agrobacterium* or for generation of cell suspensions. Do not collect roots from plants that show senescence.

Growth of *Agrobacterium*

1. Grow *Agrobacterium* to an OD_{550} of 1.5 in YEB medium supplemented with proper antibiotics to select for the maintenance of binary T-DNA vectors.
2. Pellet the cells by centrifugation (5000 g for 15 min at room temperature).
3. Resuspend cells in MSAR I liquid medium at an OD_{550} of 0.5 and distribute 50–80 mL aliquots into large petri dishes.
4. Collect *Arabidopsis* plants from the Erfenmeyer flasks, remove their roots, cut the roots in small pieces and place them into the *Agrobacterium* suspension for at least 30 min.

5. Remove the *Agrobacterium* suspension with a pipette from the petri dishes and plate the roots onto MSAR I plates.
6. Alternatively, collect the root segments in an Erlenmeyer flask containing 40 mL MSAR I medium and the place the flask on a rotary shaker. Coculture the roots with *Agrobacterium* for 36–48 h.

Selection of Transformed Plants

1. Transfer the roots from MSAR I medium containing 300 mg/L claforan and 500 mg/L tricarcillin/clavulanic acid. Alternatively, add claforan at a 500 mg/L concentration.
2. Subculture the liquid root culture every 3–4 d using the medium above.
3. Root cultures in petri dishes do not need to be subcultured. After 14 d, transfer the roots onto MSAR II plates containing claforan (300 mg/L), tricarcillin/clavulanic acid (500 mg/L), and, depending on the type of selectable marker encoded by the T-DNA, for example, 15 mg/L hygromycin or 100 mg/L kanamycin.
4. Subculture the root explants at 14-d intervals by decreasing the claforan concentration to 300 mg/L after the third subculture.
5. During the first 4 wk on selective MSAR II medium, the nontransformed tissues die and green transgenic calli appear on the plates. Within 8 wk after *Agrobacterium* infection, the transformed calli form embryo-like structures and quickly regenerate to shoots.
6. The regenerating shoots must be continuously transferred onto MSAR III medium in jars and, when they reach a size of 1–2 cm, into large test tubes (e.g., 20 × 2.5 cm diameter) containing 10–20 mL SG medium to set seed.
7. The test tubes are closed with loose cotton plugs and placed in growth chambers providing a 16 h light and 8 h dark cycle.
8. Typically, root culture from a single Erlenmeyer flask yields explants for 10–12 MSAR II plates resulting in an average of 500–100 transformed plants.

ESTABLISHMENT AND TRANSFORMATION OF *ARABIDOPSIS* CELL SUSPENSIONS

An advanced version of tissue culture transformation methods is based on the cocultivation of cell suspensions with *Agrobacterium* that yields an extremely high number of transformants (in the range of 10^6–10^7). Therefore, this method ideally suits the purpose of activation of T-DNA tagging if a proper selection for gene activation is available, for example, using inhibitors of particular enzymes or signaling factors, and plant regeneration is not immediately essential. Although embryonic cell suspensions can be established with care and transformed similarly by *Agrobacterium*, the regeneration of a large number of transformants is a very laborious process. Moreover, alternation in ploidy (from 2n to 4n or 8n) are unavoidable in cell cultures maintained in the presence of high concentrations of auxin (i.e., 2,4-D). An increase in the ploidy level reduces the recovery of recessive mutations, but permits the identification of transformants that carry dominant mutations induced by activation T-DNA tagging.

Equipment and Reagents

- *SC medium:* 4.33 g MS medium, a double concentration of B5 vitamins (2 mg nicotinic acid, 2 mg pyridoxine HCl, 20 mg thiamine HCl, and 200 mg myo inositol for 1 L), 30 g sucrose, 0.5 mg/L IPAR. The vitamins and hormones are prepared as concentrated stocks, filter sterilized, and added to SC after autoclaving.
- *Metal sieve:* pore sizes 250 and 850 μm.

Method

Initiation of *Arabidopsis* Cell Suspension

1. Place 20 seedlings into a 250 mL Erlenmeyer flask containing 35 mL liquid MSAR medium with 3.0% sucrose and fix the flasks onto a rotary shaker adjusted to 120 rpm at 25°C under an 8 h light and 16 h dark period. Within 14–20 d the flasks will be filled with roots. Transfer the roots from an Erlenmeyer flask into a petri dish, remove all green tissues.
2. Cut the roots into about 2 mm pieces, and transfer the explants into a 250 mL Erlenmeyer flask that contains 50 mL suspension culture SC medium. Cover the flasks with aluminium foil and place them on a rotary shaker (120 rpm at 24°C).
3. After 15–20 d, approximately 3–5 g roots (fresh wt.) should be available in each flask.
4. Exponential growth rate of roots in important for obtaining fine-cell suspensions. Therefore, avoid yellow-brown stationary-phase cultures.

Subculture of Cell Suspension

1. After 21 d, harvest the micro calli released from the roots by passing the root culture medium through an 850 μm sterile metal sieve into a sterile Erlenmeyer flask.
2. Pellet the cells, and replace half volume of the medium with new SC medium.
3. Add 50 mL fresh SC medium to the roots and culture them for an additional 21 d to generate more micro calli.
4. Subculture the cell suspension every 7 d by sieving and replacing half of the medium with new SC medium after pelleting the cells.
5. After the fifth subculture, a 50 mL suspension can usually be diluted with 100 mL SC and divided into 3 equal aliquots into new flasks for amplification.

Transformation of Cell Suspension with *Agrobacterium*

1. Logarthmically-proliferrating cell suspensions are usually obtained from *Arabidopsis* ecotype Col -0 after 5–6 wk.
2. The cell suspensions are then filtered through a 250 μm sieve and subcultured twice.
3. The resulting fine-cell suspension is inoculated with *Agrobacterium*. For this purpose, grow *Agrobacterium* in YEB medium and pellet the cells from

a 1.5 mL culture in a sterile microfuge tube by centrifugation for 2 min at room temperature.

4. Suspend the bacterial pellet in 1.5 mL SC medium and add to a 35 mL cell suspension immediately after subculturing.
5. Cocultivate the cells with *Agrobacterium* for 48 h, then add claforan (300 mg/L) and tricarcillin/clavulanic acid (300 mg/L).
6. Subculture the cells 7 d after the addition of *Agrobacterium* in the same SC medium.
7. After 1 wk, transfer the cells into a 50 mL Falcon tube, pellet the cells by centrifugation at 1000 g for 3 min and suspend them at a density of 10^3 cells/mL in either 0.2% phytagel dissolved in 3% sucrose or in liquid MSAR II medium.
8. Plate the cells on MSAR II medium containing 0.2% phytagel, claforan (200 mg/L), tricarcillin/clavulanic acid (200 mg/L), and suitable antibiotics (e.g., hygromycin or kanamycin).
9. If you wish to select for transformed calli without plant regeneration, use MSAR I instead of MSAR II medium for plating.
10. Transformed calli appear within 2–3 wk and yield shoots within 35–40 d following the *Agrobacterium* infection.

IN PLANTA TRANSFORMATION OF *ARABIDOPSIS* BY INFILTRATION

The first successful approach using a large scale *in situ* infection of *Arabidopsis* with *Agrobacterium* was the seed transformation method. With this meghod, a large population of transformants could be raised with less labor than using tissue culture techniques. Furthermore, seed transformation overcame the problems of recognition and removal of polyploids and aneuploids during the genetic analysis. The recognition that *Arabidopsis* survives systemic *Agrobacterium* infection led to the development of *in planta* transformation techniques using vacuum infiltration. Because the transformation process must target the reproductive organ, this technology was further perfected by dipping only the flower buds into *Agrobacterium* suspension containing 5% sucrose and Silwet L-77 surfactant.

To amplify the number of inflorescence axes produced by a plant for *Agrobacterium* infection, the primary inflorescence stem is often removed. This results in a quick outgrowth of side-inflorescence stems from the rosette, providing ample material for transformation. Most binary vectors currently used are devoid of proper partitioning functions, which would provide stable plasmid maintenance; therefore, to increase the transformation frequency, an application of *in planta* selection for the maintenance of *Agrobacterium* binary T-DNA vectors is advisable. In the case of the pPVC vector system developed in our laboratory, the T-DNA carries an ampicillin/carbenicillin-resistance gene. The media for *Agrobacterium* infiltration and floral-dip transformation can therefore be supplemented with carbenicillin. Which is not deleterious to *Arabidopsis*.

By optimization of infiltration techniques, on average, over 400 transformed seeds are routinely obtained from a single pot containing 10–15 plants. Using 500–1000 pots, one can reproducibly raise a T-DNA mutagenized population of 200,000 transformants within 3–6 mo. Based on the expectation that T-DNA integration occurs randomly and the size of the *Arabidopsis* genome is 120 Mb, such a population is predicted to carry a T-DNA tagging that may be used as an alternative to chemical mutagens to solve specific genetic problems, such as the isolation of multiple alleles and suppressors for a given mutation.

Equipment and Reagents

i. Agrobacterium YEB medium: 5 g beef extract, 1 g yeast extract, 5 g peptone, and 5 g sucrose per 1 lit. water.
ii. SG medium: regular MSAR medium, but contains half concentration of macro salts and 0.5% sucrose.
iii. Infiltration medium (1 M, pH5.7): half a concentration of Murashige and skoog basal salt mixture, 1 × B5 vitamins, 5% sucrose, and is supplemented after autoclaving with 0.05 μM BAP and 0.005% Silwet L-77.

Method

Growth of Plant Material

1. Prepare plastic pots (12 cm diameter) with wet soil and sow 10–15 seeds in each.
2. Grow seedlings in short days using an 8 h light (24°C) and 16 h (18–20°C) dark period in a growth chamber or greenhouse.
3. When the diameter of rosettes reaches a size of 3–5 cm (usually after 15–18 d). Transfer the pots to long-day conditions (16 h ligh and 8 h dark) to induce flowering.
4. Cut off the primary inflorescence stem when it reaches 1–3 cm to induce the development of secondary inflorescence from the rosette buds.
5. When the size of inflorescence reaches 5–10 cm, carrying closed flower buds and only a few open flowers, the plant material is ready for transformation.

Growth of *Agrobacterium*

1. Grow 100 mL liquid culture of *Agrobacterium* in YEB medium at 28°C for 2 d, applying a proper selection of the maintenance of binary T-DNA vectors.
2. Best transformation results were reported by using the strain GV3101 with helper Ti-plasmid pMP90 (26–28). A modified version of this *Agrobacterium* host, GV3101 (pMP90 RK), is used in combination with the pPCV-type binary T-DNA vectors that carry a carbenicillin resistance gene with their T-DNA.
3. Whe using pPCV vectors, supplement both *Agrobacterium* and *in planta* infiltration medium with 100 mg/L carbenicillin.
4. Inoculate several 2 L Erlenmeyer flasks containing 500 mL YEB medium with 10 mL starter culture, and grow the bacterial cultures to late logarithmic phase for 12–14 h. Harvest the cells by centrifugation (5000 g for 10 min

at room temperature) and resuspend the cells at an OD_{550} of 0.8–1.2 in infiltration medium. Typically, 500 mL *Agrobacterium* culture yields 1 L bacterial suspension for infiltration.

Vacuum Infiltration

1. Assemble a glass or plastic bell jar with a vacuum pump and place trays (e.g., 30 × 10 × 5 cm), or beakers (12 cm), containing the *Agrobacterium* culture into the jar.
2. Place 2 pots in inverted position into each tray, or 1 pot inverted into each beaker such that only the inflorescence shoots are submerged into the *Agrobacterium* suspension. Apply vacuum for 5 min, then release the vacuum quickly by disconnecting the fitting between the pump and the bell jar.
3. The bacterial suspension can be resued 3–4 times. Thus, an infiltration of 16 pots needs about 1–1.5 L *Agrobacterium* suspension in 1 M.
4. Cover the pots for 24 h with plastic quick-seal bags with the corners cut off. If the humidity in your greenhouse is high enough (i.e., 70%–80%), this step is not necessary.

Collection of Seeds

1. The vacuum infiltrated plants complete flowering within 10–12 d.
2. When the last flowers are fertilized, place the inflorescence stems of 5–8 plants into a translucent paper bag (i.e., 13 × 30 cm).
3. Fix the closed bags with a trap to wood or plastic sticks fitted into the pots.
4. Water the plants for an additional 10 d and then let the pots dry out.
5. Collect the seed bags and clean the seeds.
6. Take 5–10 seed aliquots derived from different pots to test the transformation efficiency in a germination assay.

Seed Sterilization

1. Sterilize seeds as described above and sow then on SG medium in petri dishes containing suitable antibiotics to select for T-DNA transformed plants.
2. Using hygromycin selection (25 mg/L), nontransformed seedlings are arrested and show bleaching after opening their cotyledons, whereas transformed plants develop leaves and log roots within 5–8 d.

Selection of Transformants

1. If at least 5–10 transformants were obtained from 500 to 1,500 seeds, bulk the seed material, sterilize it in aliquots, and plate the seeds at a high density on selective SG medium with 1000–5000 seeds per large petri dish.
2. To save greenhouse space, the transformants picked from petri dishes can be grown in jars on SG medium with hygromycin, and, when required, with claforn to eliminate possible *Agrobacterium* infection, under short day conditions for another 10–14 d.

3. Transfer the seedlings into soil by planting them individually into 5 × 5 cm quadratic pots and grow them for 10–14 d under short-day conditions to produce large rosettes.
4. Collect rosette leaves (200–500 mg) for DNA purification using a pooling strategy, and transfer the plants to long-day conditions to collect M2 progeny from each plant separately within 3–4 wk.
5. Screen mutants on the basis of phenotype.

PCR AMPLIFICATION OF T-DNA TAGGED PLANT DNA FRAGMENTS FOR AUTOMATIC DNA SEQUENCING

Direct sequencing of T-DNA insert junctions is a cumulative approach to sequence-based identification of T-DNA tagged genes. Methods using inverse PCR employ 2 T-DNA specific primers that allow the amplification and subsequent bidirectional sequencing of plant DNA fragments that are linked to either the left or right T-DNA ends, or both. Alternative techniques, such as the asymmetric interlaced PCR or amplification or insertion mutagenized sites, use either a degenerated primer or an adaptor in combination with T-DNA end primers to amplify plant DNA sequences flanking the T-DNA tags. This experiment describes the application of a log range iPCR technique that allows routine amplification and sequencing of T-DNA linked plant DNA fragments ranging in size from 300 bp to 12 kb.

Equipment and Reagents

i. SG medium: regular MSAR medium, but contains half concentration of macro salts and 0.5% sucrose
ii. DNA extraction buffer: 200 mM Tris-HCl (pH 8.5), 50 mM EDTA, 500 mM NaCl, 14 mM 2 mercaptoethanol and 1% SDS
iii. HTE buffer: 100 mM Tris-HCl (pH 8.0), 50 mM EDTA
iv. LTE buffer: 10 mM Tris-HCl (pH 7.5), 1 mM EDTA
v. RNase A: (SERVA) prepare 10 mg/mL stock solution in LTE buffer and boil for 10 min; store at 4°C
vi. Proteinase K: (Merck) prepare freshly 2 mg/mL stock solution in LTE buffer and incubate at 37°C for 10 min
vii. T_4 DNA ligase buffer: 66 mM Tris-HCl (pH 7.2), 10 mM $MgCl_2$, 5 mM dithiothreitol, 1 mM ATP
viii. Nested primers: for binary *Agrobacterium* vectors pPCV6 NFHyg and pPCV621, the following left (LB) and right (RB) T-DNA border primers are used:
LB1 (5′-CTCAGACCAATCTGAAGATGAAATGGGTATCTGGG-3′)
LB2 (5′-CTGGGAATGGCGAAATCAAGGCATCGATCGTGAAG-3′)
RB1 (5′-GTTTCGCTTGGTGGTCGAATGGGCAGGTAGGC-3′)
RB2 (5′-CAGTCATAGCCGAATAGCCTCTCCACCC-3′)
ix. Elongase PCR buffer: 60 mM Tris-SO_4 (pH 9.1), 18 mM (NH_4) SO_4, 1 mM $MgSO_4$, and 250 μM of each dATP, dCTP, dTTP, and dGTP
x. ABI Prism-Dye Terminator Cycle Sequencing kit

Method

DNA Preparation

1. Collect 2–4 leaves (50–200 mg) from M1 plants in microfuge tubes or M2 seedlings germinated on selective SG medium.
2. Snap freeze the tubes in liquid N_2 and grind the material to a fine powder with a glass or Teflon rod fitting the shape of the microfuge tube.
3. Add 750 µL freshly made DNA extraction buffer, mix the extracts well, and incubate the tubes with occasional shaking at 65°C for 10 min.
4. Add 250 µL 5 M potassium acetate (pH 6.0), mix the tubes, and place them on ice for 20 min.
5. Spin the tubes in a microfuge for 3 min, and in the meantime, prepare 1 mL pipette tips plugged with small pices of Miracloth.
6. Filter the cleared lysates through the Miracloth-plugged tips into new tubes and precipitate the DNA with 700 µL isopropanol on ice for 30 min.
7. Collect the DNA by centrifugation in a microfuge for 5 min, wash the pellet with 70% ethanol, and dissolve the DNA in 475 µL THE buffer containing 200 µg/mL heat-treated RNase A. Incubate the samples for at least 33 min at 37°C.
8. Add 125 µL proteinase K solution and incubate the samples for 1 h at 37°C.
9. Extract the samples twice with 600 µL phenol:chloroform:isoamyl alcohol (25:24:1) and with chloroform:isoamyl alcohol (24:1) at 0°C.
10. Add 60 µL 3 M sodium acetate (pH 6.0) and 400 µL cold isopropanol, mix the tubes, and precipitate the DNA on ice for 30 min.
11. Pellet the DNA by centrifugation in a microfuge for 5 min and dissolve in 100 µL LTE buffer.

DNA Digestion

1. Digest 5 µg plant DNA in 200 µL appropriate restriction-endonuclease buffer for at least 2 h using an enzyme (100 U) that has either no or only a single cleavage site within the T-DNA.
2. After resting 10 µL aliquots by gel electrophoresis, extract the samples twice with phenol:chloroform:isoamyl acohol, add 20 µL of 3 M sodium acetate (pH6.0), and precipitate the DNA with 150 µL isopropanol on ice for 30 min.
3. Pellet the DNA by centrifugation in a microfuge for 5 min, wash with 70% ethanol, dry briefly, and dissolve in 25 µL sterile water.

Ligation of DNA

1. Self ligate 1 µg digested DNA in 200 µL T_4 DNA ligase buffer containing 10% polyethylene glycol (PEG 4000) at 15°C for 8 h.
2. Extract the samples twice with phenol:chloroform:isoamyl alcohol and precipitate the DNA as described above. Digest the samples with an enzyme that cleaves once the T-DNA segment used for self ligation, but has a cleavage site frequency in the *Arabidopsis* genomic DNA less than 1 for every 50 Kb (i.e., Sma 1, Sal I, etc.).

3. Linearization of the self-ligated DNA is, however, not essential for iPCR amplification, thus this step may be omitted.

Design of PCR Primers

The primer design depends on whether the enzyme used above cleaves within the T-DNA.

If the enzyme does not cleave in the T-DNA:

1. The self-ligated plant DNA sequence can be PCR-amplified using 2 T-DNA primers that are oriented toward the left and right termini of the T-DNA.
2. Because deletions are often generated at the left border during T-DNA integration, design the left border primer such that it is located at least 100 bp upstream of the left 25 bp border repeat.
3. In contrast, deletions at the right T-DNA end are rare, thus the second primer can safely be placed 20–30 bp upstream of the right 25 bp repeat.
4. It is advisable to design nested primers placed at different distances from the T-DNA ends to ensure the success of iPCR and increase the amount of information obtained by subsequent sequencing of the rescued plant DNA segment using the T-DNA end primers.

If the enzyme cleaves the T-DNA:

1. Design 2 primers such that they face the enzyme cleavage site from btoh the 5′ and 3′ orientations.
2. Use these primers in pair-wise combinations with the T-DNA end primers for PCR amplification and then independently for PCR sequencing.

Polymerase Chain Reactions

1. This iPCR reactions are performed using either Elongase (Gibco BRI) or Takara LA-PCR enzyme mixes as recommended by the manufacturers.
2. Amplify 0.5 µg template DNA in 50 µL Elongase PCR buffer containing 1–2 SU enzyme and 0.2 µM of each primer.
3. Following denaturation at 95°C, run 35 cycles (94°C for 30 s, 65°C for 30 s, 68°C for 8 min) followed by elongation at 68°C for 10 min.
4. Test a 2 µL aliquot from each PCR reaction by gel electrophoresis.
5. In rare cases, when no amplification of defined fragments is obtained, perform a new PCR with the nested primers using either fresh template DNA or 1 µL from a 500-fold diluted PCR reaction mix.
6. Purify the PCR amplified DNA fragments from agarose gel using electroelution.
7. Extract the samples twice with 600 µL phenol:chloroform:isoamyl alcohol (25:24:1) and with chloroform:isoamyl alcohol (24:1) at 0°C.
8. Add 60 µL 3 M sodium acetate (pH 6.0) and 400 µL cold isopropanol, mix the tubes, and precipitate the DNA on ice for 30 min.
9. Pellet the DNA by centrifugation in a microfuge for 5 min and dissolve in 100 µL LTE buffer.

10. Sequence the DNA fragments using an AB1 Prism Dye Terminator Cycle Sequencing Kit according to the protocol.

Sequence Analysis

1. To find possible gene hits, analyze the obtained plant DNA sequences using the BLASTN and BLASTX programs by performing searches in the nonredundant EST and *Saccharomyces* databases (http://ncbi.nlm.nih.gov).
2. Alternatively, run searches using, for example, databases at http://www.tigr.org/tbd/.
3. After translating the DNA to protein sequences it is possible to search for motifs at http://coot.embl-heidelberg.de/SMART/ or http://www.motif.genome.ad.jp/. Other useful sequence analysis facilities are listed at http://www.columbia.edu/-ej67.dtb_lst.htm, http://www.genome.ad.jp/, http://www.infobiogen.fr/services/dbcat/, and http://www.molbio.net/genomivs.html.

DNA TRANSFECTION MEDIATED BY LIPOFECTION

Lipofection is the method of choice for DNA delivery into animal cells cultured *in vitro*. It was also one of the first methods of DNA delivery into plant protoplasts. Transformation frequencies of 4×10^{-5} have been reported. The integrated DNA apparently does not undergo rearrangements, but multiple copies may become integrated in tandem. In some cases, plasmid DNA of 9 kb and larger have been integrated intact. (Figure 3.12)

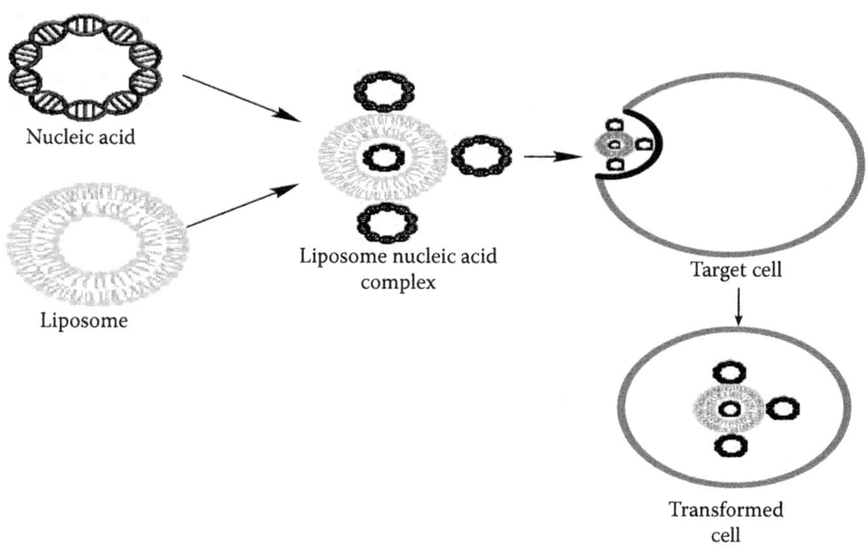

FIGURE 3.12 DNA transfection mediated by lipofection.

Chemicals

Lipofection Reagent

Several types of lipofection reagents are available, but in this experiment, 2 common lipids are used:

1. Lipofectin (N-[1-(2,3-dioleoyloxy) propyl]-N,N,N-trimethylammonium chloride [DOTMA]). It is a monocationic lipid.
2. Transfectam (Spermine-5-carboxy-glycinedioctadecyl-amide [DOGS]). Polycationic lipids such as DOGS may be substituted for lipofectin.
   a. NaCl (5 M)
   b. Sodium citrate (pH 5.5, 20 mM) containing 150 mM NaCl
   c. Nucleic scid and oligonucleotides
   d. Plasmid DNA
   e. Cell-culture growth medium

Equipment

- Test tubes, polystyrene or polypropylene
- Tissue-culture dishes (60 mm)
- Exponentially growing culture of mammalian cells

Method

1. 24 h before lipofection, harvest exponentially growing mammalian cells by trypsinization and replate them on 60 mm tissue culture dishes at a density of 10^5 cell/dish. Add 5 mL of growth medium, and incubate the culture for 20–24 h at 37°C in a humidified incubator with an atmosphere of 5%–7% CO_2.

Note: The cells should be ~75% confluent at the time of lipofection. If the cells are grown for fewer than 12 h before transfection, they will not be well anchored to the substratum and are likely to detach during exposure to lipid.

2. For each 60 mm dish of cultured cells to be transfected, dilute 1–10 μg of plasmid DNA into 100 μL of sterile deionized water (if using transfectin) or 20 mM sodium citrate containing 150 mM Nacl (pH 5.5) (if using transfectam) in a polystyrene or polypropylene test tube. In a separate tube, dilute 2–50 μL of the lipid solution to a final volume of 100 μL with sterile deionized water or 300 mM NaCl.

Note: When transfecting with lipofectin, use polystyrene test tubes; do not use polypropylene tubes, because the cationic lipid DOTMA can bind nonspecifically to polypropylene. For other cationic lipids, use the tubes recommended by the manufacturer.

3. Incubate the tubes for 10 min at room temperature.
4. Add the lipid solution to the DNA, and mix the solution by pipetting up and down several times. Incubate the mixture for 10 min at room temperature.
5. While the DNA-lipid solution is incubating, wash the cells to be transfected 3 times with serum-free medium. After the third rinse, add 0.5 mL of serum-free medium to each 60 mm dish and return the washed cells to a 37°C humidified incubator with an atmosphere of 5%–7% CO_2.

Note: It is very important to rinse the cells free of serum before the addition of the lipid-DNA liposomes. In some cases, serum is a very effective inhibitor of the transfection process.

6. After the DNA lipid solution has incubated for 10 minu, add 900 μL of serum-free medium to each tube. Mix the solution by pipetting up and down several times. Incubate the tubes for 10 min at room temperature.
7. Transfer each tube of DNA-lipid medium solution to a 60 mm dish of cells. Incubate the cells for 1–24 h at 37°C in a humidified incubator with an atmosphere of 5%–7% CO_2.
8. After the cells have been exposed to the DNA for the appropriate time, wash them 3 times with serum-free medium. Feed the cells with complete medium and return them to the incubator.
9. If the objective is stable transformation of the cells, proceed to next step. Examine the cells 24–96 h after lipofection using 1 of the following assays.
10. To isolate stable transfectants: after the cells have incubated for 48–72 h in complete medium, trypsinize the cells and replace them in the appropriate selective medium. Change this medium every 2–4 d for 2–3 wk to remove the debris of dead cells and to allow colonies of resistant cells to grow. Thereafter, individual colonies may be cloned and propagated for assay.

CALCIUM-PHOSPHATE MEDIATED TRANSFECTION OF EUKARYOTIC CELLS

This approach can be applied to virtually all mammalian cells, and a very large number of cells can be treated with little effort. In this method, the DNA preparation to be used for transfection is first mixed with a calcium chloride solution, which is then added very slowly, with mixing, to a phosphate buffer.

Calcium-phosphate coprecipitation has been a popular transfection method since its introduction in the early 1970s (Graham and van der Eb, 1973) because the components it requires are easily available and inexpensive. Furthermore, the technique is easy to master, it is effective with many types of cultured cells, and it can be used for both transient and stable transfection of a variety of cultured cell types. However, calcium-phosphate coprecipitation is prone to variability because of its sensitivity to slight changes in pH, temperature, and buffer salt concentrations, and it can be cytotoxic to many types of cell cultures, especially of primary cells. In addition, it is unsuitable for *in vivo* transfer of nucleic acids to whole animals, and it shows relatively poor transfection efficiency compared with other chemical transfection methods such as lipid-mediated transfection.

The principle of calcium-phosphate coprecipitation involves mixing DNA with calcium chloride in a buffered saline/phosphate solution to generate a calcium-phosphate–DNA coprecipitate, which is then dispersed onto cultured cells. Calcium phosphate facilitates the binding of the condensed DNA in the coprecipitate to the cell surface, and the DNA enters the cell by endocytosis. Aeration of the phosphate buffer while adding the DNA-calcium chloride solution helps to ensure that the precipitate that forms is as fine as possible, which is important because clumped DNA will not adhere to or enter the cell efficiently.

Chemicals and Reagents

- $CaCl_2$ (2.5 M)
- Chloroquine (100 mM): dissolve 52 mg of chloroquine diphosphate in 1 mL of deionized distilled water. Sterilize the solution by passing it through a 0.22 μm filter; store the filtrate in foil wrapped tubes at −20°C.
- Giemsa stain (10% w/v): the Giemsa stain should be freshly prepared in phosphate buffered saline or H_2O and filtered through Whatman No. 1 filter paper before use.
- Glycerol (15% v/v) in 1 × HEPES buffered saline: Add 15% (v/v) autoclaved glycerol to filter-sterilized HEPES buffered saline solution just before use.

2× HEPES Buffered Saline

- 140 mM NaCl
- 1.5 mM $Na_2HPO_4.2H_2O$
- 50 mM HEPES
- Dissolve 0.8 g of NaCl, 0.027 g of $Na_2HPO_4.2H_2O$, and 1.2 g of HEPES in a total volume of 90 mL of distilled water. Adjust the pH to 7.05 with 0.5 N NaOH and then adjust the volume to 100 mL with distilled water.
- Methanol
- Phosphate buffered saline
- Sodium butyrate (500 mM): In a chemical fume hood, bring an aliquot of stock butyric acid solution to a pH of 7.0 with 10 N NaOH.
- 0.1× TE (pH 7.6)
- 1 mM Tris-Cl (pH 7.6)
- mM EDTA (pH 7.6)

Nucleic Acids and Oligonucleotides

- Plasmid DNA: Dissolve the DNA in 0.1× TE (pH 7.6) at a concentration of 25 μL of plasmid solution required per mL of medium
- Cell culture growth medium
- Growing culture of mammalian cell

Equipment

- Tissue culture dishes (60 mm)
- Test tubes
- Humidified incubator

Method (Figure 3.13)

1. 24 h before transfection, harvest exponentially growing cells by trypsinization and replate them at a density of 1×10^5–4×10^5 cells/cm² in 60 mm tissue-culture dishes or 12-well plates in the appropriate complete medium.

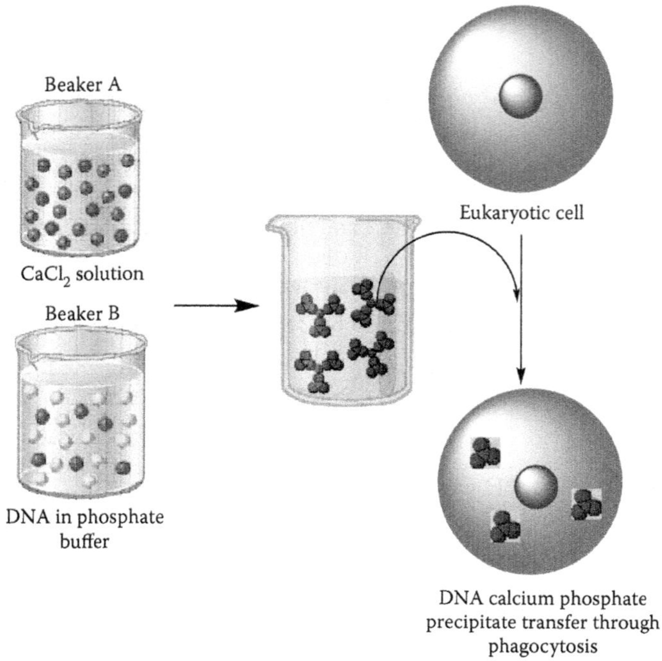

FIGURE 3.13 Calcium-phosphate mediated transfection of eukaryotic cells.

Incubate the cultures for 20–24 h at 37°C in a humidified incubator with an atmosphere of 5%–7% CO_2. Change the medium 1 h before transfection.

2. Prepare the calcium phosphate DNA coprecipitate as follows: combine 100 μL of 2.5 M $CaCl_2$ with 25 μg of plasmid DNA in a sterile 5 mL plastic tube and, if necessary, bring the final volume to 1 mL with 0.1× TE (pH 7.6). Mix 1 volume of this 2 × calcium-DNA solution with an equal volume of 2 × HEPES-buffered saline at room temperature. Quickly tap the side of the tube to mix the ingredients and allow the solution to stand for 1 min.
3. Immediately transfer the calcium phosphate-DNA suspension into the medium above the cell monolayer. Use 0.1 mL of suspension for each 1 mL of medium in a well or 60 mm dish. Rock the plate gently to mix the medium, which will become yellow-orange and turbid. Carry out this step as quickly as possible because the efficiency of transfection declines rapidly once the DNA precipitate is formed.
4. Transfected cells that will not be treated with transfection facilitators should be incubated at 37°C in a humidified incubator with an atmosphere of 5%–7% CO_2. After 2–6 h incubation, remove the medium and DNA precipitate by aspiration. Add 5 mL of warmed (37°C) complete growth medium and return the cells to the incubator for 1–6 d. Proceed to step 6, below, to assay for transient expression of the transfected DNA, or proceed directly to step 7, below, if the objective is stable transformation of the cells.

5. The uptake of DNA can be increased by treatment of the cells with chloroquine in the presence of the calcium phosphate DNA coprecipitate or exposure to glycerol and sodium butyrate following removal of the coprecipitate solution from the medium.

Treatment of Cells with Chloroquine

Chloroquine is a weak base that is postulated to act by inhibiting the intracellular degradation of the DNA by lysosomal hydrolases. The concentration of chloroquine added to the growth medium and the time of treatment are limited by the sensitivity of the cells to the toxic effect of the drug. The optimal concentration of chloroquine for the cell type used should be determined empirically.

a. Dilute 100 mM chloroquine diphosphate 1:1000 directly into the medium either before or after the addition of the calcium phosphate-DNA coprecipitate to the cells.
b. Incubate the cells for 3–5 h at 37°C in a humidified incubator with an atmosphere of 5%–7% CO_2.
c. After the treatment with DNA and chloroquine, remove the medium, wash the cells with phosphate buffered saline, and add 5 mL of warmed complete growth medium. Return the cells to the incubator for 1–6 d. Proceed to step 6, below, to assay for transient expression of the transfected DNA, or proceed directly to step 7, below, if the objective is stable transformation of the cells.

Treatment of Cells with Glycerol

This procedure may be used following treatment with chloroquine. Because cells vary widely in their sensitivity to the toxic effects of glycerol, each cell type must be tested in advance to determine the optimum time (30 sec to 3 min) of treatment.

a. After cells have been exposed for 2–6 h to the calcium phosphate-DNA coprecipitate in growth medium, remove the medium by aspiration and wash the monolayer once with phosphate buffered saline.
b. Add 1.5 mL of 15% glycerol in 1 × HEPES-buffered saline to each monolayer, and incubate the cells for the predetermined optimum length of time at 37°C.
c. Remove the glycerol by aspiration, and wash the monolayers once with phosphate buffered saline.
d. Add 5 mL of warmed complete growth medium, and incubate the cells for 1–6 d. Proceed to step 6, below, to assay for transient expression of the transfected DNA, or proceed directly to step 7, below, if the objective is stable transformation of the cells.

Treatment of Cells with Sodium Butyrate

The mechanism through which sodium butyrate acts is not known with certainty; however, the compound is an inhibitor of histone deacetylation, which suggests that treatment may leads to histone hyperacetylation and a chromatin structure on the incoming plasmid DNA that is predisposed to transcription.

a. Following the glycerol shock, dilute 500 mM sodium butyrate directly into the growth medium. Different concentrations of sodium butyrate are used depending on the cell type. For example;

CV-1	10 mM
NIH3T3	7 mM
HeLa	5 mM
CHO	2 mM

b. Incubate the cells for 1–6 d. Proceed to step 6 to assay for transient expression of the transfected DNA, or proceed directly to step 7 if the objective is stable transformation of the cells.

6. To assay the transfected cells for transient expression of the introduced DNA, harvest the cells 1–6 d after transfection. Analyze RNA or DNA using hybridization. Analyze newly synthesized protein by radioimmunoassay, by immunoblotting, by immunoprecipitation following *in vivo* metabolic labelling, or by assays of enzymatic activity in cell extracts.

Note: To minimize dish-to-dish variation in transfection efficiency, it is best to (i) transfect several dishes with each construct, (ii) trypsinize the cells after 24 h of incubation, (iii) pool the cells, and (iv) replace them on several dishes.

7. To isolate stable transfections:
   a. Incubate the cells for 24–48 h in nonselective medium to allow time for expression of the transferred genes.
   b. Either trypsinize and replace the cells in the appropriate selective medium or add the selective medium directly to the cells without further manipulation.
   c. Change the selective medium with care every 2–4 d for 2–3 wk to remove the debris of dead cells and to allow colonies of resistant cells to grow.
   d. Clone individual colonies and propagate for assay.
   e. Obtain a permanent record of the number of colonies by fixing the remaining cells with ice-cold methanol of 15 min followed by staining with 10% Giemsa for 15 min at room temperature before rinsing in tap water.

TRANSFECTION MEDIATED BY DEAE-DEXTRAN

Diethylaminoethyl (DEAE)-dextran is a polycationic derivative of the carbohydrate polymer dextran, and it is one of the first chemical reagents used to transfer nucleic acids into cultured mammalian cells (Vaheri and Pagano, 1965). The cationic DEAE-dextran molecule tightly associates with the negatively charged backbone of the nucleic acid, and the net positive charge of the resulting nucleic acid-DEAE-dextran complex allows it to adhere to the cell membrane and enter into the cytoplasm via endocytosis or osmotic shock induced by DMSO or glycerol.

The advantages of DEAE-dextran method are its relative simplicity, reproducibility, and low cost, while its disadvantages include cytotoxicity and low-transfection efficiency for a range of cell types (typically less than 10% in primary cells), as well as the requirement for reduced serum media during the transfection procedure. In addition, this method is limited to transient transfections and is not suitable for generating stable cell lines.

Materials Required

Chemicals

Buffers and Solution

- Ammonium acetate (10 M)

Dialysis Buffer

- 50 mM Tris-Cl (pH 8.0)
- 10 mM EDTA (pH 8.0)
- Ethanol

Lysis Buffer

- 10 mM Tris-Cl (pH 8.0)
- 0.1 M EDTA (pH 8.0)
- 0.5% (w/v) SDS
- 20 μg/mL DNase-free pancreatic RNase
- Phenol (pH 8.0)
- TE (pH 8.0)
- Tris-buffered saline (TBS)
- Chloroquine diphosphate (100 mM): Dissolve 60 mg of chloroquine diphasphate in 1 mL of deionized distilled water, sterilize the solution by passing it through a 0.22 μm filter. Store the filtrate in foil-warpped tubes at –20°C.
- DEAE-dextran (50 mg/mL): dissolve 100 mg of DEAE-dextran in 2 mL of distilled water. Sterilize the solution by autoclaving for 20 min at 15 psi on liquid cycle. Autoclaving also assists dissolution of the polymer.
- Phosphate buffered saline (PBS)
- Tris–buffered saline with dextrose: Immediately before use, add 20% (w/v) dextrose (prepared in H_2O and sterilized by autoclaving or filtration) to the TBS solution. The final dextrose concentration should be 0.1% (v/v).
- Plasmid DNA
- Cell-culture growth medium

Equipment

- Tissue culture dishes (60 mm or 35 mm)

Method

1. 24 h before transfection, harvest exponentially growing cells by trypsinization and transfer them to 60 mm tissue culture dishes at a density of 10^5 cells/dish. Add 5 mL of complete growth medium, and incubate the cultures for 20–24 h at 37°C in a humidified incubator with an atmosphere of 5%–7% CO_2.

2. Prepare the DNA/DEAE-dextran/TBS-D solution by mixing 0.1–4 μg of supercoiled or circular plasmid DNA into 1 mg/mL DEAE-dextran in TBS-D.
3. Remove the medium from the cell-culture dishes by aspiration, and wash the monolayers twice with warmed (37°C) PBS and once with warmed TBS-D.
4. Add the DNA/DEAE-dextran/TBS-D solution (250 μL per 60 mm dish, 150 μL per 35 mm dish). Rock the dishes gently to spread the solution evenly across the monolayer of cells. Return the cultures to the incubator for 30–90 min (the time will depend on the sensitivity of each batch of cells to the DNA/DEAEdextran/TBS-D soluion). At 15–20 min intervals, remove the dishes from the incubator, swirl them gently, and check the appearance of the cells under the microscope. If the cells are still firmly attached to the substratum, continue the incubation. Stop the incubation when the cells begin to shrink and round up.
5. Remove the DNA/DEAE-dextran/TBS-D solution by aspiration. Gently wash the monolayers once with warmed TBS-D and then once with warmed PBS, taking care not to dislodge the transfected cells.
6. Add 5 mL (per 60-mm dish) or 3 mL (per 35-mm dish) of warmed medium supplemented with serum and chloroquine (100 μM final concentration), and incubate the cultures for 3–5 h at 37°C in a humidified incubator with an atmosphere of 5%–7% CO_2.
7. Remove the medium by aspiration, and wash the monolayer 3 times with serum-free medium. Add to the cells 5 mL (per 60-mm dish) or 3 mL (per 35-mm dish) of medium supplemented with serum, and incubate the cultures for 36–60 h at 37°C in a humidified incubator with an atmosphere of 5%–7% CO_2 before assaying for transient expression of the transfected DNA.
8. To assay the transfected cells for transient expression of the introduced DNA, harvest the cells 36–60 h after transfection. Analyze RNA or DNA using hybridization. Analyze newly synthesized proteins by radioimmunoassay, by immunoblotting, by immunoprecipitation following *in vivo* metabolic labelling, or by assays of enzymatic activity in cell extracts.

TRANSFECTION OF DNA BY ELECTROPORATION

Electroporation—the use of high-voltage electric shocks to introduce DNA into cells—can be used with most cell types, yields a high frequency of both stable transformation and transient gene expression and, because it requires fewer steps, can be easier than alternate techniques. This experiment describes electroporation of mammalian cells, including ES cells for the preparation of knockout, knockin, and transgenic mice, describes protocols for using electroporation *in vivo* to perform gene therapy for cancer therapy and DNA vaccination, and outlines modifications for preparation and transfection of plant protoplasts.

Materials Required

Chemicals Required

- Giemsa stain (10% w/v)
- Methanol

- Phosphate buffered saline (PBS)
- Sodium butyrate (500 mM): In a chemical fume hood, bring an aliquot of stock butyric acid solution to a pH of 7.0 with 10 N NaOH.
- Nucleic acids
- Cell-culture growth medium

Equipment

- Centrifuges
- Rotors
- Water bath
- Electroporation device
- Tissue-culture dishes

Method (Figure 3.14)

1. Harvest the cells to be transfected from cultures in the mid-to-late logarithmic phase of growth. Use either a rubber policeman or trypsin to release adherent cells. Centrifuge at 1500 rpm in a Sorvall H 1000 B rotor for 5 min at 4°C.
2. Resuspend the cell pellet in 0.5× volume of the original growth medium and measure the cell number using a hemocytometer.
3. Collect the cells by centrifugation as described in step 1, and resuspend them in growth medium or phosphate buffered saline at room temperature at a concentration of 2.5×10^6–2.5×10^7 cells/mL.
4. Transfer 400 μL aliquots of the cell suspension (10^6–10^7 cells) into as many labelled electroporation cuvettes as needed. Place the loaded cuvettes on ice.
5. Set the parameters on the electroporation device. A typical capacitance value is 1050 μF. Voltages range from 200–350 V, depending on the cell line, but generally average 260 V. Use an infinite internal resistance value. Discharge a blank cuvette containing phosphate buffered saline at least twice before beginning electroporation of cells.
6. Add 10–30 μg of plasmid DNA in a volume of up to 40 μL to each cuvette containing cells.

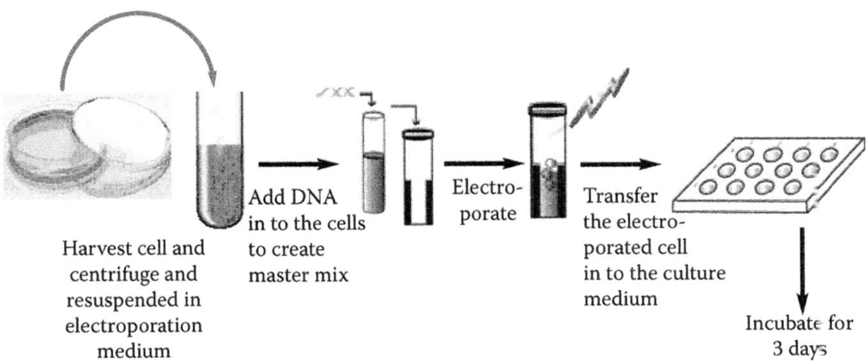

FIGURE 3.14 Electroporation techniques.

Note: Some investigations add carrier DNA to bring the total amount of DNA to 120 μg.

Gently mix the cells and DNA by pipetting the solution up and down.

7. Immediately transfer the cuvette to the electroporator and discharge the device. After 1–2 min, remove the cuvette, place it on ice, and proceed immediately to the next step.
8. Transfer the electroporated cells to a 35 mm culture dish using a micropipettor equipped with a sterile tip. Rinse out the cuvette with a fresh aliquot of growth medium, and add the washing to the culture dish. Transfer the dish to a humidified incubator at 37°C with an atmosphere of 5%–7% CO_2.
9. Repeat steps 6–8 until all of the DNA cell samples in the cuvettes are shocked. Record the actual pulse time for each cuvette to facilitate comparisons between experiments.
10. If the objective is stable transformation of the cells, proceed directly to step 11. For transient expression, examine the cells 24–96 h after electroporation using one of the following assays:
    a. If a plasmid DNA expressing *E. coli* β-galactosidase was used.
    b. If a green fluorescence protein expression vector was used, examine the cells with a microscope under 450–490 nm illumination.
    c. For other gene products, analyze the newly synthesized protein by radioimmunoassay, by immunoblotting, by immunoprecipitation following *in vivo* metabolic labelling, or by assays of appropriate enzymatic activity in cell extracts.
11. To isolate stable transfectants: After incubation for 48–72 h in complete medium, trypsinize the cells and replace them in the appropriate selective medium. The selective medium should be changed every 2–4 d for 2–3 wk to remove the debris of dead cells and to allow colonies of resistant cells to grow. Thereafter, clone individual colonies and propagate for assay.

DNA TRANSFECTION METHOD USING POLYBRENE

This method is the most commonly used method to expose cells to a coprecipitate of DNA and calcium phosphate. The method involves exposing the cells to DNA in the presence of the polycation polybrene, followed by a DMSO shock. The polybrene apparently interacts with the charge on the cell and the DNA, allowing the DNA to absorb more easily to the cell membrane.

Materials Required

Chemicals

- DMSO (30%)
- Giemsa stain (10% w/v)
- Methanol
- Polybrene (10 mg/mL)
- Sodium butyrate (500 mM)

- Plasmid DNA (1 μg/μL)
- Minimum essential medium

Equipment

- Tissue-culture dishes (90 mm)

Method

1. Harvest exponentially growing cells (e.g., CHO cells) by trypsinization, and replate them at a density of 5×10^5 cells per 90 mm tissue culture dish in 10 mL of MEM-α containing 10% fetal calf serum. Incubate the culture for 18–20 h at 37°C in a humidified incubator in an atmosphere of 5%–7% CO_2.
2. Replace the medium with 3 mL of warmed (37°C) medium containing serum, DNA (5 ng–40 μg; no carrier DNA), and 30 μg of polybrene. Mix the DNA with the medium before addition of the 10 mg/mL polybrene. Return the cells to the incubator for 6–15 h. Gently rock the dishes every 90 min during the early stages of this incubation to ensure even exposure of the cells to the DNA-polybrene mixture.
3. Remove the medium containing the DNA and polybrene by aspiration. Add 5 mL of 30% DMSO in serum containing medium. Gently swirl the DMSO medium around the dish to ensure even exposure of the cells to the solvent and place the dishes in the incubator.
4. After 4 min of incubation, remove the dishes from the incubator and immediately aspirate the DMSO solution. Wash the cells once or twice with warmed (37°C) serum-free medium, and add 10 mL of complete medium containing 10% fetal calf serum. If a sodium butyrate boost is to be included, then proceed to step 5. If not, incubate the cultures for 48 h at 37°C in a humidified incubator with an atmosphere of 5%–7% CO_2. Then proceed directly to either step 6 (to assay for transient expression) or step 7 (to establish stable transformants).
5. To facilitate the transfection of cells treated with DMSO and polybrene:
   a. Add 500 mM sodium butyrate directly to the growth medium to a final concentration of 2.5–10 mM.
   b. Incubate the cells for 20–24 h at 37°C in a humidified incubator with an atmosphere of 5%–7% CO_2.
   c. Remove the medium containing sodium butyrate, and replace it with butyrate-free medium containing 10% fetal bovine serum. Return the cells to the incubator.
6. If the objective is stable transformation of the cells, proceed directly to step 7. For transient expression, examine the cells 1–2 d after transfection using one of the following assays:
   a. If a plasmid DNA expressing *E. coli*, β-galactosidase was used to measure enzyme activity in cell lysate.
   b. If a green fluorescence protein expression vector was used, examine the cells with a microscope under 450–490 nm illumination.

c. For other gene products, analyze the newly synthesized protein by radioimmunoassay, by immunoblotting, by immunoprecipitation following *in vivo* metabolic labelling, or by assays of enzymatic activity in cell extracts.

7. To isolate stable transfectants: after the cells have incubated for 48 h in nonselective medium (to allow expression of the transferred gene to occour in step 4), either trypsinize or replate the cells in the appropriate selective medium or add the selective medium directly to the cells without further manipulation. Change this medium every 2–4 d for 2–3 wk to remove the debris of dead cells and to allow colonies of resistant cells to grow.
8. Thereafter, clone individual colonies and propagate for assay.

HYBRIDIZATION TECHNIQUES

The Hybridization Techniques is established and develop on the basis of the denaturation and renaturation of nucleic acids. Hydrogen bonds in double stranded nucleic acids can be disrupted by some physiochemical elements, and two strands of nucleic acids are separated in to single strand. Nucleic acid hybridization is a basic technique in molecular biology which involve a labelled nucleic acid probe, which is a known DNA or RNA fragment, to bind with the target nucleic acids, which is usually a poorly understood, heterogeneous population of nucleic acids. Hybridization techniques can be devided in to three types

1. For Nucleic Acid Hybridization
2. For RNA only (Northern Hybridization)
3. For DNA only (Southern Hybridization)

Nucleic Acid Hybridization

The 2 strands of DNA are held together by hydrogen bonds formed between complementary bases. When the double-helix molecule is subjected to conditions such as high temperature or alkali treatment, the 2 strands get separated resulting in denaturation of DNA. Under favourable conditions, renaturation or reannealing of the separated single strands into a double-helix state a can occur.

Reannealing of the DNA is not only confined to the single strands of the same DNA molecule, but it also occurs between DNA from different sources containing similar base sequences, and thus, has been used to study the degree of relatedness among organisms. The formation of DNA-RNA hybrids is also possible. The DNA sequences that are used for examining hybridization with the fragments of digested nucleic acids are referred to as probes.

Nucleic acid hybridization is a function of temperature, time, salt concentration, G + C content, probe length, and hybridization volume. For DNA with a G + C content of 50%, a temperature of 65°C is generally chosen, which is approximately 20°C below the actual melting temperature (T_m) of DNA at a standard salt concentration of 0.5 M NaCl. Addition of compounds such as formamide decreases the T_m of nucleic acids. So, hybridization in the presence of 50% formamide is usually done at a 42°C.

Hybridization volumes should be kept to a minimum to ensure maximum encounter rates between the probe and the bound DNA or RNA. Therefore, inert polymers such as dextran sulfate are added because they have a large hydrated volume and take up most of the space in the hybridization solution. At low ionic strength, hybridization of nucleic acids occurs slowly and can be improved by increasing the ionic strength.

Prior to hybridization, the reactive empty sites on the membrane need to be blocked by nonspecific DNA such as denatured salmon-sperm DNA and gelatin to avoid background attachment of the probe to the membrane.

A wide variety of methods are available for labelling the probes, among which nick translation and random hexamer labelling methods are most common. In nick transation, a nick is created in double-stranded DNA by brief action of DNase I; DNA polymerase-I then replaces the strand starting at the point of nick with a new strand, which gets radioactively labelled due to incorporation of ^{32}p-dNTP in the reaction mixture. In random hexamer labelling, synthetic hexa-nucleotides of random sequence are used as primers. The double-stranded DNA is denatured by boiling and primers are annealed. Klenow enzyme (a large fragment of DNA polymerase) then extends the primers and synthesizes a strand complementary to the template using 4 dNTPs, 1 of which is labelled radioactively, thus incorporating label into the probe.

At present, nonradioactive labelling and detection systems are also available and have advantages over the radioactive ones in that they are safer to use, have a much longer usable life, and can thus be stored for much longer periods of up to 12 mo. Some of the examples of nonradioactive systems include the Biotin-labelling system, Digoxignin-DUTP labelling, and the electro-chemiluminescence (ECL) system. Detection of probe nucleic acid target hybrids can be accomplished by direct or indirect labelling methods. In the former case, the label is attached directly to the nucleic acid by a covalent bond. For example, in the ECL direct-labelling system, DNA or RNA probes are labelled directly with the enzyme horseradish peroxidase. Detection is done based on generation of light in the form of enhanced chemiluminescence, which is produced during oxidation of luminal (substrate) in a reaction catalyzed by the enzyme bound to the nucleic acid probe. Chemiluminescence thus generated is recorded by exposure for a brief period (1 h) on blue-light sensitive autoradiography film. Indirect labelling employs a hapten (e.g., biotin-dUTP or digoxigenin-dUTP) attached to the nucleic acid probe. The hapten is detected using a labelled specific binding protein (e.g., avidin or antidigoxigenin), which gets bound to the labelled probe hybridized to target DNA on one end and has an enzyme such as alkaline phosphatase or horseradish peroxidase bound at the other end; the latter catalyzes a reaction resulting in generation of colored product or chemiluminescence, depending upon the substrate.

Blotting

Presence of a specific nucleic acid fragment in a digested sample is ascertained by a hybridization technique. This involves an initial blotting of electrophoretically separated pieces of nucleic acids of the digested sample. The transfer of DNA fragments onto nitrocellulose paper or a nylon membrane is called Southern blotting which has been named after its inventor. The technique enables detection of very low concentrations of specific sequences within DNA populations and has wide applications, such as identification of transformants, study of RFLPs, and DNA fingerprinting. Hybridization efficiency of a labelled probe directly to DNA fragments present in a gel is rather low and such an approach, therefore, has poor sensitivity. Hence, after separation of restriction fragments of DNA by electrophoresis on agarose gels, they are transferred and fixed onto a membrane filter (blotting) and then hybridized with a radioactivity labelled DNA or RNA probe to recognize whether any of the blotted spots contain a piece of the sequence under investigation. In a similar way, RNA molecules can be transferred to a nitrocellulose paper and the technique is known as Northern blotting (by analogy to Southern blotting). Northern blotting has been used to detect specific mRNAs within complex RNA populations and hence is extremely valuable in studying the differential expression of specific genes in response to particular physiological or environmental stimuli. DNA or RNA samples may also be applied directly as dots on the nylon membrane. This technique is called dot blotting and is often used for rapid and accurate measurement of abundance of specific RNA or DNA-specific RNA or DNA sequences.

Experimental Outline

In this technique, DNA molecules are transferred from an agarose gel to a solid support such a nitrocellulose filter or nylon membrane. DNA present in the gel is purinated in the presence of HCl followed by denaturation in alkali. Denaturation of DNA fragments prior to blotting is essential for dissociating the 2 polynucleotide chains, which can then hybridize with the probe. Denatured DNA molecules move upward by capillary action of the buffer and, on coming in contact with the nylon membrane, they get bound to it. A brief UV treatment (2–3 min) or heating at 80°C for 2 h under vacuum facilitates firm binding of DNA to the nylon membrane.

Materials

1. Agarose gel containing DNA fragments
2. Nylon membrane (Hybond N or gene screen plus)
3. Whatman 3 mm paper
4. Glass tray (5 cm deep × 20 cm × 30 cm) and a glass plate (12.5 × 25 cm)
5. Rough filter papers
6. Parafilm or used X-rays film
7. 0.25 M HCl
8. Denaturing solution: Prepare 1.5 M NaCl solution containing 0.25 M NaOH.
9. Neutralizing solution: Make 0.5 M Tris-HCl buffer containing 1.5 M NaCl and 1 mM EDTA. Adjust pH of this solution to 7.2.
10. SSC solution (2×): Dilute the below solution 5 times.
11. SSC solution (10×): Prepare a solution of 1.5 M NaCl containing (0.15 M trisodium citrate, adjust the pH to 7.0).

Method (Figure 3.15)

1. After electrophoresis, transfer the agarose gel to a glass tray containing sufficient HCl (0.25 M) to cover the gel and gently stir for 15 min at room temperature on a shaker. Unless otherwise stated, all the subsequent operations are also carried out at room temperature.

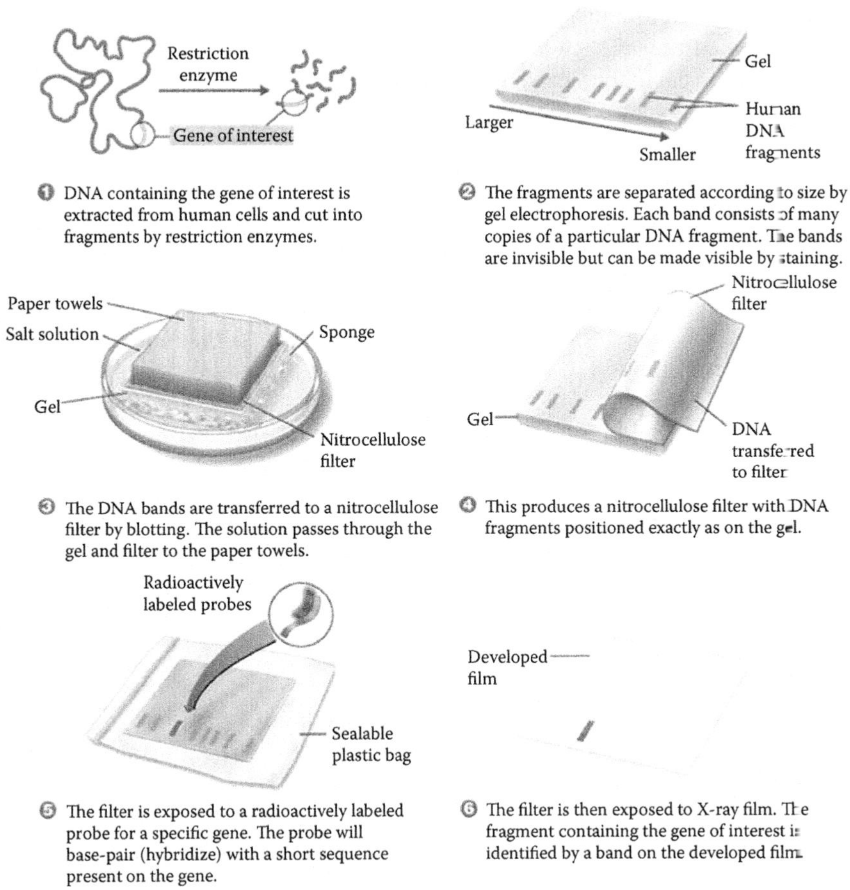

FIGURE 3.15 Nucleic acid hybridization.

2. Drain off HCl and rinse the gel with distilled water.
3. Denature the DNA by adding a sufficient volume of denaturing solution (see reagents). Gently stir for 30 min.
4. Decant the alkali and rinse with distilled water.
5. Neutralize the gel with neutralizing solution and gently stir for 30 min.
6. Remove the solution and wash the gel with 2× SSC.
7. Make a bridge on a glass tray with glass plate and place 2 layers of Whatman 3 mm paper cut to proper size on top so that the ends of the paper hang down into the tray. Wet the paper with 10× SSC and fill the tray with the same solution.
8. Place another piece of Whatman 3 mm paper, which is cut to the size of the gel and prewetted in 10× SSC, on top of the wetted filter paper and remove any entrapped air bubbles by rolling a glass pipe gently over the gel.
9. Cut a piece of nylon membrane (Hybond N or gene screen plus) of exactly the same size as that of the gel. Mark the membrane on the left hand bottom

corner with a permanent ink marker. Wet the membrane in 2× SSC and place it on top of the gel. Cover the membrane with 2 pieces of 3 mm Whatman filter paper cut to the same size and prewetted in 2× SSC. Avoid entrapment of any air bubbles between the papers.

10. Seal all the sides of the 3 mm Whatman paper with used X-ray films.
11. Place a stack of dry, rough filter papers on the top of the wet filter paper. Size the rough filter papers to the size of the wet filter paper.
12. Place a glass plate on top of the stack of filter papers, keep a weight of about 0.5 kg (500 mL water in bottle) over it and allow the transfer of DNA from gel to nylon membrane overnight.
13. Remove the weight, stack of rough filter paper, and the Whatman filter papers from the top of the membrane. Carefully remove membrane and wash it in 2× SSC.
14. Expose the membrane for not more than 2 min to UV light to achieve crosslinking and a steadfast binding of transferred DNA.

Note: Do not cros-link if 0.4 M NaOH has been used for denaturation. In this case, bake the membrane at 80°C for 2 h in a vacuum oven.

15. Air dry the membrane briefly, keep between 2 sheets of Whatman 3 mm paper and store at 4°C in a refrigerator until use.

Precautions

1. Use gloves while handling gel and nylon membrane.
2. Take care that no air bubbles are entrapped between gel, membrane, and Whatman papers.
3. Make sure that the buffer moves through the gel and membrane. The layer of papers placed on the top should not touch the layers of filter paper placed beneath the gel.

Northern Hybridization of RNA Fractionated by Agarose Formaldehyde Gel Electrophoresis

Initially, the technique of Southern blotting used for DNA transfer from gel to the membrane could not be used for blot-transfer of RNA. Instead, mRNA bands from the gel were blot-transferred into a chemically reactive paper, prepared by diazotization of amino benzyloxymethyl paper. The technique, being related to Southern blotting, was called Northern blotting (not after the name of any scientist as in Southern blotting). Later, it was shown that mRNA bands can be blotted directed onto nitrocellulose membrane, a technique which has been widely adopted. The mRNA bands blotted onto nitrocellulose membrane can be hybridized with a labelled DNA or RNA probe. The single-stranded regions of the probe are removed by nuclease (Mungbean nuclease or S-1 nuclease,) so that quantitative estimation of hybridized mRNA can also be made.

Experimental Outline

The objective of the experiment is to examine the different species of RNA on agarose gels.

Materials Required

Chemicals

- Diethylpyrocarbonate (DEPC)
- 10× and 1× MOPS running buffer
- 12.3 M (37%) formaldehyde (pH > 4.0)
- RNA sample: total cellular RNA or Poly(A)$^+$ RNA formamide formaldehyde buffer
- 0.5 M ammonium acetate and 0.8 μg/mL ethidium bromide in 0.5 M ammonium acetate or 10 mM sodium phosphate (pH 7.0) and 0.1 M formaldehyde with 10 ng/mL acridine orange
- 0.05 M NaOH/1.5 N NaCl (optional)
- 0.5 M Tris-HCl (pH 7.4) or 1.5 M NaCl (optional)
- 20 × 2 and 6× SSC.
- 0.03% (w/v) ethylene blue in 0.3 M sodium acetate (pH 5.2) (optional).
- Prehybridization or hybridization solution
- ×SSC/0.1% (w/v) SDS
- 0.2× SSC (w/v) SDS, room temperature and 42°C
- 0.1× SSC 0.1% (w/v) SDS, 68°C

Equipment

- 60°C water bath
- Oblong sponge slightly larger than the gel being blotted
- Whatman 3 MM filter paper sheets
- Nitrocellulose or nylon membrane
- UV transilluminator, calibrated
- UV transparent plastic wrap (e.g., Saran Wrap or other polyvinylindene wrap)
- Hybridization oven (e.g., Hybridiser HB-1, Techne)

Method

1. Dissolve 1.0 g agarose in 72 mL water and cool to 60°C in a water bath. When the flask has cooled to 60°C, place in a fume hood and add 10 mL of 10× MOPS running buffer and18 mL of 12.3 M formaldehyde (2.2 M final).
2. Pour the gel and allow it to set, remove the comb, place the gel in the gel tank, add sufficient 1× MOPS running buffer to cover to a depth of 1 mm. (A 1.0% gel is suitable for RNA molecules 500 bp to 10 kb in size; a 1.0%–2.0% gel should be used to resolve smaller molecules or a 0.7%–3.0% gel for longer molecules. The gel should be 2–6 mm thick after it is poured and the wells large enough to hold 60 ng/mL of sample.)
3. Adjust the volume of each RNA sample to 11 μg, then add:
   5 μL 10× MOPS running buffer
   9 μL of 3 M formaldehyde
   25 μL formamide
   Vortex and spin briefly (5–10 s) in a microcentrifuge to collect liquid. Incubate 15 min at 55°C.

4. Add 10 uL formaldehyde loading buffer, vortex, spin to collect liquid, and load 0.5–10 μL RNA into each of 2 wells arranged so that half of the gel may be stained. Run the gel at 5 V/cm until the bromophenol blue dye has migrated one-half to two-thirds the length of the gel (~3 h).
5. To stain with ethidium bromide: Remove the gel and cut off the lanes that are to be stained. Place this portion of the gel in an RNase-free glass dish, add sufficient 0.5 M ammonium acetate to cover, and soak for 20 min. Change solution and soak for an additional 20 min to remove the formaldehyde. Pour off solution, replace with 0.5 ng/mL ethidium bromide in 0.5 M ammonium acetate, and allow to stain for 40 min. If necessary, destain in 0.5 M ammonium acetate for up to 1 h.
6. Stain with acridine orange: Remove gel, cut off lanes, and stain 2 min in 1.1 M formaldehyde/l0 mM sodium phosphate containing 10 ng/mL acridine orange. If necessary, destain 20 min in the same buffer without acridine orange.
7. Examine gel on a UV transilluminator to visualize the RNA and photograph with a ruler laid alongside the gel so that band positions can later be identified on the membrane.
8. Place unstained portion of gel in an RNase-free glass dish and rinse with several changes of sufficient deionized water to cover the gel to remove the formaldehyde.
9. Partially hydrolyze the RNA by soaking the gel 20 min in 10 gel volumes of 0.05 M NaCl, followed by soaking 30 min in 10 gel volumes of 0.5 M Tris-HCl (pH 7.4)/1.5 M NaCl to neutralize (optional).
10. Replace solution with 10 gel volumes of 20× SSC and soak for 45 min (optional).
11. Place oblong sponge slightly larger than the gel in a glass or plastic dish (if necessary). Use 2 or more sponges (placed side by side). Fill the dish with enough 20× SSC to leave the soaked sponge about half-submerged in buffer.
12. Construct the transfer stack. Cut 3 pieces of Whatman 3 MM paper to the same size as the sponge. Place them on the sponge and wet them with 20× SSC. Place the gel on the filter paper and squeeze out air bubbles by rolling a glass pippet over the surface. Cut 4 strips of plastic wrap and place over the edges of the gel.
13. Cut a piece of nylon or nitrocellulose membrane just large enough to cover the exposed surface of the gel. Pour distilled water 0.5 cm deep in an RNase-free glass dish and wet the membrane by placing it on the surface of the water. Allow the membrane to submerge. For nylon membrane, leave for 5 min; for nitrocellulose membrane, replace the water with 20× SSC and leave for 10 min.
14. Place the wetted membrane on the surface of the gel. Remove air bubbles. Flood the surface of the membrane with 20× SSC. Cut 5 sheets of Whatman 3 MM paper to the same size as the membrane and place on top of the membrane. Cut paper towels to the same size as the membrane and stack on top of the Whatman 3 MM paper to a height of 4 cm.
15. Lay a glass plate on top of the stack and add a weight (0.2–0.4 kg) to hold everything in place, leave overnight.

16. Disassemble transfer stack, and recover the membrane and flatten gel. Mark the position of the wells on the membrane in pencil and mark the orientation of the gel.
17. Rinse the membrane in 2× SSC, then place it on a sheet of Whatman 3 MM paper and allow to dry completely.
18. For nitrocellulose membrane, place between 2 sheets of Whatman 3 MM filter paper and bake under vacuum 2 h at 80°C.
19. For nylon membranes, bake as described above or wrap the dry membrane in transparent plastic wrap, place RNA-side-down on a UV-transilluminator (254-nm wavelength), and irradiate for the appropriate length of time.
20. If desired, check transfer efficiency by either staining the gel in ethidium bromide or acridine orange as in step 5 (for a nylon membrane) by staining the membrane in 0.03% (w/v) methylene blue in 0.3 M sodium acetate (pH 5.2) for 45 s and destaining in water for 2 min.
21. Prepare DNA or RNA probe labelled to a specific activity of >10 s dpm/u (unit 2.10 or 3.5) and with unincorporated nucleotides removed.
22. Wet the membrane carrying the immobilized RNA in 6× SSC.
23. Place the membrane RNA-side-up in a hybridization tube and add 1 mL formamide prehybridization/hybridization solution per 10 cm^2 of membrane. Place the tube in the hybridization oven and incubate with rotation 3 at 42°C (for DNA probe) of 60°C (for RNAprobe).
24. If the probe is double-stranded, denature it by heating in a water bath or incubator for 10 min at 100°C and transfer to ice.
25. Pippete the desired volume of probe (to give 10 μg/mL if the specific activity is 10^9 dpm/μg) into the hybridization tube and continue to incubate with rotation overnight at 42°C (for DNA probe) or 60°C (for RNA probe).
26. For denatured probe, add to hybridization tube as soon after denaturation as possible. Pour off hybridization solution and wash in: 2× SSC/0.1% SDS, 5 min with rotation at room temperature; twice with 0.2× SSC/0.1% SDS, 5 min with rotation at room temperature; twice (low-stringency wash) with prewarmed 0.2× SSC, 0.1% SDS, 15 min at 42°C; twice (moderate stringency wash, optional) with prewarmed 0.1× SSC/0.1% SDS, 15 min at 68°C; and twice high-stringency wash, optional.
27. Rinse membrane in 2× SSC at room temperature. Blot excess liquid and cover in UV transparent plastic wrap. Autoradiograph.

Do not allow membrane to dry out if it is to be reprobed.

Precautions

Formaldehyde is toxic through skin contact and inhalation of vapous. All operations involving formaldehyde should be carried out in fume hood.

Southern Hybridization of DNA Fractionated by Agarose Formaldehyde Gel Electrophoresis

The Southern hybridization technique involves transfer of DNA fragments separated in electrophoretic gels to membrane filters for detection of specific base sequences by complementary probes. Prof E.M. Southern developed this technique in 1975, hence it is referred to as Southern transfer. Southern blots are used to identify and quantitate specific DNA sequences, in analysis of genome organization and expression, in the study of genetic diseases, in DNA fingerprinting, and in analysis of PCR products.

DNA molecules are resolved on the gel on the basis of their size and transferred to positively charged nylon membrane by capillary or electrophoretic transfer. The DNA is immobilized onto the membrane by UV crosslinking. The membrane is then incubated with the protein block to reduce nonspecific interactions. A hybridized probe is detected by incubating the membrane with streptavidin-HRP (horseradish peroxidase) conjugate and finally incubating with the substrate solution until sufficient color develops. The final detection is done following addition of substrate, TMB/H_2O_2 (tetramethyl benzidine/H_2O_2 sbstrate) that reacts with HRP to give a blue-colored DNA band on the nylon membrane.

In brief, the technique of Southern hybridization involves the following practicals:

1. Electrophoretic separation of DNA molecules by agarose gel electrophoresis.
2. Electrophoretic transfer of DNA from agarose gel to nylon membrane.
3. Immobilization of DNA onto nylon membrane.
4. Hybridization and nonisotope detection of DNA of interest.

Materials Required

Chemicals and Other

- DNA marker
- Biotinylated probe
- Prehybridization buffer
- Hybridization buffer
- Blocking Buffer
- 2× wash buffer (A, B, C, and D)
- Streptavidin HRP conjugate
- Conjugate dilution buffer, 10× substrate
- 50× TAE
- Blocking Powder
- Tween-20
- Agarose
- Distilled water
- Ethidium bromide
- Ice

Glassware and Other

- Filter paper
- Nylon membrane
- Petri dish
- Forceps
- Gloves
- Measuring cylinders
- Micropipette

- Blade
- Scissors
- Thermometer
- Tips
- Transparent polythene sheet

Equipment

- Hot air oven
- Incubator shaker (45°C)
- UV transilluminator

Pre-Lab Preparations

1. Preparation of prehybridization buffer: Add 1 g of blocking powder to 10 mL of prehybridization buffer.
2. Preparation of hybridization buffer: Add 1 g of blocking powder to 10 mL of hybridization buffer.
3. Preparation of blocking buffer: Add 1 g of blocking powder to 10 mL of blocking buffer.
4. Preparation of wash buffers A, B, C, and D buffer: Thaw and mix the buffer provided thoroughly. Dilute 15 mL each of 2× buffers supplied with 15 mL of distilled water to give buffers 1× concentration, 30 mL each.
5. Preparation of streptavidin HRP-conjugate buffer: Prepare 9 mL of conjugate dilution buffer by adding 9 μL of Tween-20 to the conjugate dilution buffer. Dilute 3 μL of streptavidin-HRP conjugate with 9 mL of conjugate dilution buffer for each practical just before use.
6. Preparation of substrate buffer: Dilute 0.5 mL of substrate solution with 4.5 mL of distilled water to give a final volume of 5 mL substrate solution of 1× concentration.
7. Preparation of electrotransfer buffer: Dilute 25 mL of 10× electrotransfer buffer in 225 mL distilled water to give a final volume of 250 mL electrotransfer buffer of 1× concentration.

Method

Day 1: Agarose gel electrophoresis

1. Prepare 1.0% agarose gel containing ethidium bromide.
2. Load 20 μL of ready-to-use DNA marker supplied onto the gel. Run at 50–100 V until the dye reaches 4.5 cm from the well.
3. Cut the DNA marker lane from the agarose gel as follows: Cut the gel ~3 mm above the first band and ~2 mm below the last band, ensuring the gel measures about 4–4.5 cm. Mark the appropriate position of the gel to be cut on the gel tank.
4. Cut the filter paper (Whatman No. 2) and nylon membrane exactly to the size of the cut gel. Ensure there is no protrusion of the filter paper and the membrane from the gel.
5. Wet the cut gel, nylon membrane, filter papers, and the electrotransfer cassette in 1× electrotransfer buffer for electroblotting.

Electro-blotting:
Assemble the electrotransfer apparatus. Start the arrangement by placing filter paper on the cathode cassette cover followed by the cut gel and nylon membrane. Mark and place the soft side of the nylon membrane to the cut gel. Place the wet filter paper on the nylon membrane followed by the anode cassette cover. Tighten the electrotransfer cassette tightly with the screws.

6. Insert the cassette into the apparatus filled with 250 mL of 1× electrotransfer buffer.
7. Connect the cords to a power supply and set the voltage to 50 V for 3.5 h.
8. Turn off the power supply and remove the cassette from the apparatus. Drain the buffer.

Immobilization of DNA on membrane:

9. Remove the nylon membrane gently from the cassette and place it on a thin transparent polythene sheet and place this on a UV transilluminator (expose the soft side of membrane containing transferred DNA to UV light) with UV lamps switched on for 20 min. This helps in fixing the DNA on the membrane.
10. Turn of the UV transilluminator. Place the membrane in plastic petri dish and incubate in a hot air oven at 70°C for 30 min. This ensures complete immobilization of DNA onto the membrane.

Hybridization:

11. Bring the petri dish containing the membrane to room temperature after incubation. Add 10 mL prehybridization buffer to the petri dish and incubate at 45°C in an incubator shaker with mild shaking (70–90 rpm) for 45 min.
12. After incubation, discard prehybridization buffer, taking care not to discard the membrane.
13. Add 10 mL hybridization buffer to the petri dish containing the membrane.
14. Keep 1 vial of biotinylated probe for 10 min in a boiling water bath and immediately chill by placing it on ice for 5–10 min. Add this probe to the hybridization buffer in the petri dish. (Rinse the probe vial with 300 μL of hybridization buffer and add it to the petri dish). Incubate the petri dish in a 45°C incubator shaker with mild shaking 70 rpm for 16 h.

Day 2:

15. Decant the hybridization buffer, add 10 mL of 1× wash buffer A and gently swirl the petri dish for 5 min at RT. Repeat the wash twice, each wash for 5 min. Discard the buffer after each wash.
16. Add 10 mL of 1× prewarmed (70°C) wash buffer B and gently swirl the petri dish. Incubate at 70°C for 5 min in a hot air oven and gently swirl. Repeat the wash another 2 times. Discard the buffer after each wash.
17. Add 10 mL of 1× blocking buffer to the petri dish and incubate at room temperature for 1 h with gentle rocking.

18. Discard the blocking buffer.
19. Add 9 mL of diluted HRP-streptavidin conjugate to the petri dish and incubate at RT for 20 min with gentle rocking. Discard the conjugate buffer.
20. Add 10 mL of 1× wash buffer C to the petri dish and incubate at RT for 5 min with gentle rocking. Repeat the wash 2 more times. Discard the buffer after each wash.
21. Add 10 mL of 1× wash buffer D to the petri dish and incubate at RT for 5 min with gentle rocking. Repeat the wash 2 more times. Discard the buffer after each wash.
22. Add 5 mL of 1× substrate solution and gently swirl at RT for 15–20 min until a blue-colored band develops.
23. After blue-colored band is seen, stop the reaction by placing the membrane in distilled water.

Precautions

1. Carry out the Practical by wearing gloves, especially while handling ethidium bromide agarose gel.
2. Agarose gel is to be prepared by addition of ethidium bromide at a concentration of 0.5 µL/mL from a stock of 10 mg/mL.
3. Dilute the appropriate quantities of buffers (wash buffers A, B, C, and D) to a 1× concentration with distilled water just before use.
4. Dilute 10× electrotransfer buffer to 1× with distilled water.
5. Use autoclave tips and measuring cylinders to avoid contamination.
6. All the buffers provided should be brought to room temperature 1 d prior to use, except streptavidin HRP-conjugate buffer and substrate solution.
7. Buffers can be prepared according to the required concentration on the previous day of use, except for streptavidin HRP-conjugate buffer and substrate solution, which have to be prepared just before use.
8. Ensure no air bubbles are present between any of the layers of filter paper, cut gel and the nylon membrane.

Western Hybridization of Protein Fractionated by Agarose Formaldehyde Gel Electrophoresis

Western blotting is most commonly used for identifying specific proteins recognized by antibodies. The blotting membrane is the solid phase of immunoblotting, and it serves as an immobilized matrix capable of adsorbing and retaining the transferred molecules in the original separation pattern during the subsequent detection steps. Nitrocellulose is commonly used for immobilization of transferred proteins as it is easy to handle and has a high binding capacity. For the detection of the specific proteins, blocking is essential by using either serum, casein and animal serum. Commonly used detergents are polyoxyethylene, sorbitan, alcohol (Tween 20).

Materials Required

Chemicals

- Tris

- Glycine
- Methanol
- Water

Glassware and Other

- Beaker
- Reagent bottle
- Micro tips
- Measuring cylinder

Method (Figure 3.16)

1. After completion of SDS run, remove the gel from the glass plate sandwich. Place the gel in a tray containing 200–300 mL of transfer buffer and incubate the gel for 10 min.
2. Pour about 1 in of transfer buffer in a second tray. Fill the transfer tank with 4 L of transfer buffer.
3. Place the gel on the cassette with foam and a blotter paper between the gel and the cassette. Place the nitrocellulose on top of the gel so as not to trap any air bubbles between the gel and the membrane. Now place 2 sheets of blotter paper over the gel. It is important to use 2 sheets of paper at this step in order to position the gel near the center of the cassettes where the electrical field is most uniform. Place the second half of the cassette on the top and hook it together.
4. Quickly lift the cassette out of the tray of buffer and transfer it to one of the center slots in the buffer chamber. Be certain that the cassette is oriented with the nitrocellulose membrane on the fast, or anode side, of the gel. Tap the cassette to dislodge any bubbles that may be trapped by the cassette grid.
5. Place the power lid on the unit. Turn on the power supply to 50 volts. Run the unit for 4–5 min or as required.

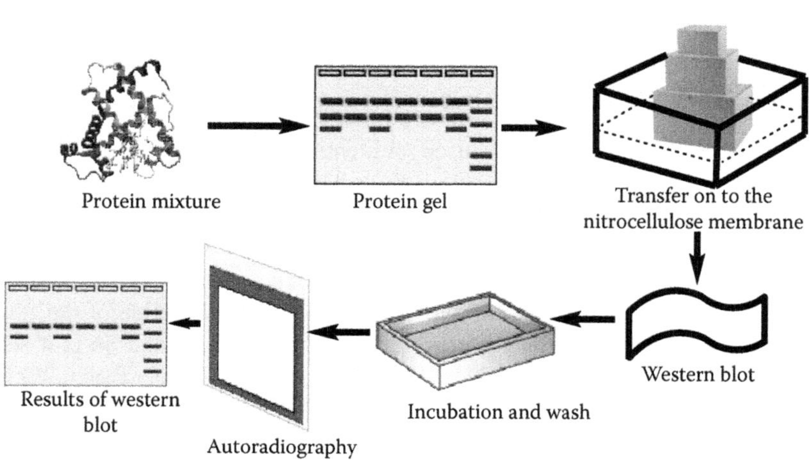

FIGURE 3.16 Western blotting.

Protein Detection by the Use of Antibodies

6. Incubate the blot in 2% bovine serum albumin (BSA) and 0.005% Tween 20 in PBS for 30 min.
7. Following blocking, the blot should be washed 2 times in PBT. First antibody incubation should be carried out either for 2 h at room temperature or overnight at 4°C.
8. Following first antibody incubation, wash the blot extensively with PBT.
9. Second antibody could either be directly conjugated to peroxidase or it could be biotinylated by application of avidin conjugatioin to peroxidase.
10. For color development, 33 mg of 4 chloronaphthol is dissolved in 10 mL of methanol and the volume made up to 50 mL with PBS. 10 μL of H_2O is added to the blots, which are incubated in this solution to develop color. Color can be intensified by addition of a pinch of imidazole. Blots can be dried and stored between 2 sheets of paper.

Anti-Phosphotyrosine Western Blotting

Western blotting is most commonly used for identifying specific proteins recognized by antibodies. The blotting membrane is the solid phase of immunoblotting, and it serves as an immobilized matrix capable of adsorbing and retaining the transferred molecules in the original separation pattern during the subsequent detection steps. Nitrocellulose is commonly used for immobilization of transferred proteins as it is easy to handle and has a high binding capacity. For the detection of the specific proteins, blocking is essential by using either serum, casein and animal serum.

Materials Required

Chemicals

- 5% milk
- 20 mM tris (pH 7.4)
- 150 mM NaCl
- 1 mM vanadate
- 0.1% Tween-20
- Antiphosphotyrosine antibody
- Anti-mouse-HRP conjugated secondary antibody

Pre-Lab Preparation

Blocking Buffer Solution for Western Blotting

S.No.	Material	Quantity
1.	20 mM Tris (pH 7.4)	2 mL 1 M Tris (pH 7.4)
2.	150 mM NaCl	3 mL 5 M NaCl
3.	10% milk	10 g skim milk powder
4.	1% BSA	1 g BSA
5.	0.1% Tween 20	100 μL Tween 20
6.	1 mM vanadate	1 mL 100 mM vanadate
7.	Double distilled water	to 100 mL

Composition of Stock solution for Western Blot

1.	1 M Tris (pH 7.4)	10 mL	20 mL
2.	5 M NaCl	15 mL	30 mL
3.	Skim milk powder	25 g	50 g
4.	Tween 20	0.5 mL	1 mL
5.	Sodium orthovanadate	0.09 g	0.18 g
6.	Double distilled water	500 mL	1 liter

Method

1. After SDS-PAGE, set up transfer as usual.
2. Block for 1 h at room temperature before probing.
3. The remaining steps are performed at room temperature in 5% milk, 20 mM Tris (pH 7.4), 150 mM NaCl, 1 mM vanadate, 0.1% Tween.
4. Blot membrane on 3 mm paper.
5. Place membrane in ECL mix for 1 min; blot dry on 3 mm paper.
6. Expose to film for 5 min and develop. Re-expose for various times as necessary.

4 Nanobiotechnology and Metabolome Profiling

OBJECTIVE: TO PERFORM METABOLOME PROFILING BY GC/MS ANALYSIS

PRINCIPLE

Identifying a wide range of primary metabolites (often after derivatization) is greatly facilitated by the high resolution of capillary GC, the reproducible fragmentation of metabolites in the mass spectrometer, and the ready availability of large mass spectral libraries. Recent developments in GC-MS have resulted in improvements in both the GC and MS capabilities. The analysis of biomolecules by MS initially was hindered by difficulties in volatizing the molecules, because, although small organic molecules could be readily vaporized by heating in a vacuum, proteins, oligo nucleotides, etc., were destroyed under these conditions. Thus, a combination of both makes this process applicable.

MATERIALS REQUIRED

Chemicals

- Liquid nitrogen
- Isopropanol
- Acetonitrile
- HPLC-grade water
- Ribitol
- Methoxy amine hydrochloride
- Pyridine
- N-methyl-N-(trimethylsilyl)trifluoroacetamide (MSTFA)

Glassware and Other

- Pestle and mortar
- GC-MS vials

Equipment

- Centrifuge
- Vortex
- Vacuum concentrator
- GC-MS

Pre-Lab Preparations

Preparation of sample: The analysis of biomolecules by MS initially was hindered by difficulties in volatizing the molecules, because, although small organic molecules could be readily vaporized by heating in a vacuum, proteins, oligo nucleotides, etc. were destroyed under these conditions. Thus, sample preparation should be done under mild alcohol grinding.

Method

Without Solvent

- Add 0.1–0.5 mL of MSTFA, weigh 1–10 mg of sample into 5 mL reaction vessel.
- Mix well and let the solution stand at room temperature until the sample has dissolved (5–10 min) or heat at 60°C for 15 min.
- Allow to cool at room temperature and analyze the sample.

With Solvent

- Dissolve 1–10 mg of sample in pyridine, dimethyl sulfoxide, dimethyl formamide plus hydrofluoric acid or acetonitrile.
- Add 0.1–0.5 mL of MSTFA.
- Cap vial, mix well, and let the solution stand at room temperature until the sample has dissolved (5–10 min) or heat at 60°C for 15 min. Cool to room temperature and analyze the sample.

Sample Analysis

- Perform the analysis of components of extracts using gas chromatography 7890A with mass selective detector MS-5975C (Agilent Technologies, USA) controlled by Chem-Station software fitted with a DB-1I MS fused silica column (25 m × 0.2 mm; 0.33 μm film thickness), which is operated in El mode.
- Maintain ion source temperature at 250°C; obtain the mass spectrum of compounds in samples by electron ionization at 70 eV.
- Collect data at a rate of 1 MS spectrum per s in positive ion modes from m/z 50 to 950.
- Use helium (99.99%) as the carrier gas with flow rate of 1.5 mL/min in the split mode (25:1).
- Start GC oven temperature at 40°C (for 5 min), raising to 150°C at a 5°C/min rate without holding. Allow the temperature to increase to 250°C at a 20°C/min rate, and maintain at 250° for 2 min; finally, ramp the oven temperature to 290°C at a rate of 20°C/min, and hold for 5 min at this temperature.
- Inject an aliquot of 1 μL of extract of the plant genotypes into the column with injector temperature at 250°C.

Data Analysis

The retention time/mass pairs generated by the Chem-Station software are then exported for subsequent analysis in Agilent Gene-Spring MS software/XCms/MetaboAnalyst, according to the following workflow:

1. Check alignment and normalization of features
2. Check data quality with hierarchical clustering
3. Identify features with differential abundances across classes using 1-way and 2-way analysis of variance (ANOVA)
4. Perform principle component analysis (PCA) to show discriminating classes
5. Visualize fold changes

The result of the analysis to this point is a mass list of metabolites that shows statistically significant variations in abundance between experimental classes. Search of the mass list against the NIST library.

OBJECTIVE: TO ANALYZE THE PRESENCE OF KNOWN METABOLITES IN PLANT EXTRACT BY HPLC

PRINCIPLE

High performance liquid chromatography (HPLC) is a form of column chromatography that pumps a sample mixture or analyte in a solvent (known as the mobile phase) at high pressure through a column with chromatographic packing material (stationary phase).

MATERIALS REQUIRED

Chemicals

- HPLC grade methanol
- Acetonitrile
- Formic acid
- Water, standard

Glassware and Other

- Volumetric flasks

Equipment

- HPLC instruments
- Sonicator
- Filtration assembly syringe filter (0.45 μm)

METHOD

Preparation of Standard

- Prepare stock solution (1 ppm) of eugenol in volumetric flasks.
- Filter the solutions by a 0.45 μm syringe filter.

Preparation of Sample

- Through drying and grinding process, the whole leaf is obtained from powder for further steps
- Weigh the dried powder (10 g) and keep in 250 mL conical flask. Add 100 mL of autoclaved distilled water, hot water, or absolute ethanol. Seal the opening with parafilm and aluminum foil and mix gently.

- Keep the solution in an incubator shaker at 37°C with gentle swirling at 120 rpm.
- Use a muslin cloth to filter the solution. Using Whatman filter paper No.1 to doubly filter the filtrate.
- Collect the supernatant of cold/hot water, while using a large glass plate already kept inside a fan equipped incubator at 37°C to keep it dry; keep the supernatant at –20°C for freezing. Collect the supernatant of ethanolic extract.
- Finally, lyophilize the sample in a vacuum.
- Stock solution of 1.0 mg/mL. is obtained by dissolving in distilled water to the lyophilized plant extracts (aqueous and ethanolic).
- Filter all the solutions by a 0.45 μm syringe filter.

Preparation of Mobile Phase

- Mix HPLC-grade water:methanol:acetonitrile in 45:35:20 v/v ratio.
- Filter the mixture with filtration assembly having 0.45 μm membrane filter.
- Keep the filtered mixture in sonicator for 10 min for degassing.

Instrumentation and Chromatographic Conditions

- Analysis is carried out with an Agilent 1120 HPLC system controlled by EZChrom Elite.
- Separation is carried out in reverse phase Zorbax C18 (4.6 × 250 mm particle size 5 μm) analytical column (Agilent) with 20 μL sample.
- Use a flow rate of 1.0 mL/min and the run time of 15 min.
- The detection of components is carried out at 210 nm using diode array detector.
- Peak identification is based on the retention time.

Quantification of Components

- The content of compounds in extracts is determined by extrapolating standard concentration vs peak area.
- Stock solution of eugenol (1000 μL mL^{-1} each) is prepared in mobile phase.
- Appropriate dilution of stock solution of standard is made in respective mobile phase to obtain working standard solutions of eugenol in the concentration ranges of 0.02–0.20 μg and used to draw calibration curve.
- Triplicate determinations of sample are carried out and the average taken in drawing the calibration curve.

OBJECTIVE: TO PERFORM THIN LAYER CHROMATOGRAPHY (TLC) FOR METABOLOME PROFILING

Principle

Thin layer chromatography, or TLC, is a method for analyzing mixtures by separating the compounds in the mixture. TLC can be used to help determine the number of components in a mixture, the identity of compounds, and the purity of a compound.

Metabolomics based on liquid thin layer chromatography (TLC)) is a powerful tool for studying static responses of biological systems to different physiological or pathological conditions. With few variations within and between batches, unwanted and uncontrolled data variation is removed, allowing useful information to be extracted, making the profiling reliable.

Requirements

- TLC plates (Silica gel 60)
- Rice varieties for extract preparation
- Glass chamber
- Mobile phase DCM:Methanol (8:2)
- Spotting capillaries
- Standard 2-AP
- Visualizing reagent

Procedure

1. With the help of capillary action, add an extract of each variety (30 μL) 1 cm above the bottom of the activated plate.
2. Perform chromatography in a glass TLC tank saturated with mobile phase of dichloromethane (DCM) and methanol in ratio 8:2.
3. Remove the plate from the TLC chamber when the solvent reaches up 2/3 of the plate.
4. Prepare visualizing reagent (2, 4-dinitrophenyl hydrazine) by dissolving 2 g of 2, 4-dinitrophenyl hydrazine in 15 mL concentrated sulphuric acid, followed by adding 15 mL of 95% ethanol and diluting the solution to 1 50 mL with distilled water.
5. Spray the visualizing reagent after drying the TLC plate and observe the spot.
6. Scan the TLC plates by TLC scanner.
7. Generate a chromatogram of each plate using the win CATS software available in the Camag TLC scanner.
8. Create a 2-dimensional area of interest and a table for each spot at the TLC plate and obtain the R_f-value for each spot through the corresponding table, designated as Retardation Factor Value.

OBJECTIVE: TO SYNTHESIZE AND CHARACTERIZATION OF SILVER NANO-PARTICLES

Principle

Synthesis and characterization of silver nanoparticles is an eco-friendly, cost efficient, rapid, and easy method for synthesis of silver nanoparticles. It is basically nanoparticles of silver between 1 nm and 100 nm in size. While frequently described as being "silver," some are composed of a large percentage of silver oxide because of their large ratio of surface-to-bulk silver atoms. Nanoparticle synthesis refers to methods for creating nanoparticles. Nanoparticles can be derived from larger molecules, or synthesized by "bottom-up" methods that, for example, nucleate and grow particles from fine molecular distributions in liquid or vapor phase. This, for characterization of the synthesized nanoparticles, utilizes UV-visible spectroscopy, scanning electron microscopy (SEM), energy dispersive X-ray spectroscopy (EDX), field emission scanning electron microscope (FESEM), transmission electron microscope (TEM), X-ray diffraction (XRD), dynamic light scattering (DLS), and Fourier transform infrared spectroscopy (FT-IR) analysis.

Requirements

- Conical flask
- Trisodium citrate
- Silver nitrate
- Magnetic stirrer with hot plate
- Magnetic beads
- Weighing balance
- pH meter
- UV-Vis spectrophotometer.

Stock Solutions

- Silver Nitrate 100 mM
- Trisodium citrate 5%

Method

1. Boil 0.5 mM AgNO3; to this solution, add 1/10th volume of 1% trisodium citrate, drop by drop; use a magnetic stirrer vigorously. Heat the solution until a pale yellow color forms.
2. Allow the solution to cool at RT with constant stirring.
3. *Observation*: Use TEM analysis of silver nanoparticles and spectrophotometric analysis to check the status of synthesized particles.

OBJECTIVE: TO STUDY THE EFFECT OF SILVER NANOPARTICLES ON PLANT MAPK EXPRESSION

Principle

In commonly known abiotic stresses, the understanding of AgNP stress or silver nanotoxicity in plants is still under study. Gene expression, protein function, and pathways are used to elucidate similarities and differences in different abiotic stresses. MAPK (mitogen activated protein kinases) are involved in signal transduction for regulating plant growth or biotic stress and various other processes in a plant cell. Different signaling pathways often cross talk through the MAPK cascade. Because they are involved in plant growth, under both normal and stressful conditions, they can be used as a tool to elucidate the effects of nanoparticles at the molecular level.

Requirements

- DEPC-treated pestle mortar
- Tips and Eppendorf tubes
- Liquid nitrogen
- RNA Xpress reagent
- Chloroform
- Isopropanol
- DEPC-treated water
- 75% ethanol
- MAPK specific primers
- RT-PCR kit (Fermentas)
- DNase-I agarose
- Ethidium bromide
- Electrophoresis assembly
- Gel documentation system

Method

RNA Extraction

Sample Preparation

Use 1 mL for 50–100 mg of RNA-Xpress Reagent and homogenize plant tissue samples in a DEPC-treated pestle and mortar.

Phase Separation

1. In-order to permit the complete dissociation of nucleoprotein complexes, incubate the homogenized samples for 5 min at room temperature (15–25°C).
2. Add 200 mL chloroform per mL of RNA-Xpress reagent used. Cover the sample tightly, shake vigorously for 15 s and allow to stand for10 min at room temperature (15–25°C).
3. Centrifuge the resulting mixture at 12,000 × *g* (13,000 rpm) for 15 minutes at 4°C.
4. Following centrifugation, mixture separates into lower deep red organic phase (containing protein), an interphase (containing DNA) and a colorless upper aqueous phase containing RNA.

RNA Precipitation

5. Transfer the aqueous phase containing RNA to a fresh tube and add 500 mL of isopropyl alcohol. Allow the sample to stand for 5–10 minutes at room temperature (15–25°C).
6. Centrifuge at 12,000 × *g* (13,000 rpm) for 10 minutes at 4°C. The RNA precipitate, often invisible before centrifugation, forms a gel-like pellet on the side and bottom of the tube.

RNA Wash

Remove the supernatant and wash the RNA pellet by adding 1 mL (minimum) of 75% ethanol. Vortex the sample and then centrifuge at 7,500 × *g* (10,500 rpm) for 5 min at 4°C.

Note: If the RNA pellets form a froth, wash in 75% ethanol at 12,000 × *g* (13,000 rpm).

DNase Treatment

Purify the RNA sample by treating the DNase treatment.

cDNA Preparation

Follow the protocol of the Fermentas cDNA synthesis kit.

Expression Profiling or MAPK3

Design the MAPK3 gene specific primers using online tool *Primer3 or NCBI primer tool.* Run PCR with the MAPK3 gene specific primers, using cDNA as a template.

Analyze the PCR product via agarose gel electrophoresis, visualization.

OBJECTIVE: TO DEMONSTRATE COMPETITIVE FORMAT BASED IMMUNO-CHROMATOGRAPHIC STRIP USING GOLD NANOPARTICLES FOR ON-SITE DETECTION OF TELIOSPORES OF KARNAL BUNT (TILLETIAINDICA) IN PLANT

PRINCIPLE

Recent progress in nanotechnology and analyte-nanoparticles complex (DNA-nanoparticle and antibody-nanoparticle conjugate systems) have emerged as novel and extremely powerful tools for plant disease diagnostics, nanoparticles with the ability to bind to their target with high affinity and improve cell penetration given by the antibodies. This specialized vehicle, which can encapsulate several chemotherapeutic agents, can be engineered to possess the desirable properties, allowing overcoming the successive physiological conditions and to cross biological barriers and reach a specific tissue or cell. Moreover, antibody-conjugated nanoparticles have shown the ability to be internalized through receptor-mediated endocytosis and accumulate in cells without being recognized by the P-glycoprotein.

REQUIREMENTS

- Sample pad
- Adsorbent pad
- Conjugate pad
- Nitrocellulose membrane
- Phosphate buffer saline (pH 7.4)
- Goat anti-rabbit antibody
- Anti-teliosporic antibody
- Extraction buffer

Procedure

1. Extract teliospores from the bunted seeds or infected plant samples.
2. Transfer the teliospores (about 10 mg) into the mortar pestle and add 1 mL of extraction buffer.
3. Extract out the teliosporic protein by crushing the teliospores with pestle in extraction buffer.
4. Transfer 3–5 μL of teliosporic protein into the vial containing 100 μL of buffer.
5. Dip the strip into the vial containing buffer + teliosporic protein.
6. Wait until the colloidal gold (brilliant red) travels to the adsorbent pad. This should takes 7–8 min.
7. If 1 band develops, it indicates the presence of teliospores and, ultimately, Karnal bunt in plant.
8. If 2 bands develop, it indicates the sample is negative and plant samples are not infected with Karnal bunt.

METABOLIC PROFILING USING BODY FLUIDS BY NMR SPECTROSCOPY

Principle

NMR is widely used in metabolic profiling to identify and quantify metabolites. This is because it allows nondestructive measurement of a wide array of molecules having different structures. Moreover, quantification is easily detected based on the peak area of the specific metabolite related to the peak area of a known concentration of the internal standard. External reference compounds can also be used for the quantification of metabolites by inserting a coaxial capillary containing the solution of the reference compound into the NMR tube during measurement. Moreover, NMR exhibits high reproducibility of measurement and requires minimum sample preparation as compared with other techniques, such as mass spectrometry (MS).

Method

1. Biofluids such as serum, plasma, urine, amniotic fluid, cerebrospinal fluid, saliva, and other digestive fluids can be analyzed by liquid-state NMR spectroscopy.
2. Solvent suppression is required during the measurement because of the large water signal dominating the spectrum.
3. Water pre-saturation is done by applying a weak radiofrequency (RF) during the relaxation/mixing time and WATERGATE (WATER suppression by GrAdient-Tailored Excitation) is done.
4. Sample preparation such as deproteinization by reaction with organic solvents or ultrafiltration is used and the resulting metabolite extract measured is by a single pulse NMR experiment.
5. Freeze the tissues in liquid nitrogen to quench the enzymatic reactions, followed by grinding the tissue in a cooled mortar or with an electric homogenizer

6. Then take the tissue sample and pack into the rotor with the addition of D_2O for signal locking during NMR measurement.
7. Insert the sample rotor into the NMR probe at an angle of 54.7° relative to the external magnetic field B_0 and spinning the rotor will result in narrow line width due to the cancelation of the orientation dependent chemical shift anisotropy (CSA) and dipolar coupling effects in the solid samples.
8. The acquired free induction decay (FID) data can be processed using ACD/NMR Processor (ACD/Labs, Toronto, Canada) in which 2x zero-filling and an exponential window function equivalent to 0.3 Hz line broadening is usually applied before Fourier transformation. The spectra will be phased, baseline corrected, referenced to standard compound signal of a specific chemical shift, and exported as an ASCII file.
9. The ASCII file is imported into Excel and the spectral region from 0.02 ppm to 10 ppm is chosen for subsequent statistical analysis.
10. Software such as topspin, NMRlab, ProMetab, matNMR, and MetaboLab can also be applied to process the raw FID data to visualize the spectra.
11. Spectral resonance assignment is usually performed by searching chemical shift lists with the help of MetaboID2 (a graphical user interface that offers comparison of experimental spectra with authentic metabolite spectra to guide assignment).

Note: For samples containing minimum proteins and lipids (e.g. urine), phosphate buffer made in D_2O can be added to the sample and measured by a single pulse NMR experiment with water suppression.

Appendix I: Preparation of Buffers

ACETATE BUFFER

Stock Solution

A: 0.2 M solution of acetic acid (11.55 mL in 1000 mL)
B: 0.2 M solution of sodium acetate (16.4 g of C_2H_2Na or 27.2 g of $C_2H_3O_2Na.3H_2O$ in 1000 mL)
x mL of A, y mL of B, diluted to a total of 100 mL

x	y	pH
46.3	3.7	3.6
44.0	6.0	3.8
41.0	9.0	4.0
36.8	13.2	4.2
30.5	19.5	4.4
25.5	24.5	4.6
20.0	30.0	4.8
14.8	35.2	5.0
10.5	39.5	5.2
8.8	41.2	5.4
4.8	45.2	5.6

BORIC ACID: BORAX BUFFER

Stock Solution

A: 0.2 M solution of boric acid (12.4 g in 1000 mL)
B: 0.05 M solution of Borax (19.5 g in 1000 mL; 0.2 M in terms of sodium borate)
50 mL of A, x mL of B, diluted to a total of 200 mL

x	pH
2.0	7.6
3.1	7.8
4.9	8.0
7.3	8.2
11.5	8.4
17.5	8.6
22.5	8.7
30.0	8.8
42.5	8.9
59.0	9.0
83.0	9.1
115.0	9.2

CARBONATE: BICARBONATE BUFFER

Stock Solution

A: 0.2 M solution of anhydrous sodium carbonate (21.2 g in 1000 mL)
B: 0.2 M solution of sodium bicarbonate (16.8 g in 1000 mL) x mL of A, y mL of B, diluted to a total of 200 mL

x	y	pH
4.0	46.0	9.0
7.5	42.5	9.3
9.5	40.5	9.4
13.0	37.0	9.5
16.0	34.0	9.6
19.5	30.5	9.7
22.0	28.0	9.8
25.0	25.0	9.9
27.5	22.5	10.0
30.0	20.0	10.1
33.0	17.0	10.2
35.5	14.5	10.3
38.5	11.5	10.4
40.5	9.5	10.5
42.5	7.5	10.6
45.5	5.0	10.7

CITRATE BUFFER

Stock Solution

A: 0.1 M solution of citric acid (21.01 g in 1000 mL)
B: 0.1 M solution of sodium citrate (29.41 g $C_6H_5O_7Na_3.2H_2O$ in 1000 mL) x mL of A, y mL of B, diluted to a total of 100 mL

x	y	pH
46.5	3.5	3.0
43.7	6.3	3.2
40.0	10.0	3.4
37.0	13.0	3.6
35.0	15.0	3.8
33.0	17.0	4.0
31.5	18.5	4.2
28.0	22.0	4.4
25.5	24.5	4.6
23.0	27.0	4.8
20.5	29.5	5.0
18.0	32.0	5.2
16.0	34.0	5.4
13.7	36.3	5.6
11.8	38.2	5.8
9.5	40.5	6.0
7.2	42.8	6.2

GLYCIN: HCl BUFFER

Stock Solutions

A: 0.2 M glycine (15.01 g in 1000 mL)
B: 0.2 N HCl
25 mL of A, x mL of B diluted to a total of 100 mL

x	pH
22.0	2.2
16.2	2.4
12.1	2.6
8.4	2.8
5.7	3.0
4.1	3.2
3.2	3.4
2.5	3.6

PHOSPHATE BUFFER

Stock Solutions

A: 0.2 M solution of monobasic phosphate (27.8 g in 1000 mL)
B: 0.2 M solution of dibasic sodium phosphate (53.65 g of $Na_2HPO_4.7H_2O$ or 71.7 g of $Na_2HPO_4.12H_2O$ in 1000 mL)
x mL of A, y mL of B, diluted to a total of 200 mL

x	y	pH
93.5	6.5	5.7
92.0	8.0	5.8
90.0	10.0	5.9
87.7	12.3	6.0
85.0	15.0	6.1
81.5	18.5	6.2
77.5	22.5	6.3
73.5	26.5	6.4
68.5	31.5	6.5
62.5	37.5	6.6
45.0	55.0	6.9
39.0	61.0	7.0
33.0	67.0	7.1
28.0	72.0	7.2
23.0	77.0	7.3
19.0	81.0	7.4
16.0	84.0	7.5
13.0	87.0	7.6
10.5	89.5	7.7
8.5	91.5	7.8

GENERAL PROTEIN STAINS FOR BLOT TRANSFERS ARRANGED IN ORDER OR INCREASING SENSITIVITY

Detection Reagent	Approximate Sensitivity	Matrix
Fast Green FCF	–	NC
Ponceau S	–	NC
Coomassie Brilliant Blue R-250	1.5 μg	NC
Amido black 10B	1.5 μg	NC
India ink	100 μg	NC + nylon
Colloidal iron	30 μg	NC + nylon
In situ biotinylation + HPR avidin	4 μg	NC
Colloidal gold		NC

SPECIFIC DETECTION METHODS OF BLOT TRANSFERS ARRANGED IN ORDER OF INCREASING SENSITIVITY

Method	Approximate Sensitivity
Peroxidase-protein A	2.0
Peroxidase-second antibody	1.5
Gold second antibody	1.5
^{125}1-second antibody	1.0
Peroxidase double sandwich	0.8
Peroxidase-anti peroxidase	0.5
Avidin-Biotin-Peroxidase Complex (ABC)	0.5
Gold-second antibody with silver enhancement	0.1

SUMMARY OF STAINING PROCEDURES FOR PROTEIN

Staining Mixture	Staining Time (h)	Destaining Mixture	Components
Amido black 10B 1–10 g dissolved in 250 mL methanol; then add 100 mL acetic acid + 650 mL H_2O	0.5–2.0	10% acetic acid	Very widely used simple; prefixing in 12.5% TCS May be advantageous
Coomassie Brilliant Blue R-250 1 g dissolved in 500 mL methanol; then add 100 mL acetic acid + 400 mL H_20	1.0	500 mL methanol + 100 mL acetic acid + 400 mL H_2O (store in 7% acetic acid)	Very sensitive
0.5 g in 50 mL H_2O mixed with 950 mL 12.5% TCA	0.5–1.0	10% TCA for 20–30 min (store in 7% acetic acid)	Before staining, gel can be prefixed for 0.5 h in 12.5% TCA (or overnight if desired); very sensitive

(Continued)

Staining Mixture	Staining Time (h)	Destaining Mixture	Components
Xylene Cyanide Brilliant G 2.5 g dissolved in 500 mL methanol; then add 500 mL 25% TCA or 0.2 g dissolved in 100 mL H_2O add 100 mL to H_2SO_4; stand for atleast 3 h and filter; add 22.2 mL 10 N KOH to filtrate, add 28.7 g TCA Anilinonaphthalene sulphonate 30 mg in 1L of 0.1 N Sodium phosphate pH 6.8	0.5–1.0	5% TCA for 10–60 min (store in 7% acetic acid) Store in staining mixture, or better H_2O	Complicated to make up but no destaining required; rapid and sensitive Fluorescent levelling, less sensitive than staining methods but faster and preserves bio-logical/enzymatic activity
O-phthaladehyde (OPA) Soak gels for 10 min in solution of 2-mercaptoethanol (70–80 μL per 100 mL) in 0.1 M barbiturate phosphate or hydrogen carbonate Add OPA (20 mg) in 2 mL methanol, mix and allow to stand in dark for 15 min Examine under U.V. illumination at 365 nm.	25 min		Sensitive fluorescent staining; very rapid, no destaining step sensitivity varies with enzymatic and biological activities retained.

SAFETY PROCEDURES

Chemicals

A number of chemicals used in any molecular biology laboratory are hazardous. All manufacturers of hazardous materials are required by law to supply the user with pertinent information on any hazards associated with their chemicals. This information is supplied in the form of a Material Safety Data Sheet or MSDS. This information contains the chemical name, CAS#, health hazard data, including first aid treatment, physical data, fire and explosion hazard data, reactivity data, spill or leak procedures, and any special precautions needed when handling this chemical. A file containing MSDS information on the hazardous substances should be kept in the lab. In addition, MSDS information can be accessed on World Wide Web. You are strongly urged to make use of this information prior to using a new chemical and certainly in the case of any accidental exposure or spill. The instructor/lab manager must be notified immediately in the case of an accident involving any potentially hazardous reagents.

All of the following listed chemicals are highly toxic and hazardous.

- *Acrylamide:* It is a neurotoxin and possible carcinogen. It is most dangerous when dry (inhalation hazard) so wipe any spilled liquid acrylamide off the bench before it has a chance to dry.
- *Chloroform:* It is a hazardous carcinogenic and toxic solvent, readily absorbed through the skin, degreases and deproteinates the skin; dispense in the fume hood. It readily melts some plastics like polystyrene. Avoid chloroform on outside of tubes as it erases labels.
- *Alcohol:* Highly flammable.
- *Phenol:* It is quite toxic and is a suspected carcinogen. It is rapidly absorbed through skin. Small quantities are metabolized by the body: so, it is not accumulated poison. If phenol is spilled on body or clothes, immediately remove the contaminated clothing and wash small spills under running water for at least 5 min. Do not rinse contaminated skin with alcohol to dissolve the phenol: it increases the surface area and can cause accelerated absorption through the skin.
- *Ethidium bromide:* It is carcinogenic hence may cause heritable genetic damage.
- *Liquid nitrogen:* It has a greater risk of cryogenic burns, always handle in a well aerated and open area
- *Silver nitrate:* Possible carcinogen, causes burns absorbed through the skin and stains the skin.
- *Sodium thiosulphate:* Irritant, dispense stock in the fume hood.
- *TEMED:* Highly flammable, harmful by inhalation; causes burns.
- *All other chemicals in powder or crystal* form that you are likely to encounter on this course will generally be irritants in one way or another, possibly causing irreversible effects on contact with the skin or on inhalation, always wear gloves when weighing, dispensing and handling solutions.

These chemicals are not harmful if used properly. Always wear a lab coat, gloves, and safety glasses. Never mouth-pipet them. If accidentally splash any of these chemicals on skin, immediately wash the area thoroughly with water and inform the instructor. Discard the waste in appropriate containers.

Ultraviolet Light

Exposure to ultraviolet light can cause acute eye irritation. Since the retina cannot detect UV light, you can have serious eye damage and not realize it until 30 min to

24 hours after exposure. Therefore, always wear appropriate eye protection when using UV lamps.

Electricity

The voltages used for electrophoresis are sufficient to cause electrocution. Cover the buffer reservoirs during electrophoresis. Always turn off the power supply and unplug the leads before removing a gel.

General Housekeeping

All common areas should be kept free of clutter and all dirty dishes, electrophoresis equipment, etc. should be dealt with appropriately. Since you have only a limited amount of space to call your own, it is to your advantage to keep your own area clean. Since you will use common facilities, all solutions and everything stored in an incubator, refrigerator, etc. *must be labelled.* In order to limit confusion, each person should use his initials or other unique designation for labelling plates, etc. Unlabelled material found in the refrigerators, incubators, or freezers may be destroyed. Always mark the backs of the plates with your initials, the date, and relevant experimental data, for example, strain numbers.

Appendix II: Definitions

1. *Atomic weight:* Atomic weight of an element is the relative weight of the atom on the basis of oxygen as 16.

   For example, Atomic weight of sodium is 23.
2. *Molecular weight:* The sum of the atomic weights of all the atoms in a molecule is its molecular weight.

   For example, Molecular weight of H_2SO_4 is 98, since

Hydrogen	2×1	=	2
Sulphur	1×32	=	32
Oxygen	4×16	=	64
			98

3. *Equivalent weight:* Equivalent weight of a substance is the number of grams of the substance required to react with, replace or furnish one mole of H_2O^+ or OH^-.

   The equivalent weight of an acid is the weight that contains one atomic weight of acidic hydrogen, that is, the hydrogen that reacts during neutralization of acid with base.

   For example, the equivalent weight of H_2SO_4 is 49. Since H_2SO_4 contains two replaceable hydrogens, equivalent weight is molecular weight/2, that is, $98/2 = 49$.
4. *Percent solution (w/v):* One percent solution of a substance contains one gram of the substance in 100 mL of the solvent. If v/v is given, it means 1 mL in 100 mL of solvent.
5. *Molar solution (M):* One molar solution of a substance contains one mole or one gram molecular weight of the substance in one litre of solution.

   For example, 1 M NaOH contains 40 g sodium hydroxide in one litre solution. Likewise, one millimolar solution of a substance contains.
6. *Normal solution (N):* One normal solution of a substance contains on equivalent or one gram equivalent weight of the substance in one litre solution (i.e., molecular weight divided by the hydrogen equivalent of the substance).

   For example, 1 N H_2SO_4 contains 49 g H_2SO_4 in one liter solution.
7. *Buffer:* A solution containing both a weak acid and its conjugate weak base whose pH changes only slightly on the addition of acid or alkali.
8. *pH:* pH is a value taken to represent the acidity or alkalinity of an aqueous solution. It is defined as logarithm of the reciprocal of the hydrogen ion concentration of the solution, that is, $\text{pH} = \log \frac{1}{[H^+]}$
9. *Dilute acids:* While preparing dilute acids, add slowly to water preferable under cooling.

   Alkali solutions: When preparing concentrated alkali solutions (e.g., 40% NaOH) dissolve the alkali in distilled water under cooling.

10. *Water:* In biochemical experiments, it is advised to use distilled water for all purposes. So, water means distilled water in the methodologies given.

CONCENTRATION OF SOLUTIONS

Molarity (M):	The number of moles of solute per litre of solution (The term molar is equivalent to mol l^{-1})
Molality (m):	The number of moles of solute per kg of solvent
% (w/v):	The weight in grams of a solute per 100 mL of solution
% (w/w)	The weight in grams of a solute per 100 g of solution

UNITS OF LENGTH

1 micron = 1 μ = 1 μm = 1×10^{-6} m = 1×10^{3} nm = 1×10^{4} Å
1 Å = 0.1 nm = 1×10^{-4} μm = 1×10^{-10} m
1 nm = 10 Å = 1×10^{-3} μm = 1×10^{-9} m

CONVERSION UNITS OF TEMPERATURE

Temp. in °F = (Temp. in °C × 1.8) + 32
Temp. in °C = (Temp. in °F–32) × 5/9
Temp. in K = (Temp. in °F + 459.67) × 5/9

UNITS OF RADIOACTIVITY

Curie (Ci) = That quantity of a radioactive substances in which 3.7×10^{10} atoms disintegrate per second. Thus 1Ci = 3.7×10^{10} dps = 2.22×10^{12} dpm.

Becquerel (Bq) = That quantity of a radioactive substance in which one atom disintegrate per second.
Thus 1 Ci = 3.7×10^{10} Bq
1 Bq = 1 dps = 60 dpm

SOME ISOTOPES AND THEIR PROPERTIES

Isotope	Type of Decay	Half-Life	Average Energy
^{3}H	β^-	12.3 *y*	0.0055
^{14}C	β^-	5570 *y*	0.05
^{32}p	β^-	14.5 *d*	0.70
^{35}p	β^-	87.4 *d*	0.049
^{36}Cl	β^-	3.07×10^{5} *y*	0.30

y = year; d = day

Appendix III: Preparation of Buffers

ACETATE BUFFER

Stock Solutions

A: 0.2 M solution of acetic acid (1.5 mL of Glacial Acetic Acid in 100 mL distilled water)

B: 0.2 M solution of sodium acetate (0.64 g of sodium acetate or 2.72 g of sodium acetate trihydrate in 100 mL distilled water).

x mL of A, y mL of B, diluted to a total of 100 mL

x	y	pH
46.3	3.7	3.6
44.0	6.0	3.8
41.0	9.0	4.0
36.8	13.2	4.2
30.5	19.5	4.4
25.5	24.5	4.6
20.0	30.0	4.8
14.8	35.2	5.0
10.5	39.5	5.2
8.8	41.2	5.4
4.8	45.2	5.6

BORIC ACID-BORAX BUFFER

Stock Solutions

A: 0.2 M solution of boric acid (12.4 g in 1000 mL)

B: 0.05 M solution of borax (19.5 g in 1000 mL; 0.2 M in terms of sodium borate)

50 mL of A, x mL of B, diluted to a total of 200 mL

x	pH
2.0	7.6
3.1	7.8
4.9	8.0
7.3	8.2
11.5	8.4
17.5	8.6
22.5	8.7
30.0	8.8
42.5	8.9
59.0	9.0
83.0	9.1
115.0	9.2

CARBONATE-BICARBONATE BUFFER

Stock Solutions

A: 0.2 M solution of anhydrous sodium carbonate (21.2 g in 1000 mL)
B: 0.2 M solution of sodium bicarbonate (16.8 g in 1000 mL) x mL of A, y mL of B, diluted to a total of 200 mL

x	y	pH
4.0	46.0	9.0
7.5	42.5	9.3
9.5	40.5	9.4
13.0	37.0	9.5
16.0	34.0	9.6
19.5	30.5	9.7
22.0	28.0	9.8
25.0	25.0	9.9
27.5	22.5	10.0
30.0	20.0	10.1
33.0	17.0	10.2
35.5	14.5	10.3
38.5	11.5	10.4
40.5	9.5	10.5
42.5	7.5	10.6
45.0	5.0	10.7

CITRATE BUFFER

Stock Solutions

A: 0.1 M solution of citric acid (21.01 g in 1000 mL)
B: 0.1 M solution of sodium citrate (29.41 g $C_6H_5O_7Na_3.2H_2O$ in 1000 mL) x mL of A, y mL of B, diluted to a total of 100 mL.

x	y	pH	x	y	pH	x	y	pH
46.5	3.5	3.0	33.0	17.0	4.0	18.0	32.0	5.2
43.7	6.3	3.2	31.5	18.5	4.2	16.0	34.0	5.4
40.0	10.0	3.4	28.0	22.0	4.4	13.7	36.3	5.6
37.0	13.0	3.6	25.5	24.5	4.6	11.8	38.2	5.8
35.0	15.0	3.8	23.0	27.0	4.8	9.5	40.5	6.0
			20.5	29.5	5.0	7.2	42.8	6.2

GLYCINE-HCl BUFFER

Stock Solutions

A: 0.2 M glycine (15.01 g in 1000 mL)
B: 0.2 N HCl

25 mL of A, x mL of B diluted to a total of 100 mL

x	pH
22.0	2.2
16.2	2.4
12.1	2.6
8.4	2.8
5.7	3.0
4.1	3.2
3.2	3.4
2.5	3.6

PHOSPHATE BUFFER

Stock Solutions

A: 0.2 M solution of monobasic phosphate (27.8 g in 1000 mL)
B: 0.2 M solution of dibasic sodium phosphate (53.65 g of $Na_2HPO_4.7H_2O$ or 71.7 g of $Na_2HPO_4.12H_2O$ in 1000 mL)
x mL of A, y mL of B, diluted to a total of 200 mL

x	Y	pH	x	y	pH
93.5	6.5	5.7	45.0	55.0	6.9
92.0	8.0	5.8	39.0	61.0	7.0
90.0	10.0	5.9	33.0	67.0	7.1
87.7	12.3	6.0	28.0	72.0	7.2
85.0	15.0	6.1	23.0	77.0	7.3
81.5	18.5	6.2	19.0	81.0	7.4
77.5	22.5	6.3	16.0	84.0	7.5
73.5	26.5	6.4	13.0	87.0	7.6
68.5	31.5	6.5	10.5	89.5	7.7
62.5	37.5	6.6	8.5	91.5	7.8
56.5	43.5	6.7	7.0	93.0	7.9
51.0	49.0	6.8	5.3	94.7	8.0

TRIS (HYDROXYMETHYL) AMINOMETHANE (TRIS HCl) BUFFER

Stock Solutions

A: 0.2 M solution of Tris (hydroxymethyl) aminomethane (24.2 g in 1000 mL)
B: 0.2 N HCl
50 mL of A, x mL of B, diluted to a total of 200 mL

x	pH
5.0	9.0
8.1	8.8
12.2	8.6
16.5	8.4
21.9	8.2
26.8	8.0
32.5	7.8
38.4	7.6
41.4	7.4
44.2	7.2

GLYCINE-NaOH BUFFER

Stock Solutions

A: 0.2 M solution of glycine (15.01 g in 1L)
B: 0.2 M NaOH
50 mL of A + x mL of B, diluted to a total of 200 mL

x	pH
4.0	8.6
6.0	8.8
8.8	9.0
12.0	9.2
16.8	9.4
22.4	9.6
27.2	9.8
32.0	10.0
38.6	10.4
45.6	10.6

Appendix IV: Preparation of Standard Solutions

STANDARD NaOH SOLUTION (0.1 M/0.1 N)

Reagents

1. Approx 0.1 N NaOH solution: Prepare a saturated solution of sodium hydroxide as follows. Add in small portions about 100 g of AR Grade NaOH to 100 mL of water in a flask, stopper and allow to stand for a few days or until the Na_2CO_3 sttles to the bottom. The clear solution which is carbonate-free contains about 50 g of NaOH/100 mL (approx 50% solution or 12.5 M). From this solution, pipette out 5.3 mL of clear supernatant liquid and transfer to a 1.0 litre rubber stoppered bottle and dilute up to the mark.
2. Standard HCl solution (0.1 N): Prepare and standardize as described before.
3. Phenolphthalein solution: Dissolve 0.5 g of phenolphthalein in 100 mL of 95% alcohol.

Method

Using a pipette, transfer 20 mL of standard HCl solution into 150 mL beaker. Add 25 mL of water and one drop of phenolphthalein indicator. Titrate with the sodium hydroxide (approx 0.1 N) until a faint pink color is obtained. Repeat titration 2–3 times. Calculation the exact normality of NaOH using the relationship $N_1 V_1 = N_2 V_2$.

STANDARD SODIUM THIOSULPHATE SOLUTION (0.1 N)

Reagents

1. Approx 0.1 N $Na_2S_2O_3.5H_2O$ solution: Dissolve 24.82 g of sodium thiosulphate in freshly boiled and cooled water and dilute to 1.0 L.
2. 0.1 N potassium dichromate: Dissolve 4.903 g of potassium dichromate in water and dilute to 1.0 L.
3. Starch indicator: To 1.0 g of starch add 10 mL of water and pour the suspension into 90 ml of boiling water. boil for 1–2 min and cool.
4. 15% Kl solution: Dissolve 15 g of Kl in water and dilute to 100 mL.
5. HCl solution.

Method

Place in glass-stopped flask 20 mL of 0.1 N potassium and 10 mL of 15% Kl solution. Add 5 mL of HCl. Dilute with 100 mL of freshly boiled and cooled water and titrate against the 0.1N $Na_2S_2O_3.5H_2O$ until the yellow color has almost disappeared. Add a few drops of starch indicator and continue to add NaS_2O_3 until the blue color just disappears. Calculate the exact normality of $Na_2S_2O_3$ using the relationship $N_1 V_1 = N_2 V_2$.

STANDARD H_2SO_4 SOLUTION (0.05M/0.1N)

REAGENTS

1. Approx 0.05 M/0.1 N H_2SO_4 solution: Dilute 3.0 mL of pure concentrated H_2SO_4 solution (sp. Gr. 1.84, 96%) to 1.0 L with water in a volumetric flask and mix thoroughly.
2. Anhydrous Na_2CO_3 solution (0.1 N): Prepare as described under standard HCl.
3. Methyl orange indicator: Prepare as described under standard HCl.

METHOD

Fill a clean, dry burette with H_2SO_4 solution (0.05 M/0.1 N) and titrate against a known volume (10 mL) of anhydrous Na_2CO_3 solution using 2 drops of methyl orange as an indicator until an orange or faint pink color end point is obtained. Repeat titration 2–3 times.

CALCULATION

Calculation the exact normality of H_2SO_4 using the relationship, $N_1V_1 = N_2V_2$ as described before.

STANDARD HCl SOLUTION (0.1 M OR 0.1 N)

REAGENTS

1. Approx 0.1 M/0.1 N HCl solution: Dilute 9.0 mL (sp. Gr. 1.18, 35%) or 8.5 mL (sp. gr. 1.19, 37%) of pure concentrated HCl to 1.0 litre with distilled water in a volumetric flask. Invert several times and transfer to a clean, dry bottle.
2. Anhydrous Na_2CO_3 solution (0.1 N): Dry the pure anhydrous Na_2CO_3 on a watch glass in the oven at 110°C for an quantitatively to a beaker and dissolve in about 20 mL of water. Transfer to 100 mL volumetric flask with washings, up to the mark and mix.
3. Methyl orange indicator solution: Dissolve 0.5 g of methyl orange in 1.0 L of water.

METHOD

Fill a clean, dry burette with HCl solution (approx 0.1 M/0.1 N) and titrate against a known volume (10 mL) of sodium carbonate (0.1 N) using 2 drops of methyl orange indicator until an orange or faint pink color end point is obtained. Note the titre value and repeat titration 2–3 times.

Calculation

Using $N_1V_1 = N_2V_2$ relationship, the exact normality of HCl can be calculated.
Where,

N_1—Normality of HCl to be calculated
N_2—Normality of $Na_2CO_3 = 0.1$ N
V_1—Volume of HCl (titer value)
V_2—Volume of Na_2CO_3 (10 mL)

Note

Methyl red may also be used in place of methyl orange as an indicator provided the CO_2 in solution is expelled by boiling before the end point is reached.

STANDARD HYDROGEN PEROXIDE SOLUTION (0.1 N)

Reagents

1. 0.1 N H_2O_2: Take about 7.0 g of chemically pure H_2O_2 solution (ca. 7.0 P.C.) in a 500 mL measuring flask and dilute to the mark with water. Shake, keep in a dark, cool place.
2. Standard $Na_2S_2O_3$ solution (0.1 N): Prepare and standardize as described before.
3. Standard potassium permanganate solution: Prepare and standardize as described before.

Method

1. Standardization with thiosulphate solution
   Take 25 mL of H_2O_2 in a 250 mL flask, treat with 10 mL of 4 N H_2SO_4, 1.0 g of Kl and 4 drops of 3% ammonium molybdate to catalyze the reaction. Titrate against standard sodium thiosulphate using starch as an indicator until the blue color just disappears.
2. Standardization with potassium permanganate
   Take 25 mL of H_2O_2 in a 250 mL flask, titrate with 0.1 N permanganate solution in presence of 10 mL of dilute H_2SO_4 (1:5) until a light permanent pink color is obtained.

STANDARD POTASSIUM PERMANGANATE SOLUTION (0.1 N)

Reagents

1. Potassium permanganate solution (0.1 N approximate): Dissolve about 3.2 g of pure crystals of potassium permanganate in 200 mL of hot water and allow to stand for 60 min at about 100°C. The mouth of the container is well closed with a glass-stopper to disallow any addition of dust particles

into the permanganate solution. The solution is then filtered after 2 days standing through either a sintered filter or a filter of glass wool but never filter through a filter paper because the latter reduces the strength of permanganate considerably. The solution is kept in dark because of its sensitivity to light.
2. Oxalic acid ($H_2C_2O_4.2H_2O$) or sodium oxalate $(COONa)_2$ (0.1 N): Dissolve 6.3 g of pure crystals of oxalic acid in 200 mL of water and transfer to a litre flask. Make up with water or slowly heat 6.7 g of pure sodium oxalate salt in a platinum crucible until the flame of carbon monoxide formed disappears. Dissolve the contents of crucible with water and make up to 1.0 L.
3. 4 N H_2SO_4.

METHOD

Dilute 10 mL of standard oxalic acid or oxalate solution to 20 mL water and treat with 15 mL of 4 N H_2SO_4. Heat the solution to 70–80°C and titrate against permanganate solution until the solution becomes slightly pink. The color is permanent. Calculate the normality of permanganate solution using the relationship $N_1V_1 = N_2V_2$.

STANDARD IODINE SOLUTION (0.1 N)

REAGENTS

1. Iodine solution (0.1 N): Dissolve 12.7 g of iodine (A.R.) and 40 g of iodate-free potassium iodide in 50 mL of water. Stir the mixture until the whole of iodine dissolves completely. Transfer it to a liter flask and make up to the mark with water. Keep in a dark and cool place.
2. Standard arsenious oxide solution (0.1 N): Dry about 8.0 g of arsenious oxide (As_2O_3) for 2 h at 105–110°C in a platinum crucible and cool in a desiccator. Weigh accurately 4.945 g of dried sample and dissolve in 10 mL of 4 N NaOH solution. The solution may be heated if required. Transfer to a litre flask and treat with 4 drops of phenolphthalein. The alkalinity of medium is neutralized by carefully adding just enough dilute HCl (disappearance of pink color). Then mix the solution with another 50 mL of dil. HCl and make up the volume to the liter mark.
3. Starch solution: Dissolve 1.0 g of starch and 5 g of $HgCl_2$ in 1.0 mL of water and add the suspension to 500 mL of boiling water. Boil until the clear solution is obtained and store in a glass-stoppered bottle.

METHOD

Take 25 mL of standard arsenious oxide solution in 250 mL of flask and mix with 50 mL of water, 1.0 g of sodium bicarbonate and 1.0 mL of starch solution. Titrate against prepared I_2 solution until blue color appears. Calculate the normality of I_2 using the relationship $N_1V_1 = N_2V_2$.

Note

Best results are obtained if the starch solution is added just before the end point is reached.

STANDARDIZATION OF THE FEHLING'S SOLUTION

Reagents

1. Fehling's solution A (copper sulphate solution): Dissolve 34.639 g of $CuSO_4.5H_2O$ (pure recrystallized) in water, dilute to 500 mL and filter.
2. Fehling's solution B (alkaline tartrate solution): Dissolve 173 g of Rochelle salt (sodium potassium tartrate) and 50 g of NaOH in water, dilute to 500 mL. Allow to stand for 2 days and filter. Mix equal volumes (5 mL each) of both immediately before use.
3. Standard glucose solution (0.5%): Dissolve accurately 0.5 g of pure anhydrous glucose powder in water and dilute to 100 mL with water.
4. 1.0% Methylene blue: Dissolve 1.0 g in 100 mL 95% alcohol.

Method

Into a conical flask, pipette out 5.0 mL of each of Fehling's solutions A and B. Dilute with about 10 mL of water and add 3 glass beads to it. Keep the flask over a wire gauze or hot plate and bring to boil. Add from the burelle the standard glucose solution slowly, 1 mL at a time till the color just turns from blue to red. Then add 3–4 drops of 1.0% methylene blue indicator which gives a blue color to the solution and continue to add the glucose solution drop-wise till a brick red color end point of solution is obtained.

Again in another conical flask, take 5.0 mL of each of the Fehling's solutions A and B, dilute with about 10 mL of water and add 3 glass beads into it. Boil the contents as above and add from the burette the standard glucose solution 2.0 mL less than the quantity required in the first titration. Then add 3–4 drops of methylene blue indicator and continue the titration till the brick red colored end point is obtained. Repeat the whole process of titration 3–4 times and calculate the mean of the last 3 readings (excluding the first reading). From the mean titration reading, calculate the amount of anhydrous glucose corresponding to 10 mL of the mixed Fehling's solution A and B.

Note

The titration should be completed within total boiling time of about 3 min. Note down the first burette reading.

Calculation

Let 'a' mL be the mean titration reading.

10 mL of Fehling's A & B together = 'a' mL of std. glucose solution.

Now, 100 mL of std. glucose soln. = 0.5 g of glucose

$\therefore$ 'a' mL of std. glucose soln. $= 0.5 \times \frac{a}{100}$ g of glucose

$\therefore$ 10 mL of Fehling's solutions $= 0.5 \times \frac{a}{100}$ g of glucose

A and B contain

STANDARD SODIUM NITRITE SOLUTION (0.1N)

METHOD

50 mL of 0.1 N permanganate solution and 5 mL of dilute H_2SO_4 (1:5) are mixed together and treated with 25 mL of approx 0.1 N $NaNO_2$ solution. The mouth of flask is closed with a glass stopper and allowed to stand for 15 min. Then the excess of permanganate is titrated back by procedure described under preparation of standard potassium permanganate solution or iodometrically after adding 2.0 g of Kl and then titrating with standard thiosulfate solution.

CALCULATION

Note

Volume of 0.1 N $KMnO_4$ consumed by nitrite
= (50-volume of 0.1 N standard solution used for back titration)
= actual volume of permanganate = V_1
Calculate normality of $NaNO_2$ using the relationship $N_1V_1 = N_2V_2$

Appendix V: Common Stock Solutions

10 M Ammonium Acetate

To prepare a 10 M solution in 100 mL, dissolve 77 g of ammonium acetate in 70 mL of H_2O at room temperature. To prepare a 5 M solution in 100 mL, dissolve 38.5 g in 70 mL of H_2O. Adjust the volume to 100 mL with H_2O. Sterilize the solution by passing it through a 0.22 μm filter. Store the solution in tightly sealed bottles at 4°C or at room temperature. Ammonium acetate decomposes in hot H_2O and solutions containing it should not be autoclaved.

Ampicillin

Prepare a stock of 100 mg/mL in water. Sterilize by filtration. Store at −20°C but avoid repeated freeze/thaw cycles. Use at a final concentration of 100 μg/mL.

Cresol Red Loading Dye—2.5X—for PCR Reactions

1 M sucrose, 0.02% cresol. Prepare 1% cresol red in water (0.5 g/50 mL). To prepare loading dye, dissolve 17 g sucrose in a total volume of 49 mL of water. Add 1 mL 1% cresol red.

EB buffer (Qiagen Recipe)

10 mM Tris-Cl pH 8.3

EDTA Stock

To prepare 1 liter, 0.5 M EDTA pH 8.0: Add 186.1 g of disodium EDTA-2 H_2O to 800 mL of H_2O. Stir vigorously on a magnetic stirrer. Adjust the pH to 8.0 with NaOH (approx. 20 g of NaOH pellets). Dispense into aliquots and sterilize by autoclaving. The disodium salt of EDTA will not go into solution until the pH of the solution is adjusted to approx. 8.0 by the addition of NaOH. For tetrasodium EDTA, use 226.1 g of EDTA and adjust pH with HCl.

6x Gel Loading Buffer

0.25% Bromophenol blue
0.25%Xylene cyanol FF
15% Ficoll Type 4000
120 mM EDTA

IPTG

IPTG is isopropylthio-b-D-galactoside. Make a 20% (w/v, 0.8 M) solution of IPTG by dissolving 2 g of IPTG in 8 mL of distilled H_2O. Adjust the volume of the solution to 10 mL with H_2O and sterilize by passing it through a 0.22 μm disposable filter. Dispense the solution into 1 mL aliquots and store them at –20°C.

LB Medium

To make 1 liter, use 10 g tryptone, 5 g yeast extract, 10 g NaCl. Adjust pH to 7.0. Sterilize by autoclaving.

LB Agar

Dispense 15 g per liter of agar directly into final vessel. Prepare LB medium as above and add to agar. NOTE: Agar will not go into solution until it is autoclaved (or boiled). If adding antibiotics, autoclave medium first and allow to cool until warm to the touch, then add the antibiotic. Dispense about 30 mL per plate. Allow plates to dry either at 37°C overnight or 20 minutes in a laminar flow hood (lids removed). Store in original Petri plate bags, inverted, at 4°C for up to 2 weeks.

NaCl

To prepare 1 liter of a 5 M solution: Dissolve 292 g of NaCl in 800 mL of H_2O. Adjust the volume to 1 liter with H_2O. Dispense into aliquots and sterilize by autoclaving. Store the NaCl solution at room temperature.

NaOH

The preparation of 10 N NaOH involves a highly exothermic reaction, which can cause breakage of glass containers. Prepare this solution with extreme care in plastic beakers. To 800 mL of H_2O, slowly add 400 g of NaOH pellets, stirring continuously. As an added precaution, place the beaker on ice. When the pellets have dissolved completely, adjust the volume to 1 liter with H_2O. Store the solution in a plastic container at room temperature. Sterilization is not necessary.

20X SB (Electrophoresis Buffer)

(Buffer diluted to 1X should be 10 mM Sodium hydroxide and pH 8.5) for 1 liter, weigh out 8 g NaOH and ~40 g boric acid—add water, dissolve and add additional boric acid until pH = 8.0; bring final volume to 1 liter.

SDS Stock

10% or 20% (w/v) SDS. Also called sodium lauryl (or dodecyl) sulfate. To prepare a 20% (w/v) solution, dissolve 200 g of electrophoresis-grade SDS in 900 mL of H_2O.

Heat to 68°C and stir with a magnetic stirrer to assist dissolution. If necessary, adjust the pH to 7.2 by adding a few drops of concentrated HCl. Adjust the volume to 1 liter with H_2O. Store at room temperature. Sterilization is not necessary. Do not autoclave.

Use a mask when weighing this out.

20x SSC:

0.3 M Na(3) citrate

3 M NaCl

SOB Medium: per liter:

Bacto-tryptone 20 g

Yeast extract 5 g

NaCl 0.584 g

KCl 0.186 g Mix components and adjust pH to 7.0 with NaOH and autoclave.

2 M Mg^{++} stock:

MgCl 2-6H2O 20.33 g

MgSO 4 -7H2O 24.65 g

Distilled water to 100 mL. Autoclave or filter sterilize.

2 M Glucose:

Glucose 36.04 g

Distilled water to 100 mL. Filter sterilize.

For SOB Medium + magnesium: Add 1 mL of 2 M Mg ++ stock to 99 mL SOB Medium.

For SOCMedium: Add 1 mL of 2 M Mg ++ stock and 1 mL of 2 M Glucose to 98 mL of SOB Medium.

3 M Sodium Acetate—pH 5.2

To prepare a 3 M solution: Dissolve 408.3 g of sodium acetate-$3H_2O$ in 800 mL of H_2O. Adjust the pH to 5.2 with glacial acetic acid. Adjust the volume to 1 liter with H_2O. Dispense into aliquots and sterilize by autoclaving.

Southern Solutions:

Depurination solution (for Southern blotting)

250 mM HCl

Denaturation solution (for Southern blotting)

1.5 M NaCl

0.5 M NaOH

Neutralization solution (for Southern blotting)

1.5 M NaCl

0.5 M Tris-HCl, pH adjusted to 7.5

50x TAE

Prepare a 50x stock solution in 1 liter of H_2O:

242 g of Tris base

57.1 mL of glacial acetic acid

100 mL of 0.5 M EDTA (pH 8.0)

The 1x working solution is 40 mM Tris-acetate/1 mM EDTA.

5X (or 10X) TBE

Prepare a 5x stock solution in 1 liter of H_2O:

54 g of Tris base

27.5 g of boric acid

20 mL of 0.5 M EDTA (pH 8.0)

The pH of the concentrated stock buffer should be approx. 8.3. Some investigators prefer to use more concentrated stock solutions of TBE (10x as opposed to 5x). However, 5x stock solutions is more stable because the solutes do not precipitate during storage. Passing the 5x or 10x buffer stocks through a 0.22 μm filter can prevent or delay formation of precipitates.

TE buffer:

10 mM Tris-Cl (pH, usually 7.6 or 8.0)

1 mM EDTA (pH 8.0)

Use concentrated stock solutions to prepare. If sterile water and sterile stocks are used, there is no need to autoclave. Otherwise, sterilize solutions by autoclaving for 20 minutes. Store the buffer at room temperature.

1M TRIS-CL: USED AT VARIOUS pHs

Using Tris base: To make 1 liter, dissolve 121 g Tris Base in 800 mL of water. Adjust pH to the desired value by adding approximately the following:

pH = 7.4 about 70 mL of concentrated HCl

pH = 7.6 about 60 mL of concentrated HCl

pH = 8.0 about 42 mL of concentrated HCl

Make sure solution is at room temperature before making final pH adjustments. Bring final volume to 1 liter. Sterilize by autoclaving.

Using Trizma tables: An alternate procedure for preparing Tris solutions is to combine the proper amount of Tris Base and Tris Hydrochloride to achieve the desired value using Sigma's Tris tables.

WB (10% redistilled glycerol, 90% distilled water, v/v)

In a 1-liter graduated cylinder, add 100 mL of glycerol and 900 mL of distilled water. Cover with parafilm and mix thoroughly. Sterilized by autoclaving, and chill to 4°C.

Western Blotting Solutions:

1X Transfer buffer 1: 25 mM Tris, 192 mM Glycine, pH 8.3

Mix 3.03 g Tris, 14.4 g glycine; add dd water to 1 liter—do not adjust pH.

1X Transfer buffer 2: 25 mM Tris, 192 mM Glycine, pH 8.3, 20% methanol

Mix 3.03 g Tris, 14.4 g glycine; add 200 mL methanol; add dd water to 1 liter—do not adjust pH. (NOTE: methanol is not needed for PVDF membranes)

10X Western Buffer: 200 mM Tris pH = 7.5; 1.5 M NaCl

To prepare 1X Western Buffer, dilute 10X buffer to 1X, adding Tween-20 to 0.1%. Remove 50 mL and set aside for the last two washes. To the remainder, add I-Block to 0.2%, heating gently with constant stirring until dissolved. Bring to room temperature before using.

X-gal 5-bromo-4-chloro-3-indolyl-b-D-galactoside (same recipe for X-phosphate)

Make a 2% (w/v) stock solution by dissolving X-gal in dimethylformamide at a concentration of 20 mg/mL solution. Use a glass or polypropylene tube. Wrap the tube containing the solution in aluminum foil to prevent damage by light and store at −20°C. It is not necessary to sterilize X-gal solutions.

Solution and stuff

A. Solutions

1. QB

Stock	For 100 mL	Final []
2 M KPO_4 (PH 7.8)	5 mL	100 mM
0.5 M EDTA	200 μL	mM
Triton X-100	1 mL	1%
80% Glycerol	12.5 mL	10%
dH_2O	81.1 mL	
Store RT		
DTT (1.0 M)	100 μL	1 mM
(Alternatively directly add 15.4 mg DTT per 100 mL)		

2. 2 M KPO_4 (pH 7.8)

Stock	For 200 mL	Final []
K_2HPO_4	63.2 g	
KH_2PO_4	5.0 g	
pH should be		~7.8
If not, adjust pH to 7.8		
F, a/c		

3. 0.5 M EDTA (pH 8.0)

Stock	For 250 mL	Final []
EDTA	46.52 g	0.5 M
H_2O to	250 mL	
pH w/ 10N NaOH to 8.0		
(Alternatively use ~5 pellets of NaOH.)		
f, a/c.		
Store RT.		

Note: EDTA will not completely go into solution until the pH approaches 8.0 and the H_2O is almost at final volume. Essentially, the pH needs to be continuously adjusted as the EDTA dissolves.

4. 10 N NaOH

Stock	For 250 mL	Final
NaOH	100 g	10 N
dH_2O to 250 mL		
Store at RT in a PLASTIC bottle. (NaOH will react with glass.)		

5. 80% Glycerol

Stock	For 100 mL	Final
100% Glycerol	80 mL	80%
dH_2O	20 mL	
a/c		

6. 1 M DTT

Stock	For 10 mL	Final
DTT	1.545 g	1 M
0.01 M NaOAc to 10 mL		
(pH 5.2)		
Filter sterilize		
Aliquot into portions		
Store at –80°C		
0.01 M NaOAc is 33 μL of 3 M NaOAc pH~5.2 in 9.67 mL dH_2O.)		
Solutions for Yeast DNA Extraction		

- 40 mM EDTA/90 mM 2-Mercaptoethanol (2-ME)

40.0 mL	0.5 M EDTA, pH8.0
3.8 mL	2-Mercaptoethanol (12 M stock)
456.2 mL	sterile ddH_2O
500 mL	Store at room temperature

- SCE

		Final Concentration
2 M Sorbitol	50 mL	1.0 M
1 M Sodium citrate	10 mL	0.1 M
0.25 M EDTA, pH7.0	24 mL	60 mM
Sterile ddH_2O	16 mL	
	100 mL	

- Filter sterilize and store at room temperature.
- SCE/2-ME/Lyticase (2 mL): 2 mL SCE + 16 μL 2-ME + 0.2 mg Lyticase

- Miniprep Yeast Lysis Buffer

Miniprep Lysis Buffer		**Final Concentration**
1 M Tris.HCl, pH 7.4	5 mL	50 mM
0.5 M EDTA, pH8.0	5 mL	25 mM
5 M NaCl	10 mL	500 mM
1 M MgCl2	300 μL	3 mM
12 M 2-Mercaptoethanol	25 μL	3 mM
Nonidet P-40	100 μL	0.1%
10% SDS	10 mL	1%
bring volume to	100 mL with sterile ddH_2O.	

- Filter sterilizes and store at room temperature.
  York Modified Glycerol
    - 20 mM Na-phosphate buffer, pH 7.0
    - 50% glycerol
    - 0.5% Triton X-100
    - 2 mM $MgCl_2$
    - 1 mM NaN_3
    - 1 mM DTT

  Rigor Buffer
    - 10 mM Na-phosphate buffer, pH 7.0
    - 100 mM NaCl
    - 2 mM $MgCl_2$
    - 2 mM EGTA
    - 0.1 mg/ml soybean trypsin inhibitor
    - 1 mM DTT
    - with and without 0.5% Triton X-100

  TM buffer
    - 50 mM Tris-Cl, pH 7.5
    - 10 mM $MgSO_4$

  STET
    - 8% sucrose
    - 5% Triton X-100
    - 50 mM Tris-HCl, pH 8.0
    - 50 mM EDTA, pH 8.0
    - store at 4°C

  Lysozyme Stock Solution (10 mg/mL)
    - 150 mg lysozyme
    - 3.75 mls 1M Tris-HCl, pH 8.0
    - 11.25 mls H2O
    - store at –20°C.

Appendix VI: Abbreviations/ Acronyms

Amp	ampicillin	ng	nanogram(s) = 10^{-9} gram
APS	ammonium persulfate	nm	nanometer(s) = 10^{-9} meter
BME	β-mercaptoethanol	OD	optical density
BPB	bromophenol blue	ODx	optical density at x nm
BSA	bovine serum albumine	PCR	polymerase chain reaction
	CSPD, Disodium 3-(4 methoxyspiro {1,2- dioxetane-3, 2′(5′-chloro) tricycle [3.3.1.13,7]decan}-4-yl) phenyl phosphate		
CTAB	mixed alkyltrimethyl-ammonium	RFLPs	restriction fragment length polymorphisms
dATP	deoxyadenosine 5′- triphosphate	RNA	ribonucleic acid
dCTP	deoxycytidine 5′-triphosphate	RT	room temperature
ddH_2O	double-distilled water	RXN	reaction(s)
dGTP	deoxyguanosine 5′-triphosphate	SDS	sodium dodecyl sulphate
dH_2O	distilled water	s	second(s)
Dig	digoxigenin	SGB	sample gel buffer
Dig-dUTP	digoxigenin-11-dUTP	SS DNA	salmon sperm DNA
DNA	deoxyribose nucleic acid	SSC	saline sodium citrate
dNTPs	deoxynucleoside 5′triphosphates	STE	sodium Tris-EDTA (also TEN)
DTT	dithiothreitol	TAE	Tris-acetate EDTA (buffer)
dUTP	deoxyuridine 5′-triphosphate	TBE	Tris-borate EDTA
EDTA	ethylenediaminetetraacetate	TE	Tris-EDTA (buffer)
EtBr	ethidium bromide	TEMED	N,N,N′,N′-Tetramethylethylenediamine
EtOH	ethanol	TNE	Tris Sodium (Na) EDTA (buffer)
g	gram(s)	Tris	Tris(hydroxymethyl)amino-methane
h	hour(s)	TTE	Triton Tris-EDTA (buffer)
kb	Kilobases	TTP	thymidine 5′-triphosphate
KOAc	potassium acetate	U	unit(s) of enzyme
LMP	low melting point	UV	ultraviolet
mA	milli Amperes	V	volts
min	minute(s)	XC	xylene cyanole
mL	milliliter(s)	°C	degree Celsius
MSI	Micron Separations Inc.	µg	microgram(s) = 10^{-6} gram
MW	molecular weight	µL	microliter(s) = 10^{-6} liter
NaOAc	sodium acetate		

Appendix VII: Molecular Weights of Selected Laboratory Chemicals

Acrylamide	71.08	Phenol	94.11
Ammonium Acetate	77.08	Phenylmethylsulfonyl fluoride	174.2
Ammonium Bicarbonate	79.06	Polyethylene glycol	c. 8000
Ammonium Chloride	53.49	Polyvinylpyrrolidone	40,000
Ammonium Persulfate	228.2	Potassium Acetate	98.14
Ammonium Sulfate	132.14	Potassium Chloride	74.55
Ampicillin	371.4	Potassium Phosphate, anhydrous, monobasic	136.1
ATP, disodium salt	605.19	Potassium Phosphate, trihydrate, dibasic	228.2
dATP, disodium salt	535.1	Sodium Bicarbonate	84.01
Bis	154.17	Sodium Azide	65.01
Boric Acid	61.83	Sodium Bicarbonate	84.01
Calcium Chloride	147.0	Sodium chloride	58.45
Cesium Chloride	168.37	Sodium hydroxide	40.0
Chloramphenicol	323.1	Sodium iodide	149.89
Citric Acid	294.1	Sodium Dodecyl Sulfate	288.38
dCTP, sodium salt	467.2	Sodium Phosphate, monohydrate	137.99
Dithiothreitol	154.24	Sodium Phosphate, dibasic	142.0
EDTA, disodium salt	372.24	Spermidine, trihydrochloride	254.6
Ficoll type 400	400,000	Spermine, tetrachloride	348.2
Glucose	180.2	Sucrose	342.3
Glutamic acid	147.1	TEMED	116.2
dGTP, sodium salt	507.2	Thiamine	337.3
Glycine	75.07	Thymidine	242.2
Hydroxyquinoline	145.2	Tris	121.1
IPTG	238.31	Tris-HCl	157.6
2-mercaptoethanol	14.26 M	dTTP, sodium salt	482.2
Magnesium Chloride, hexahydrate	203.30	Urea	60.06
Magnesium Sulfate, anhydrous	120.39	X-gal	408.64

Appendix VIII: SI Units of Mass

The SI unit of mass (weight) is given below:

Weight	Symbol	Multiple of Gram	Lower Equivalent
megagram	M	10^6	1000 kilogram
kilogram	K	10^3	1000 gram
gram	g	1	1000 milligram
milligram	m	10^{-3}	1000 microgram
microgram	μ	10^{-6}	1000 nanogram
nanogram	n	10	1000 picogram
picogram	p	10	1000 femtogram

Appendix IX: Constants of Acids and Bases

			Commercial Concentrated Reagent		
Acid or Base	**Formula**	**Molecular wt.**	**Specific Gravity**	**% by Weight**	**Molarity (M)**
Acetic acid	CH_3COOH	60.1	1.05	99.5	17.4
Ammonium Hydroxide	NH_4OH	35.0	0.89	28	14.8
Formic acid	$HCOOH$	46.0	1.20	90	23.4
Hydrochloric acid	HCl	36.5	1.18	36	11.6
Nitric acid	HNO_3	63.0	1.42	71	16.0
Perchloric acid	$HClO_4$	100.5	1.67	70	11.6
Phosphoric acid	H_3PO_4	80.0	1.70	85	18.1
Sulphuric acid	H_2SO_4	98.1	1.84	96	18.0

Appendix X: Densities of Liquids

If the density of the solution is known, it is possible to calculate the water volume (weight) that is necessary for its preparation. It is very convenient, because otherwise phrase "add water to the final volume…" forces you to use the graduated cylinder for the preparation of solution (calibration of all other laboratory glassware is only approximate).

A list of densities of some chemical substances is given below.

Substance	Density [g/mL]	Substance	Density [g/mL]
Acetic acid glacial	1.049	Isopropanol	0.785
Acetic acid 36%	1.049	KCl, 0.25 M	1.044
Acetone	0.788	KOH, 5.0 M	1.217
AcOK, 5.0 M	1.218	LB	1.024
1.0 M	1.046		
AcONa, 3.0 M	1.121	LiCl, 9 M	1.194
2.5 M	1.147	8 M	1.172
$AcONH_4$, 10 M	1.073	Mercaptoethanol (ß)	1.114
Acrylamid:BIS, 40%	1.044	$MgCl_2$, 1 M	1.073
Ammonium hydroxide, NH_4OH, 28%	0.89	$MgSO_4$, 1 M	1.112
Butanol (n)	0.81	$MnCl_2$, 1 M	1.099
$CaCl_2$, 1 M	1.086	Nitric acid, $HNO_3$71%	1.42
Chloroform	1.48	NaCl, 5 M	1.186
Denhardt sol., 50X	1.004	NaOH, 50%	1.53
DMSO	1.100	NaOH, 10 M	1.329
EDTA, 0.5 M	1.096	$NaxHPO_4$, 1 M, pH~7	1.104
Ethanol, anhydr	0.79	Nonident P-40	1.064
96%	0.80		
70%	0.868		
Formaldehyde, 37%	1.09	Octane	0.71
Formamid	1.133	PEG 6000, 20%	1.041
		40%	1.075
Glucose, 10%(w/v)	1.040	PEG/TAE, (15%/1X)	1.032
2 M	1.13		
Glycerol	1.25	Phosphoric acid H_3PO_4, 85%	1.70
HCl, 37% (12.07 M)	1.19	Phenol	1.054
Isoamyl alcohol	0.813	Sulfuric acid H_2SO_4, 96%	1.84
Isobutanol	0.871	Sarcosyl, 10% (w/v)	1.017
		30% (w/v)	1.027

(Continued)

Substance	Density [g/mL]	Substance	Density [g/mL]
SDS, 10%	1.018	Tris.Cl, 1 M, pH 7-8	1.041
		2M, pH 7-9	1.10
SSC, 20X	1.163	TritonX-100, 10%	1.016
30X	1.245		
SSPE, 20X	1.136	TritonX-100, 100%	1.06
TAE, 50X	1.088	Tween 20, 100%	1.108 (1.101)
TBE, 10X	1.070	$ZnCl_2$, 1 M	1.111

Appendix XI: Weights and Measures

Decimal fractions

Fraction	Symbol	Prefix	Abbreviation
Thousandth	10^{-3}	Milli	M
Millionth	10^{-6}	Micro	μ
Billionth	10^{-6}	Millimicro (or nano)	mμ
Trillionth	10^{-12}	Micromicro (or pico)	μμ

Molar concentration

Mole Fraction	Name	Abbreviation	Amount/L	Amount/ml
	Molar	M	1 mole/L	1 mmole/ml
10^{-3} M	Millimolar	mM	1 millimole/L	1 μmole/ml
10^{-6} M	Micromolar	μM	1 μmole/L	1 mμmole/ml
10^{-9} M	Nanomolar	nM	1 mμmole/L	1 μμmole/ml

Weights

Name	Abbreviation	Fraction of Gram (g)
Gram	g	
Milligram	mg	10^{-3}
Microgram (gamma)	μg	10^{-6}
Nanogram	ng	10^{-9}
Picogram	pg	10^{-12}

Volume

Name	Abbreviation	Fraction of Litre
Liter	L	
Milliliter	mL	10^{-3}
Microliter	μL	10^{-6}
(lambda)	(λ)	

Length

Name	Abbreviation	Fraction of a Meter
Centimeter	cm	10^{-2}
Millimeter	mm	10^{-3}
Micron	μ	10^{-6}
Millimicron	mμ	10^{-9}
Angstrom unit	°A	10^{-10}

Temperature

Temp. in °F = (Temp. in °C × 1.8) + 32
Temp. in °C = (Temp. in °F–32) × 5/9
Temp. in K = (Temp. in °F + 459.67) × 5/9

Appendix XII: pK Values of Acids/Bases

Acids/Bases	pK Values
Acetic acid	4.76
Ammonia	4.74
Barbitone	7.98
Barbituric acid	4.04
Boric acid	9.23
Carbonic acid	6.37 [pK_1]
	10.25 [pK_2]
Phosphoric acid	2.12 [pK_1]
	7.21 [pK_2]
	12.32 [pK_3]
Pyridine	8.64
Sulphuric acid	0.4 [pK_1]
	1.92 [pK_2]

Appendix XIII: Preparation of Indicators

1. 0.04% Bromophenol blue: Grind 0.1 g of dry bromophenol blue powder in a mortar with 14.9 mL of 0.01 M NaOH. Dilute to 250 mL with water.
2. 0.04% Bromothymol blue: Grind 0.1 g of bromothymol blue dry powder in a mortar with 16 mL NaOH. Dilute to 250 mL with water.
3. 0.5% Methyl red: Dissolve 0.5 g of methyl red in 100 mL of 95% ethyl alcohol.
4. Methylene blue-methyl red mixed indicator: This solution consists of 1 part of 0.2% methyl red in alcohol and 1 part of 0.1% methylene blue in alcohol. The pH is at 5.4. The color is green in alkaline (pH 5.6), colorless at pH 5.4 (end point) and reddish pink in acid solution (pH 5.2).
5. 0.02% Methylene blue: Dissolve 0.2 g of "certified" methylene blue in 1L of water.
6. 0.04% Phenol red: Grind 0.1 g of dry phenol red powder in a mortar with 28.2 mL of 0.01 M NaOH. Dilute to 250 mL with water.
7. 0.5% Phenolphthalein solution: Dissolve 0.5 g phenolphthalein in 100 mL of 95% alcohol.
8. 1% Starch: Mix 1 g of soluble starch with a little water, stir and add to 100 mL of boiling water.
9. 0.1% Bromocresol green: Dissolve 20 mg in 20 mL water.
10. 0.5% Congo red: Dissolve 5 g of congo red in 900ml of water and dilute to a litre with 95% alcohol.
11. Methyl orange: Dissolve 0.5 g of methyl orange in 1.0 L of water.
12. Litmus: Dissolve 1.0 g of azolitmin in 1.0 L of water.
13. Potassium chromate: Dissolve 5.0 g of potassium chromate in 100 mL of water.

Appendix XIV: pH Ranges of Indicators

Indicator	pH Range	Color Change
Acid cresol red	0.2–1.8	Red-yellow
Acid metacresol purple	1.2–2.8	Red-yellow
Thymol blue	1.2–2.8	Red-yellow
Benzo yellow	2.4–4.0	Red-yellow
Bromocresol green	3.8–5.4	Yellow-blue
Bromophenol blue	3.0–4.6	Yellow-blue
Methyl red	4.4–6.0	Red-yellow
Chlorophenol red	5.2–6.8	Yellow-red
Bromocresol purple	5.2–6.8	Yellow-purple
Bromothymol blue	6.0–7.6	Yellow-blue
Phenol red	6.8–8.4	Yellow-red
Cresol red	7.2–8.8	Yellow-red
Meta cresol purple	7.6–9.2	Yellow-purple
Thymol blue	8.0–9.6	Yellow-blue
Phthalein red	8.6–10.2	Yellow-red
Tolyl red	10.0–11.6	Red-yellow
Parazo orange	11.0–12.6	Yellow-orange
Acyl blue	12.0–13.6	Red-blue

Appendix XV: Use of Equipment

GENERAL COMMENTS

It is to everyone's advantage to keep the equipment in good working condition. As a rule of thumb, don't use anything unless you have been instructed in the proper use. This is true not only for equipment in the lab but also departmental equipment. Report any malfunction immediately. Rinse out all centrifuge rotors after use and in particular if anything spills. Please do not waste supplies - use only what you need. If the supply is running low, please notify either the instructor/lab manager *before* the supply is completely exhausted. Occasionally, it is necessary to borrow a reagent or a piece of equipment *from another lab. Except in an emergency, notify the instructor.*

MICROPIPETTORS

Most of the experiments you will conduct in this laboratory will depend on your ability to accurately measure volumes of solutions using micropipettors. The accuracy of your pipetting can only be as accurate as your pipettor and several steps should be taken to insure that your pipettes are accurate and are maintained in good working order. Each pair of students will be assigned a set of pipettors and upon receipt, they should be labeled with the students' name. They should then be checked for accuracy following the instructions given by the instructor. If they need to be recalibrated, do so. We have two different types of pipettors, Rainin pipetmen, and Oxford benchmates. Since the pipettors will use different pipet tips, make sure that the pipet tip you are using is designed for your pipettor. DO NOT DROP IT ON THE FLOOR. If you suspect that something is wrong with your pipettor, first check the calibration to see if your suspicions were correct, then notify the instructor.

USING A pH METER

Biological functions are very sensitive to changes in pH and hence, buffers are used to stabilize the pH. A pH meter is an instrument that measures the potential difference between a reference electrode and a glass electrode, often combined into one combination electrode. The reference electrode is often AgCl 2. An accurate pH reading depends on standardization, the degree of static charge, and the temperature of the solution.

AUTOCLAVE OPERATING PROCEDURES

Place all material to be autoclaved in an autoclavable tray. All items should have indicator tape. Separate liquids from solids and autoclave separately. Make sure lids on all bottles are loose. Does not crowd large number of items in tray- in order for all items to reach the appropriate temperature, one must allow sufficient air/steam circulation.

1. Make sure chamber pressure is at 0 before opening the door.
2. Place items to be autoclaved in the autoclave and close the door. Some autoclaves require that you also lock the door after it's closed.
3. Set time—typically 20 minutes.
4. Temperature should be set at 121° C already, but double-check and change if necessary.
5. Set cycle: If liquid, set "liquid cycle" or "slow exhaust." If dry, set "dry cycle" or "fast exhaust" + dry time.
6. Start the cycle. On some autoclaves, the cycle starts automatically at step 5. On others, turn to "sterilize."
7. At the end of the cycle, check that: a. the chamber pressure is at 0; b. the temp is <100°C.
8. Open door.
9. Remove contents using gloves and immediately tighten all caps.

OPERATING INSTRUCTIONS FOR SPECTROPHOTOMETER

- To measure the absorbance of a solution in the short-wave range (<300 nM) use the quartz cuvettes. Disposable plastic cuvettes are available for reading in the visible range.
- Turn the spectrophotometer on—the switch is on the right in the back.
- Allow the instrument to calibrate. Do not open the chamber during this time. The deuterium lamp is OFF by default. To read absorbance in the UV range, turn the deuterium lamp on as follows after the machine has completed its calibration: Depress the function key until Fn5 is displayed. Press the mode key until d2on is displayed. Press enter. For best accuracy, the deuterium lamp should be warmed up for 20 minutes.
- Press the function key until Fn0 is displayed. Press enter. Using the up or down arrow keys, enter in the desired wavelength.
- Prepare a reference cuvette containing the same diluent as your sample. Prepare your sample.
- Place the reference cuvette in cell #1 and place your samples in cells #2-6.
- Press the cell key until cell #1 is in position. Press the Set Reference key to blank against the appropriate buffer. Press the cell key to advance to read the next sample.

USE OF BALANCE IN LABORATORY

We prefer to use a balance for preparing:

1. Agarose gels (weighing of agarose and buffer)
2. AA gels (weighing of all components)

3. 1x electrophoresis buffers from the stocks (weighing of water and stock solution)
4. Salt solutions
5. In the cases of viscous liquids, large volumes etc.

It is really convenient if there are 2 balances in the laboratory. One (that is used more often) is with the 0.01g gauge and the range to 1.5–3 kg. It is used for preparation of large volumes of buffers, balancing of centrifuge tubes etc. The other balance has 0.0001 g gauge and the range of 50–100 g. It is used for preparation of small volumes of buffers (it is sufficient if the neighboring laboratory has such a scale).

Appendix XVI: Disposal of Buffers and Chemicals

1. Any uncontaminated, solidified agar or agarose should be discarded in the trash, not in the sink, and the bottles rinsed well.
2. Any media that becomes contaminated should be promptly autoclaved before discarding it. Petri dishes and other biological waste should be discarded in Biohazard containers which will be autoclaved prior to disposal.
3. Organic reagents, for example, phenol, should be used in a fume hood and all organic waste should be disposed of in a labeled container, not in the trash or the sink.
4. Ethidium bromide is a mutagenic substance that should be treated before disposal and should be handled only with gloves. Ethidium bromide should be disposed of in a labeled container. Decontamination of EtBr includes:
   a. When EtBr conc. is >0.5 mg/mL:
      i. Add sufficient water to reduce the concentration of EtBr to <0.5 mg/mL.
      ii. Add 1 volume of 0.5 M KMnO4. Mix, add 1 volume of 2.5 N HCl (210 mL HCl in 1 L water) . Mix and allow the solution to stand for several hr at room temperature.
      iii. Add 1 volume of 2.5 N NaOH (100 g NaOH in 1 L water). Mix, and discard the solution.
   b. When EtBr conc. is <0.5 µg/mL:
      * electrophoresis buffer contains 0.5 µg/mL EtBr
      i. Add 29 g Amberlite XAD-16 for 1 L of solution.
      ii. Store the solution for 12 hr at room temperature, shaking it intermittently.
      iii. Filter the solution through a Whatman No. 1 filter, and discard the filtrate.
      iv. Seal the filter and Amberlite resin in a plastic bag, and dispose of the bag in the hazardous waste.
5. Dirty glassware should be rinsed, all traces of agar or other substance that will not come clean in a dishwasher should be removed, all labels should be removed (if possible), and the glassware should be placed in the dirty dish bin. Bottle caps, stir bars and spatulas should not be placed in the bins but should be washed with hot soapy water, rinsed well with hot water, and rinsed 3 times with distilled water.

Appendix XVII: Equilibration of Phenol

Phenol dissolves and denatures proteins that are tightly associated to DNA. Its equilibration is must for separating DNA from protein.

1. Melt the phenol at 65°C if required.
2. Add 1 g of hydroxyquinoline to 1 L phenol; stir it in the bottle until dissolved.
3. Pour into a 2 L separatory funnel, add 5 mL saturated NaCl, 900 mL 1 M Tris-Base (121 g Tris-Base/1L H_2O, reserve 100 mL), mix and allow to separate.
4. Keep lower (yellow) phenol phase and discard H_2O, add 1 L 0.1 M Tris-Base, mix, allow to separate.
5. Keep lower (yellow) phenol phase and discard H_2O phase.
6. Repeat the process until the pH of phenolic phase is >7.8.
7. Add 1 L $CHCl_3$: Isoamyl alcohol (96:4, v/v), mix, and store in brown bottle at 4°C.

Appendix XVIII: Sterile Technique

1. All media, including plates, liquid media and top agar must be autoclaved immediately after it is prepared. It is best to prepare media in several small bottles, only opening one at a time. Check the bottle for contamination before you use it by gently swirling it and looking for cloudy material in the center. Always grow up a small amount of broth alone when growing cells overnight. A small amount of contamination is not always evident until the media is incubated at 37°C.
2. Use a flame on inoculating loops and on the lips of media bottles before and after pipetting from them. Never leave a media or agar bottle open on the bench and don's take an individually-wrapped pipet out of its protective wrapper until you are ready to use it (i.e., don't walk across the room with an unwrapped pipet). Always use a fresh, sterile pipet or pipet tip when pipetting culture media, and never go back into a media bottle or cell culture with a used pipet.
3. To prevent wide-scale, untraceable contamination, each person should have his own stock of liquid culture media, top agar, plates, 100% glycerol, glycerol stocks of cells, etc. and don't share.
4. Overnight cultures should be grown only from a single colony on a fresh plate or from a previously-tested glycerol stock that was grown from a single colony. To prepare an overnight culture from a glycerol stock, take an individually-wrapped 1-ml pipet and a culture tube of media to the –80°C freezer. Quickly remove the cap from the freezer vial containing the glycerol stock, scrap a small amount of ice from the surface of the culture, replace the cap on the freezer vial, and place the pipet into the culture tube. Sufficient numbers of bacteria are present in the ice in order for the culture to grow to saturation in 16 hours. Never let the glycerol stock thaw.
5. Think about what you are doing. The best defense is common sense.

Appendix XIX: Some Facts and Units

- Avogadro's Number: 6.02 × 10e23
- 1 bp DNA = 660 daltons
- 1 Kb duplex DNA = 6.6 × 10e5 daltons
- 1 Kb duplex DNA = 333 amino acid coding capacity = 37,000 dalton protein
- 1 uM = 1 pmole/ul
- pM = 10e-12 M
- 1 ug/mL of a 1 Kb nucleic acid fragment = 3 nM fragment ends
- A 1 × 10e6 dalton restriction fragment = 1 ug/pmole of fragment
- Bacteriophage lambda = 30 × 10e6 daltons, = 30 ug/ pmole lambda
- 1 bacteriophage lambda plaque contains ~10e6 phage particles
- A lambda lysate has approximately 2 × 10e10 pfu/mL
- 1 O.D. of duplex DNA; A260 of 1.0 = 50 ug/mL
- 1 O.D. single-stranded RNA; A260 of 1.0 = 40 ug/mL
- 1 O.D. single-stranded DNA; A260 of 1.0 = 33 ug/mL
- O.D.660 *E. coli* cells x 5 = cells/mL × 10e8
- Haploid human genome = 3 × 10e9 bp
- Haploid human genome = 33 Morgans in length (3300 cM)
- Haploid yeast (*S. cerevisiae*) genome = 1.5 × 10e7 bp
- *E. coli* genome = 4.72 × 10e6 bp
- Wild type lambda genome = 4.85 × 10e4 bp
- Density of duplex DNA = ([molar fraction G + C] + 14.19)/8.63 ~1.7 g/mL
- Density of RNA ~1.9 g/mL
- Density of lambda phage ~1.5 g/mL
- Density of protein ~1.3 g/mL
- Density of carbohydrate ~1.5–2.0 g/mL
- 1 uCi = 7 x 10e6 cpm (Cherenkov radiation) = 2.2 × 10[6] dpm
- e32P half life = 14.3 days
- e35S half life = 87.4 days

Appendix XX: Calculating Amounts of Nucleic Acids

O.D. Units: The "absorbance" of a chemical is a product of its (concentration) x (optical path length) x (extinction coefficient, E). Nucleic acids have a peak absorbance in the ultraviolet range at about 260 nm. When the spectrophotometer has a path length of 1 cm, absorbance = "optical density" (O.D.), and O.D. = E x concentration. Extinction coefficients vary with the type of nucleic acid. Double stranded DNA (dsDNA) has an E = 20 g^{-1} cm^{-1} L). Depending on the reference you read, the E for single stranded DNA (ssDNA) is 20 or 30 g^{-1} cm^{-1} L), while E for RNA is 25 g^{-1} cm^{-1} L).

Extinction coefficients can be used to estimate the concentration of a sample dissolved in a known aqueous volume, or to calculate the number of grams or moles of nucleic acid according to the following formulas:

Determining Concentrations:

1 A260 O.D. Unit for dsDNA = 50 μg/mL

1 A260 O.D. Unit for ssDNA = 33 or 50 μg/mL

1 A260 O.D. Unit for RNA = 40 μg/mL

Examples: A 1:50 dilution of dsDNA gives an A260 = 0.063. [DNA] = 0.063x50/20 = 0.16 mg/mL. This is for a 1 cm path length. The path length of the 5 μl cuvette is only 0.5 mm = 1/20 cm. Therefore multiplication by 20 to give a 1 cm path length and then dividing by 20 for the extinction coefficient cancel each other. So for that cuvette, [DNA] = (A260–A320) x dilution factor. The A320 is used to subtract absorbance due to particles in suspension in this small chamber.

Determining Moles: For ssDNA oligonucleotides, an estimate of the number of moles can be obtained by using an approximate mw of 350 Daltons/nucleotide. Accordingly, 1 O.D. Unit ssDNA = 33 × 10^{-6}/(350 × length) moles = 94.3/length nmoles

An O.D. of 1.0 corresponds roughly to:

10-mer	10 nmoles
20-mer	5 nmoles
50-mer	2 nmoles
100-mer	1 nmole

Also, since 1 O.D. Unit = 33 μg, then 1 μg = 2.86/length nmoles.

A precise estimate of the molecular weight of dephosphorylated oligos can be calculated using oligos:

mw = [(#A's × 312.2) + (#G's × 328.2) + (#C's × 288.2) + (#T's × the formula)]

mw = [(#A's × 312.2) + (#G's × 328.2) + (#C's × 288.2) + (#T's × 302.2) – 61.0][1]

For phosphorylated 302.2) + 17.0][1]
For long (at least 30bp) dsDNA
approximate mw = 700 Daltons/base pair; mw of 1 kb is about 7×10^5 Daltons
(649 Daltons/bp is more accurate, but harder to remember)
μg in a pmole = length × 700 g/mole × 10^{-12} moles/pmole × $10^6$1;g/g
= length × (7×10^{-4}) μg/pmole
Multiply pmoles by 2 for pmoles 5′ ends for dsDNA.
ex: 1 kb DNA is about 0.7 μg/pmole
1 μg of a 1 kb DNA = 1.52 pmol; 3.03 (pmol ends)
1 pmole of 1 kb DNA = 0.66 μg

MW RNA: length x 350 (341 is more accurate)

Other Useful Conversions for Oligonucleotides

Oligonucleotide length to molecular weight
350 × length = molecular weight
Mass to μmoles
μg/(350 × length) = μmoles
(μg × 10^6)/(350 × length) = pmoles
pMoles to μg
pmoles x length x 350 /10^6 = μg
μg to pmoles
μg x 10^6/(length x 350)

MW of Oligonucleotide	pMoles 3′ Ends per Microgram of DNA
10×10^6	0.05
1×10^6	0.50
0.1×10^6	5.00
0.01×10^6	50.0

Melting Temperature Formulas (14-70 bases):
Tm = 81.5°C + 16.6(log[Na]) + 0.41(fraction G + C)–600/L
where
Na = [monovalent cations] usually 50–60 mM
L = primer length
fraction G + C = G + C content of primer
Also, Tm (approx) = 4°C x(fraction G + C) + 2°C x(fraction A + T)

Oligonucleotide Stock Concen trations for Amplifications:[2]
mls to dissolve oligo = Total ODs (or A260) / (E x molar concentration needed)
Count the numbers of each base in the oligo. Compute E
= [#A's × (16,000)] + [#G's × (12,000)] + [#C's x (7000)] + [#T's × (9600)] + [#I's × (12,000)]
Make your oligonucleotide solution more concentrated than needed so that when you add it to a working solution, the final concentration will be correct.

Ex: 18-mer with 2(A) + 6(G) + 8(C) + 2(T).
E = 2(16,000) + 6(12,000) + 8(7000) + 2(9600) = 179,200

Total ODs in tube = 11
Concentrations needed = 5 × 10–4 = 0.5 mM
mls needed to dissolve = $11/(179{,}600 \times 5 \times 10^{-4}) = 0.123$ *mL*

Glossary

A: Adenine residue in either DNA or RNA.

A Domain: The conserved 11 bp sequence of A-T base pairs in the yeast ARS element that comprises the replication origin.

A-DNA: One of the many possible double-helical structures of DNA. It is a right-handed double helix, fairly similar to the more common and well-known B-DNA form, but with a shorter, more compact, helical structure. A-DNA is thought to be 1 of 3 biologically active double-helical structures, along with B- and Z-DNA. It appears likely that it occurs only in dehydrated samples of DNA, such as those used in crystallographic experiments and possibly is also assumed by DNA-RNA hybrid helices and by regions of double-stranded RNA.

A Site (of ribosome): The site that an aminoacyl-tRNA enters to base with the codon.

Abduction: A method of reasoning in which one chooses the hypothesis that would, if true, best explain the relevant evidence. It is also used to just mean the generation of hypothesis to explain observations or conclusions.

***ab initio*:** A method that derives a 3D structure from initial physical forces and interactions.

Abiogenesis: An early theory that held some organisms originated from nonliving material.

Abiotic elicitor: An elicitor that is not of biological origin; for example, UV light, salts of heavy metals, etc.

Abiotic stress: Outside (nonliving) factors, such as soil conditions, drought, and extreme temperatures, which can cause harmful effects to plants.

Abortive transduction: Failure of incorporation of transducing DNA into the recipient chromosome.

Abscisic acid: A plant growth regulator involved in abscission, dormancy and inhibition of seed germination. It also affects somatic embryogenesis in some plant species.

Abscissa: The horizontal axis of a graph.

Absorb (Ab, away + sorbere, to suck in): To suck up, or to take in; in the cell, materials are taken in (absorbed) from a solution.

Absorption: (1) In general: The process of absorbing or taking up of water and nutrients by assimilation or imbibition. The taking up by capillary, osmotic, chemical, or solvent action, such as the taking up of a gas by a solid or liquid, or taking up of a liquid by a solid. (2) In biology: The movement of a fluid or a dissolved substance across a cell membrane. In plants: Water and mineral salts are absorbed from the soil by roots. In animals: Soluble food material is absorbed into the circulatory system through the cells lining the alimentary canal.

Absorption spectrum: A graph depicting the ability of a substance to absorb lights of various wavelengths.

Abundance (of an mRNA): The average number of molecules per cell.

Abundant mRNAs: Consist of a small number of individual species, each present in a large number of copies per cell.

Abzymes: A common name for catalytic antibodies that have been engineered to catalyze a chemical reaction.

Acaricide: A pesticide used to kill or control mites or ticks.

Acceptor arm (of tRNA): A short duplex that terminates in the CCA sequence to which an amino acid is linked.

Acceptor site: The splice site at the 3′ end of an intron.

Accessible surface: The surface that is traced by the center of a probe molecule (usually water) as it rolls on the van der Waals surface of a molecule.

Accession number: An identifier supplied by the curators of the major biological databases upon submission of a novel entry that uniquely identifies that sequence (or other) entry.

Accessory bud: Lateral bud occurring at the base of a terminal bud or at the side of an axillary bud.

Acclimatization: The adaptation of a living organism (plant, animal, or microorganism) to a changed environment that subjects it to physiological stress.

Acellular: Describing tissues or organisms that are not made up of separate cells but often have more than 1 nucleus.

Acentric chromosome: A chromosome without a centromere.

Acentric fragment: Segment of chromosome, generated by breakage, tha lacks a centromere and is lost at cell division.

Acetosyringone: A phenolic compound produced by dicot plant cells at wound sites; acts as signal molecule for the expression of vir regulon of *Agrobacterium tumefaciens.*

Acetyl co-enzyme: A compound formed in the mitochondria when an acetyl group (CH_3CO-)—derived from break-down of fats, proteins, or carbohydrates—combines with the thiol group (-SH) of co-enzyme A.

Acid: Any compound that releases hydrogen ions when dissolved in water.

Acid-base and redox catalysis: A mode of enzyme action; electron transfer generally involves transition metals or coenzymes present as a component of such enzymes.

Acid break: In case of fermentation, a rapid conversion of acids into acetone butanol leading to a fast decline in the acidity.

Acidic domain: A type of activation domain.

Acid precipitation: It includes acid rain, acid fog, acid snow, and any other form of precipitation that is more acidic (less than pH 5.6) than normal. Excess acidity is derived from certain air pollutants, namely, sulphur dioxide and oxides of nitrogen.

Acoustic wave biosensor: Device that detects a change in the mass of the biological component of the biosensor.

Acquired: Developed in response to the environment (not inherited), such as a character trait (acquired characteristic) resulting from environmental effect(s). cf acclimatization.

Acquired immunity: Another term for adaptive immunity.

Acquired immunodeficiency syndrome (AIDS): A syndrome caused by infection by the human immunodeficiency virus (HIV).

Acquision of totipotency: Formation of proembryogenic cell masses in cultured cells/tissues, usually in response to a high auxin concentration.

Acridines: Mutagens that act on DNA to cause the insertion or deletion of a single base pair. They were useful in defining the triplet nature of the genetic code.

Acridine dye: A chemical compound that intercalates between adjacent base pairs of the double helix, which causes a frameshift mutation.

Acrocentric: A chromosome in which the centromere lies very near to one end.

Acropetal: (1) Developing or blooming in succession toward the apex, such as leaves or flowers developing acropetally. (2) Transport or movement of substances toward the apex, such as the movement of water through the plant. The opposite tendency is termed basipetal.

Acrosome: An optical organelle in the head of a spermatozoon.

Actin: One of the 2 contractile proteins in muscle (the other being myosin). Actin is also found in the microfilaments that form part of the cytoskeleton of all cells.

Action Letter: An official communication that informs a sponsor of a decision by the agency. An approval letter allows commercial marketing of the product.

Action spectrum: A graph relating the degree of physiological response caused by different wavelengths of light.

Activated charcoal; activated carbon: Charcoal that has been treated to remove hydrocarbons to increase its adsorptive properties. The inhibitory substances in nutrient medium get adsorbed to charcoal when present in the medium. Sometimes phenolamines present as contaminants in charcoal may stimulate root growth *in vitro*.

Activated sludge: The sediment from biologically treated sewage; it contains the microbes for anaerobic digestion of sewage.

Activated sludge process: It uses large aerated vessels for oxidation of liquid waste, which continuously flows into the vessels and treated water flows out of them at a predetermined rate to obtain optimal digestion of the dissolved organic matter.

Activation: The time-varying value that is the output of a neuron.

Activation domain: The part of a transcription factor that makes contact with the initiation complex.

Activation energy: The free energy needed to elevate a molecule from its stable ground state to the unstable transition state.

Activation function: A function that translates a neuron's net input to an activation value.

Activator: A protein that stimulates the expression of a gene, typically by acting at a promoter to stimulate RNA polymerase. In eukaryotes, the sequence to which it binds in the promoter is called a response element.

Active collection: A collection which complements a base collection from which seed samples are drawn for distribution, exchange, and other purposes such as multiplication and evaluation.

Active immunity: (1) A type of acquired immunity whereby resistance to a disease is built up by either having the disease or having a vaccine to it. (2) Immunity acquired after stimulation with an antigen by natural infection or other exposure.

Active initiation: A process in which RNA polymerase starts transcription but terminates before it has left the promoter. It then reinitiates. Several cycles may occur before the elongation stage begins.

Active site: The restricted part of an enzyme to which a substrate binds.

Active transport: An energy-consuming process that moves molecules against an electrochemical gradient. Energy for the movement is provided by hydrolysis of ATP.

Acute transforming virus: Carries a gene(s) that originated in a cellular genome. Its transforming capacity is the result of expression of that gene. Because the gene replaced viral sequences, the virus does not have the capacity to replicate independently.

Acylation: The attachment of a lipid side chain to a polypeptide molecule.

Ada enzyme: An enzyme in *Escherichia coli* that is involved in the direct repair of alkylation mutations.

Adaptability: The potential or ability of a population to adapt to changes in the environmental condition through changes in the genetic structure.

Adaptation: An internal change in a system in response to an external event in the system's environment.

Adaptation traits: The complex of traits related to reproduction and survival of the individual in a particular production environment. Adaptation traits contribute to individual fitness. They are the traits subjected to selection during the evolution of animal genetic resources. By definition, these traits are also important to the ability of the animal genetic resource to be sustained in the production environment.

Adaptin: A subunit of the cytosolic adaptor proteins that mediate formation of clathrin-coated vesicles. There are several types of adaptin subunits.

Adaptive: Subject to adaptation, that is, able to change over time to improve fitness.

Adaptive immunity: The response mediated by lymphocytes that are activated by their specific interaction with antigen. The adaptive immune response develops over several days as lymphocytes with antigen-specific receptors are stimulated to proliferate and become effectors cells. It is responsible for immunological memory.

Adaptive radiation: Evolutionary diversification of a generalized ancestral from with the production of a number of specialized forms by adaptation.

Adaptive surface: A surface plotted in 3 dimensions with all possible combinations of allele frequencies for different loci plotted along the plane, and the mean fitness for each combination plotted as the height of the surface.

Adaptor: A synthetic single-strand of oligonucleotides that, after self-hybridization, produces a molecule with cohesive ends and an internal restriction endonuclease site. When the adaptor is inserted into a cloning vector by means of cohesive ends, the internal sequence provides a new restriction endonuclease site.

Adaptor protein: Proteins bind to signals in the cytoplasmic tails of transmembrane cargo proteins and recruit clathrin molecules in the assembly of clathrin-coated pits and vesicles. Different types of adaptor proteins function at different compartments. Each adaptor protein contains four different subunits.

Addendum (pl: addenda): In formulation of tissue culture media, an item or a constituent substance to be added.

Addiction system: A survival mechanism used by plasmids. The mechanism kills the bacterium upon loss of the plasmid.

Additive allelic effects: Effects of alleles at a locus, where the heterozygote is exactly intermediate between the 2 homozygotes.

Additive gene effects: Additive allelic effects summed across all the loci that contribute to genetic variation in a quantitative trait.

Adaptive genetic variance: The proportion of genetic variation that is the summation of the effect of all individual genes influencing a trait. It is the average effect of substituting 1 allele for another.

Adenine (A): One of the 2 purines (9-membered double-ringed nitrogenous base) in DNA and RNA.

Adenocarcinoma: Cancer that originates in cells that line some of the internal organs.

Adenosine: The nucleotide containing adenine as its base.

Adenosine deaminase (ADA) deficiency: A severe immunodeficiency disease that results from a lack of the enzyme adenosine deaminase. It usually leads to death within the first few months of life.

Adenosine diphosphate (ADP): Lower energy form of ATP, having 2 (instead of 3 in ATP) phosphate groups attached to the adenine base and ribose sugar.

Adenosine triphosphate (ATP): A common form in which energy is stored in living systems. It consists of a nucleotide (with ribose sugar) with 3 phosphate groups.

Adenovirus: A group of DNA viruses that cause diseases in animals. In man, they produce acute respiratory tract infections with symptoms resembling the common cold. They are used in gene cloning, as vectors for expressing large amounts of recombinant proteins in animal cells. They are also used to make live- virus vaccines against more dangerous pathogens.

Adenovirus-associated virus (AAV): A small virus that requires adenovirus for growth. This virus causes no known human disease, and hence recombinant AAV is used as a gene therapy vector.

Adenylate cyclase: An enzyme that uses ATP as a substrate to generate cyclic AMP, in which 5′ and 3′ positions of the sugar ring are connected via a phosphate group.

Adept (antibody-directed enzyme pro-drug therapy): A way to target a drug to a specific tissue. The drug is administered as an inactive pro-drug, and then converted into an active drug by an enzyme administered with a second infection. The enzyme is coupled to an antibody that concentrates it in the target tissue. When the enzyme arrives at the target tissue, the pro-drug is activated to form the active drug, while elsewhere it remains inactive.

Adherent cells: Immune cells (such as dendritic cells and macrophages) with the property of adhering to tissue culture, flasks. This property allows the cells to be separated and purified.

Adhesion (1. adhaerere, to stick to): The attraction of dissimilar for each other. A sticking together of unlike substances, such as oil and water.

Adjacent segregation: Segregation of a heterozygous reciprocal translation in which a translocated chromosome and a normal chromosome segregate together, producing an aneuploid gamete.

Adjacent-1 segregation: Segregation from a heterozygous reciprocal translocation in which homologous centromeres go to opposite poles of the first-division spindle.

Adjacent-2 segregation: Segregation from a heterozygous reciprocal translocation in which homologous centromeres go to same pole of the first-division spindle.

Adjuncts: Starch rich materials, such as wheat, rice, maize, etc., that are added to barley malt during beer production.

Adjuvant: (1) Insoluble material that increases the formation and persistence of antibodies when injected with an antigen. (2) A substance that nonspecifically activates the immune system. Many therapeutic molecules such as vaccines are provided with an adjuvant to enhance their effect.

Adoptive transfer: The transfer of immune cells from a primed donor to an unprimed recipient.

Adsorbent: A substance to which compounds adhere; for example, activated charcoal.

Adsorption: The formation of a layer of gas, liquid, or solid on the surface of a solid.

Adsorption chromatography: Separation of molecules because of their differential adsorption onto the solid matrix surface.

Adult cloning: The creation of identical copies of an adult animal by nuclear transfer from differentiated adult tissue.

Advanced: Applied to an organism or a part thereof, implying considerable development from the ancestral stage or from the explants stage.

Advantageous mutation: A mutation that increases the fitness of the organism carrying it.

Adventitious: Describes organs developing from positions on the plant from which they would not normally be derived; for example, shoots from, leaf, root, petiole, cotyledon, or embryos, etc. from any cell other than a zygote.

Aerate: To supply with or mix with air or gas. The process is aeration.

Aerobe: A microorganism that grows in the presence of oxygen.

Aerobic: (1) Needing oxygen for growth. (2) Active in the presence of free oxygen.

Aerobic bacteria: Bacteria that can live in the presence of oxygen.

Aerobic fermentation: Requires an efficient air supply, usually by sparging and stirring; 2 types: stirred tank and airlift.

Affinity: A measure of the strength of binding of 1 molecule to another, for example, of a ligand to a receptor or of a substrate to an enzyme.

Affinity biosensor: The biological material simply binds to the analyte molecules that leads to a change in its mass.

Affinity chromatography: A method for separating molecules by exploiting their ability to bind specifically to other molecules. There are several types of biological affinity chromatography. A biological molecule can be immobilized and a smaller molecule to which it is to bind can be stuck to it, or the smaller ligand can be immobilized and the macromolecule stuck to it. A variant is to use an antibody as the immobilized molecule and use it to "capture" its antigen: this is often called immune-affinity chromatography. A variation is pseudo-affinity chromatography, in which a compound that is like a biological ligand is immobilized on a solid material, and enzymes or other proteins are bound to it. Other techniques include metal affinity chromatography, in which a metal ion is immobilized on a solid support; metal ions bind tightly and specifically to many biomolecules. The metal ion is bound to a chelator or chelating group, a chemical group that binds specifically and extremely tightly to that metal.

Affinity tag/purification tag: An amino acid sequence that has been engineered into a protein to make its purification easier. These can work in a number of ways. The tag could be another protein, which binds to some other material very tightly, thus allowing the protein to be purified by affinity chromatography. The tag could be a short amino acid sequence, which is recognized by an antibody. The antibody would then bind to the protein, whereas it would not have done so before. One such short peptide, called FLAG, has been designed so that it is particularly easy to make antibodies against it. The tag could be a few amino acids, which are then used as a chemical tag on the protein. For example, a string of positively charged amino acids will bind very strongly to a negatively charged filter. This could be used as the basis of a separation system. Some amino acids bind metals very strongly, especially in pairs; this chemical property can be exploited by using a filter with metal atoms chemically linked onto it to pull a protein out of a mixture of proteins.

Afflatoxin: Toxic compounds, produced by molds (fungi) of the *Aspergillus flavus* group, that bind to DNA and prevent replication and transcription. Aflatoxins can cause acute liver damage and cancer. Animals may be poisoned by eating stored food or feed contaminated with the mold.

Agar: A polysaccharide solidifying agent used in nutrient media obtained from certain types of red algae (Rhodophyta).

Agarose: The main constituent of agar, a purified form.

Agarose gel electrophoresis: A process in which a matrix composed of a highly purified form of agar is used to separate large DNA and RNA molecules.

Agents: Independent, autonomous, software modules that can search the internet for data or content pertinent to a particular application, such as a gene, protein, or biological system.

Aggregate: (1) A clump or mass formed by gathering or collecting units. (2) A body of loosely associated cells, such as a friable callus or cell suspension. (3) Coarse inert material, such as gravel, that is mixed with soil to increase its porosity. (4) A serological reaction (aggregation) in which the antibody and antigen react and precipitate out of solution.

Agonist: A drug, hormone, or transmitter substance that forms a complex with a receptor site that is capable of triggering an active response from a cell.

***Agrobacterium*:** A genius of bacteria that includes several plant pathogenic species that cause tumor-like symptoms.

***Agrobacterium tumefaciens*:** *Agrobacterium* that causes crown gall disease. The cells form a crown gall due to Ti plasmid present in these bacteria.

***Agrobacterium tumefaciens*–mediated transformation:** A naturally occurring process of DNA transfer from the bacterium *A. tumefaciens* to plants.

Agroinfection: Introduction of a viral genome into plant cells using *Agrobacterium*; usually, a tandem dimer of viral genome is placed within the T-DNA of *Agrobacterium.*

Agropine: Plasmids carry genes coding for the synthesis of opines of the agropine type. The tumors usually die early.

Alanine (Ala, A): One of the 20 amino acids; its codons are GCA, GCC, or GCU. In mammals, it is a nonessential amino acid with a molecular formula CH_3-$CH(NH_2)$-COOH.

Alarmone: A small molecule in bacteria that is produced as a result of stress and which acts to alter the state of gene expression. The unusual nucleotides ppGpp and pppGpp are examples.

Albinism: Hereditary absence of pigment in an organism. Albino animals have no color in their skin, hair, and eyes. The term is also used for absence of chlorophyll in plants.

Albino: (1) An organism lacking pigmentation because of genetic factors. The condition is albinism. (2) A conspicuous plastome (plastid) mutant involving loss of chlorophyll.

Aleurone: The outermost layer of the endosperm in a seed.

Algal biomass: Single-celled plants, such as *Chlorella* spp. and *Spirulina* spp., are grown commercially in ponds to make feed materials. *Chlorella* is grown commercially to make into fish food; it is fed to zooplankton, and these in turn are harvested as feed for fish farms. This is a means of converting sunlight into food in a way that is more convenient and controllable than normal farming.

Algal bloom: Sudden heavy growth of algae; generally occurs due to addition of nutrients, whose scarcity is normally limiting.

Algalization: The process of application of blue-green algae (cyanobacteria) in fields as biofertilizer.

Alginate: A polysaccharide gelling agent.

Alignment: The result of a comparison of 2 or more gene or protein sequences in order to determine their degree of base or amino acid similarity. Sequence alignments are used to determine the similarity, homology, function, or other degree of relatedness between two or more genes or gene products.

Alkaline: A substance that releases hydroxyl ions or uses up hydrogen ions.

Alkylating agents: Chemical mutagens (acridine orange, proflavin) that wedge between nitrogenous bases and cause frameshift mutations.

Allele: (Gr. Allelon, of one another, mutually each other); allelomorph (adjallelic, allelomorphic). One of a pair, or series, of variant forms of a gene that occur at a given locus in a chromosome.

Allele frequency: The number of copies of an allele in a population, expressed as a proportion of the total number of copies of all alleles at a locus in a population.

Alleles: The different forms of a gene. Y and y are different alleles of the gene that determines seed color. Alleles occupy the same locus, or position, on chromosomes. An abbreviation of allelomorph, meaning 1 form or the other, distinguished from other alleles by phenotypic effects.

Allele-specific ligation: Two short, synthetic oligonucleotides that base-pair adjacent to each other on a DNA fragment and are ligated by DNA ligase. If 1 of the alleles contains a mutation overlapped by the 3′-end of 1 oligonucleotide, ligation with the other oligonucleotide will be prevented.

Allele-specific oligonucleotide hybridization (ASO hybridization): The use of an oligonucleotide probe to determine which of 2 alternative nucleotide sequences is contained in a DNA molecule.

Allelic exclusion: Describes the expression in any particular lymphocyte of only 1 allele coding for the expressed immunoglobulin. This is caused by feedback from the first immunoglobulin allele to be expressed that prevents activation of a copy on the other chromosome.

Allelopathy: The phenomenon by which the secretion of chemicals, such as phenolic and terpenoid compounds, by a plant inhibits the growth or reproduction of other plant species with which it is competing.

Allergen: (1) A substance that causes an allergic reaction. (2) An antigen that provokes an immune response.

Allergy: An unusual sensitivity to certain foods, pollens, insect bites, etc; appears on skin.

Allogamy: Cross fertilization in plants.

Allogenic: Of the same species, but with a different genotype.

Allograft: Graft obtained from genetically different individuals of the same species.

Allohaploid: A haploid obtained from an allopolyploid.

Allosteric: Regulation describes the ability of a protein to change its conformation (and therefore activity) at 1 site as the result of binding a small molecule to a second site located elsewhere on the protein.

Allosteric enzyme: An enzyme that has 2 structurally distinct forms, 1 of which is active and the other inactive. Active forms of allosteric enzymes tend to catalyze the initial step in a pathway, leading to the synthesis of molecules. The end product of this synthesis can act as a feedback inhibitor, converting the enzyme to the inactive form, thus controlling the amount of product synthesized.

Allosteric regulation: A catalysis-regulating process in which the binding of a small effector molecule to 1 site on an enzyme affects the activity at another site.

Allosteric transition: A reversible interaction of a small molecule with a protein molecule, resulting in a change in the shape of the protein and a consequent alternation of the interaction of that protein with a third molecule.

Allotetraploid: An organism with 4 genomes derived from hybridization of different species. Usually, in forms that become established, 2 of the 4 genomes are from 1 species and 2 are from another species.

Allozygote: A diploid individual that is homozygous at a locus in which the 2 genes are not identical by descent from a common ancestor.

Allozymes: Different molecular forms of an enzyme coded by different allelic forms of a gene.

Alphabet: The set of all possible symbols in an application.

Alpha helix: A helical structure formed by polypeptide that is one of the principal secondary structural forms found in proteins.

Alphalactalbumin: Protein component of milk.

Alphoid DNA: The tandemly repeated nucleotide sequences located in the centromeric regions of human chromosomes.

Alternative splicing: Describes the production of different RNA products from a single product by changes in the usage of splicing junctions.

Alu: A short, repetitive string of about 300 DNA bases that are scattered throughout the genome, often in regions that have many genes.

Alu domain: Comprises the parts of the 7S RNA of the SRP that are related to Alu RNA.

Alu family: A set of dispersed, related sequences, each ~300 bp long, in the human genome. The individual members have Alu cleavage sites at each end (hence the name).

Alu PCR: A clone fingerprinting technique that uses PCR to detect the relative positions of Alu sequences in cloned DNA fragments.

Alu sequences: A family of 300-bp sequences occurring nearly a million times in the human genome.

Amanitin (more fully α-amanitin): A bicyclic octapeptide derived from the poisonous mushroom *Amanita phalloides*; it inhibits transcription by certain eukaryotic RNA polymerases, especially RNA polymerase II.

Amber: The triplet codon UAG, 1 of the 3 termination codons that end protein synthesis.

Amber Codon: UAG-One of the nonsense or stop or chain termination codons that ends protein synthesis.

Amber mutation: Mutation of the UAG termination codon.

Ambient standards: It is an air-quality standard, set by the U.S. Environmental Protection Agency (EPA) that limits certain levels of pollutants that should not be exceeded in order to maintain environmental and human health.

Ambient temperature: Air temperature at a given time and place, not the radiant temperature.

Amino acid: An organic compound containing an amino (-NH_2) and an acid carboxyl group (-COOH).

Amino group: One of the 20 chemical building blocks that are joined by amide (peptide) linkages to form a polypeptide chain of a protein.

Amino terminus: The –NH_2 (amino) end of a polypeptide.

Aminoacylation: Attachment of an amino acid to the acceptor arm of a tRNA molecule.

Aminoacyl site (a-site): One of 2 sites on a ribosome to which the incoming aminoacyl tRNA binds.

Aminoacyl-tRNA: A tRNA linked to an amino acid. The COOH group of the amino acid is linked to the 3′- or 2′-OH group of the terminal base of the tRNA.

Aminoacyl-tRNA synthetases: Enzymes responsible for covalently linking amino acids to the 2′- or 3′-OH position of tRNA.

Amitosis: Cell division (cytokinesis), including nuclear division through constriction of the nucleus, without chromosome differentiation as in mitosis. The maintenance of genetic integrity and diploidy during amitosis is uncertain. This process occurs in the endosperm of flowering plans.

Amorph null mutation: A mutation that obliterates gene function.

Amperometric biosensor: Measures the transfer of electrons to O_2 (present in the biosensor) due to redox reactions.

Amphidiploid: A species derived by doubling of chromosomes in the F_1 hybrid of 2 species.

Amphimixis: True sexual reproduction involving the fusion of male and female gametes and the formation of a zygote.

Amphipathic: Structures have 2 surfaces, 1 hydrophilic and 1 hydrophobic. Lipids are amphipathic; and some protein regions may form amphipathic helices, with 1 charged face and 1 neutral face.

Amphiphilic: Having a polar water-soluble group attached to a water-insoluble hydrocarbon chain.

Ampholyte: A trade name for a mixture of substances with a range of isoelectric points that have high buffering capacity at their isoelectric points.

Amphotropic virus: Recombinant virus that has the capacity to infect a wide variety of mammalian (human) cells.

Ampicillin: A penicillin derived antibiotic that prevents bacterial growth by interfering with synthesis of the cell wall.

Amplicon: The DNA sequence/fragment that is amplified by PCR.

Amplification: (1) Replication of a gene library in bulk. (2) Duplication of gene(s) within a chromosomal segment. (3) Creation of many copies of a segment of DNA by polymerase chain reaction.

Amplification refractive mutation system (ARMS) PCR: This PCR was devised to achieved sequence-specific amplification.

Amplified fragment length polymorphism (AFLP): A combination of RFLP and RAPD very sensitive in detecting polymorphisms throughout the genome.

Amplified genomic library: The library that consists of recombinant phage lysates or bacterial clones of a genomic library.

Amplified ribosomal DNA restriction analysis (ARDRA): The amplification product of the gene encoding the 16S ribosomal RNA is subjected to restriction fragment length polymorphism. This analysis is important forbacterial identification/taxonomy.

Amplify: To increase the number of copies of a DNA sequence, *in vivo* by inserting into a cloning vector that replicates within a host cell, or *in vitro* by polymerase chain reaction (PCR).

Amylase: A group of enzymes that degrade starch, glycogen, and other polysaccharides, producing a mixture of glucose and maltose. Plants have both a- and b-amylase and animals have only a-amylase.

Amylolytic: The capability of breaking down starch into sugars.

Amylopectin: A polysaccharide comprising highly branched chains of glucose molecules. The water-in-soluble portion of starch.

Amylose: A polysaccharide consisting of linear chains of 100–1,000 glucose molecules. The water-soluble portion of starch.

Anabolic pathway: A pathway by which a metabolite is synthesized; a biosynthetic pathway.

Anabolism: Formation of macromolecules from smaller molecules; requires an input of energy.

Analog: biological structures which perform similar functions by similar mechanisms but evolved separately, for example, a chemical so similar to one of the normal bases that it can be incorporated into DNA.

Analogous: Feature of organisms which are functionally similar but have evolved in different ways.

Analogous structures: Body parts that serve the same function in different organisms, but differ in structure and embryological development, for example, the wings of insects and birds.

Analogy: Reasoning by which the function of a novel gene or protein sequence may be deduced from comparisons with other gene or protein sequences of known function. Identifying analogous or homologous genes via similarity searching and alignment is one of the chief uses of Bioinformatics.

Analyte: The substance or component in a sample that is being measured.

Analytical breeding: In this strategy, dihaploids are isolated from autotetraploids, and breeding work is carried out at the dihaploid level; the chromosome number of the selected dihaploid lines is then doubled to obtain normal tetraploid lines.

Anaphase: (1) The stage in nuclear division when the chromosomes move to opposite poles of the mitotic spindle within a cell. (2) The stage of mitosis or meiosis during which the daughter chromosomes (sister chromatids) pass from the equatorial plate to opposite poles of the cell (toward the ends of the spindle). Anaphase follows metaphase and precedes telophase.

Anaphase promoting complex (APC): A set of proteins that triggers proteolysis or targets required to allow chromosomes to separate.

Anatomic barriers: In case of innate immunity; these barriers include skin and surfaces of mucous membranes; together, they prevent the entry of most microorganisms.

Ancestral character state: A character state possessed by a remote common ancestor of a group of organisms.

Anchor (stop-transfer) (often referred to as a "transmembrane anchor"): A segment of a transmembrane protein that resides in the membrane.

Anchor gene: A gene that has been positioned on both the physical map and the linkage map of a chromosome.

Anchorage dependence: Describes the need of normal eukaryotic cells for a surface to attach in order to grow in culture.

Anchored PCR: When the sequence of only one end of a DNA segment is known, the primer specific to the known 3′ end is used to obtain copies of this strand; an oligo-G tail is then added at the 3′ ends of the single-stranded copies, and oligo-C is used as primer to obtain double-stranded copies.

Ancient DNA: DNA preserved in ancient biological material.

Androgen: Any hormone that stimulates the development of male secondary sexual characteristics and contributes to the control of sexual activity in vertebrate animals. Usually synthesized by the testes.

Androgenesis: Development of haploid plants from the male gametophyte following a developmental pattern resembling embryogenesis, resulting from the culture of anthers or microspores.

Androgenic grains: Pollen grains, which exhibit androgenesis; believed to be s-grains.

Anergy: The absence of an expected cell-mediated immune reaction in sensitized organisms, caused by inactivated B and T cells. Also used to describe inactivated B or T cells.

Aneuploid: Set of chromosomes differs from the usual diploid constitution by loss or duplication of chromosomes or chromosomal segments.

Aneuploidy: The loss or gain of chromosomes from the normal euploid number by various processes.

Aneusomaty: Presence of aneuploid and euploid cells in the same tissue, especially *in vivo.*

Angstrom (Å): Unit of length equal to 10^{-10} meter.

Animal biotechnology: Use of animal cells, tissues, organs, or their components to generate useful products or services; e.g., monoclonal antibodies, cell culture-derived vaccines, transgenic animals, etc.

Animal cell culture: Culture of dispersed animal cells either as a monolayer or as a suspension of cells.

Animal cell immobilization: Entrapment of animal cells in some solid material in order to produce some natural product or genetically engineered protein. Animal cells have the advantage that they already produce many proteins of pharmacological interest, and that genetically engineered proteins are produced by them with the post-translation modifications normal to animals. However, because animal cells are much more fragile than bacterial ones, they cannot tolerate a commercial fermentation process.

Animal cloning: The process of producing a group of complete normal animals from somatic cells of an adult animal; a somatic cell is fused with an enucleated fertilized egg and the egg is implanted in the uterus of a surrogate mother.

Animal genetic resources databank: A databank that contains inventories of farm animal genetic resources and their immediate wild relatives, including any information that helps to characterize these resources.

Animal genome (gene) bank: A planned and managed repository containing animal genetic resources. Repositories include the environment in which the genetic resource has developed, or is now normally found (*in situ*) or facilities elsewhere (*ex situ: in vivo* or *in vitro*). For *in vitro, ex situ* genome bank facilities, germplasm is stored in the form of 1 or more of the following: semen, ova, embryos, and tissue samples.

Anion: A negatively charged ion; opposite is cation.

Anneal: The pairing of complementary DNA or RNA sequences, via hydrogen bonding, to form a double-standard polynucleotide. Most often used to describe the binding of a short primer or probe.

Annealing: The process of heating (denaturing step) and slowly cooling (renaturing step) double-stranded DNA to allow the formation of hybrid DNA or complementary strands of DNA or of DNA and RNA.

Annotation: A combination of comments, notations, references, and citations, either in free format or utilizing a controlled vocabulary, that together describe all the experimental and inferred information about a gene or protein. Annotations can also be applied to the description of other biological systems. Batch, automated annotation of bulk biological sequences is one of the key uses of bioinformatics tools.

Annual: (1) Taking 1 year, or occurring at intervals of 1 year. (2) In botany, a plant that completes its life cycle within 1 year. During this time, the plant germinates, grows, flowers, produces seeds, and dies.

Anonymous DNA marker: A DNA marker detectable by virtue of variation in its sequence, irrespective of whether or not it actually occurs in or near a coding sequence. Microsatellites are typical anonymous DNA markers.

Anti–Sm: An autoimmune antiserum that defines the Sm epitope that is common to a group of proteins found in snRNPs that are involved in RNA splicing.

Antibiotic: Classes of natural and synthetic compounds that inhibit the growth or kill some microorganisms.

Antibiotic resistance: The ability of a microorganism to produce a protein that disables an antibiotic or prevents transport of the antibiotic into the cell.

Antibody: A protein (immunoglobulin) molecule produced by the immune system that recognizes a particular substance (antigen) and binds to it.

Antibody engineering: Refers to the use of recombinant DNA technology to modify or design antibodies.

Antibody microarray: Consists of spots of different antibodies and is used to measure the abundance of thousands of different proteins in the samples.

Anticaking agents: Substances that keep powders, such as milk powder, from caking and sticking.

Anticipation: The process whereby some genetic diseases get more severe in each successive generation.

Anticoding strand: The strand of DNA double helix that is actually transcribed. Also known as antisense or template strand. The strand has $3' \rightarrow 5'$ direction, which directs the synthesis of mRNA in the $5' \rightarrow 3'$ direction.

Anticodon: The trinucleotide sequence in anticodon loop in a tRNA which pairs with a triplet (codon) by complementary base pairing.

Anticodon arm (of tRNA): A stem loop structure that express the anticodon triplet at 1 end.

Antifoaming agents: Substances that reduce or prevent foaming in foods/fermenters.

Antigen: A molecule that is capable of stimulating the production of neutralizing antibody proteins when injected into an organism. Any foreign substance for which entry into an organism provokes an immune response by stimulating the synthesis of an antibody (an immunoglobulin protein that can bind to the antigen).

Antigen-antibody interaction: Highly specific binding of an antibody to an antigen epitope, which induced its production.

Antigen-binding domain: In antibodies, the area that binds to the antigen-epitope.

Antigen presentation: The antigenic peptides bind within the cleft of MHC I (major histocompatibility complex I) or MHC II molecules and are displayed on the surface of antigen presenting cells (APCs).

Antigen presenting cells (APCs): These cells have on their surfaces class II MHC molecules, which bind to antigen-derived peptides and present them to a group of lymphocytes, which are then activated to mount an immune response.

Antigen processing: A foreign antigen is degraded into short peptides of 8–18 amino acids by antigen presenting cells.

Antigenic determinant: The portion of an antigen that is recognized by the antigen receptor on lymphocytes. It is also called an epitope.

Antigenic index: A prediction of the antigenicity of a sequence based on its hydrophilicity, predicted side chain flexibility, surface probability, and the predicted turns.

Antigenic specificity: Ability of antibodies to differentiate between antigen molecules differing by a single amino acid.

Antigenic variation: The ability of a trypanosome to change its surface protein, so that the host is challenged with a different antigen.

Antigenically Committed: The mature B cell is said to be antigenically committed since it is predetermined in its ability to recognize and respond to a specific antigen.

Antigenicity: Ability of an antigen to interact with an antibody, which is specific to it.

Antimicrobial agent: Any chemical or biological agent that harms the growth of microorganisms.

Antinutrients: Compounds that inhibit normal uptake of nutrients.

Anti-immunoglobulin: An antibody specific to another antibody; binds to the constant region of all antibodies of the class.

Anti-oncogene: A gene for which the product prevents the normal growth of tissue.

Antioxidants: A group of chemicals that prevent oxidation, for example, ascorbic acid, citric acid. They retard senescence and browning of tissue.

Antiparallel: The 2 DNA strands are organized in opposite directions, the 5′ end of 1 strand aligns with the 3′ end of other strand.

Antiporter: A type of carrier protein that simultaneously moves 2 different types of solutes in opposite directions across the plasma membrane.

Antisense DNA: The sequence of chromosomal DNA that is transcribed.

Antisense gene: A gene that produces a transcript (mRNA) that is complementary to the pre-mRNA or mRNA of a normal gene (unusually constructed by inverting the coding region relative to the promoter).

Antisense orientation: Reverse orientation of a gene in relation to its promoter; the promoter is now located at the 5′ end of the antisense strand so that natural sense strand is now transcribed.

Antisense RNA: RNA produced by transcription of the sense strand; usually, in the case of antisense gene constructs; complementary to antisense strand of the gene and to mRNA; pairs with mRNA. RNA with sequence complementary to mRNA.

Antisense RNA technology: An antisense construct of the target gene is integrated into the genome to suppress the concerned endogenous genes.

Antisense strand: Acts as the template for the synthesis of mRNA.

Antisense therapy: *In vivo* treatment of a genetic disease by blocking translation of a protein with a DNA or RNA sequence that is complementary to a specific mRNA.

Antiseptic: Any substance that kills or inhibits the growth of disease-causing microorganisms (a microorganism capable of causing sepsis), but is essentially nontoxic to cells of the body.

Antiserum: (1) Blood serum containing specific antibodies against an antigen. Antisera are used to confer passive immunity to many diseases. (2) The fluid portion of the blood of an animal (after coagulation of the blood), containing antibodies. (3) Serum from an animal that contains antibodies against a particular antigen.

Antiterminator: A type of protein that enables RNA polymerase to ignore certain transcriptional stop or termination signals and read through them to produce longer mRNA transcripts.

Antitermination: A mechanism of transcriptional control in which termination is prevented at a specific terminator site, allowing RNA polymerase to read into the genes beyond it.

Antitermination proteins: Proteins that allow RNA polymerase to transcribe through certain terminator sites.

Antitranspirant: A compound designed to reduce transpiration when sprayed or painted on leaves of newly transplanted trees, shrubs, or vines, or used as a dip for cuttings in lieu of misting; may interfere with photosynthesis and respiration if the coating is too thick or unbroken.

Anucleate: Bacteria that lack nuclei but are of similar shape to wild-type bacteria.

AP endonuclease: An enzyme that is involved in base excision repair.

Apoenzyme: Inactive enzyme that has to be associated with a specific organic molecule called a coenzyme in order to function. The apoenzyme/coenzyme complex is called a holoenzyme.

Apolipoprotein E: Certain alleles of the gene that encodes the protein apolipoprotein E have been associated with the development of heart disease and Alzheimer's disease.

Apomixis: The asexual production of diploid offspring without the fusion of gametes. The embryo develops by mitotic division of the maternal or paternal gamete, or, in the case of plants, by mitotic division of a diploid cell of the ovule.

Apomorphic character state: A character state that evolved in a recent ancestor of a subset of organisms in a group being studied.

Apoprotein: A protein without its coenzymes, cofactors, and prosthetic groups that are required for its functionality.

Apoptosis: Genetically controlled programmed cell death. For example, the resorption of the tadpole tail at the time of its metamorphosis into a frog; removal of tissue between fingers and toes of the fetus.

Aporepressor: A protein changed into a repressor when a corepressor binds to it.

AP site: A nucleotide position in a DNA molecule where the nitrogenous base component is missing.

Aquaculture: The cultivation of plants using water as the support medium. It can also mean rearing marine life under controlled conditions in water, such as fish farming.

Aqueous biphasic (2-phase) systems: Both phases are aqueous, but they contain different polymers, which do not mix together.

Aqueous-Organic biphasic (2-phase) systems: In these systems, the enzyme molecules are suspended in the organic solvent, while the second phase is aqueous; constituted by a thin layer of water surrounding each enzyme molecule.

***Arabidopsis*:** A genus of flowering plants in the Cruciferae family. *A. thaliana* is used in research as a model plant because it has a small genome (5 pairs of chromosomes 2n = 10) and can be cultured easily with a generation time of 2 mo.

Arabinogalactan protein (AGP): A group of glycoproteins; induce somatic embryogenesis in otherwise nonembryogenic cells, for example, of carrot (cultured *in vitro*).

Archaea: One of the 2 main groups of prokaryotes, mostly found in extreme environments.

Arginine (Arg, R): An amino acid.

Arm: One of the 4 (or, in some cases, 5) arms of tRNA stem-loop structures that make up the secondary structure. The arms of a lambda phage attachment site are the sequences flanking the core region where the recombination event occurs.

Arms race: A phenomenon of evolution in which 2 or more species adapt to one another in a co-evolutionary manner.

Arrayed library: Individual primary recombinant clones (hosted in phage, cosmid, YAC, or other vector) that are placed in 2-dimensional arrays in microtiter dishes.

Array of hairpins: An assembly of alpha-helices that cannot be described as a bundle or a folded leaf.

ARS (autonomously replicating sequence): An origin for replication in yeast. The common feature among different ARS sequences is a conserved 11 bp sequence called the A-domain.

ARS vector: A yeast vector having ARS (autonomously replicating sequence) module; it is quickly lost if selection pressure is absent.

Artificial chromosome: A circular or linear vector that is stably maintained in, usually 1–2 copies per cell; several types: bacterial artificial chromosome (BAC), PI-derived artificial chromosome (PAC), yeast artificial chromosome (YAC), mammalian artificial chromosome (MAC).

Artificial intelligence (AI): The study or application of techniques to emulate "intelligent" behaviors, such as planning, vision, or language processing.

Artificial intelligence in medicine (AIM): The application of artificial intelligence methods to solve problems in medicine, such as diagnosis or therapy planning.

Artificial life: The study of biological processes within the confines of a computer.

Artificial medium: In plant tissue culture: a chemically defined medium, having inorganic salts, vitamins, sucrose, agar, and the required growth regulators. In animal cell and tissue culture: medium having various organic and inorganic nutrients; may or may not contain serum, etc.

Artificial neural network (ANN): A network of neurons that are connected graphically through synapses or weights.

Artificial seeds: A synthetic seed consisting of a somatic embryo surrounded by nutrient medium that is protected by a thin membrane of chemical.

Artificial selection: The practice of choosing individuals from a population for reproduction, usually because these individuals possess 1 or more desirable traits.

Asbestos fibers: Crystals of asbestos, a natural mineral, that have the form of minute strands; asbestos is a serious health hazard in indoor spaces.

Ascertainment: In scientific research, ascertainment bias occurs when false results are produced by nonrandom sampling, and conclusions made about an entire group are based on a distorted or nontypical sample. Ascertainment bias may be easy to recognize or difficult to detect.

Ascites: The fluids that accumulate in the peritoneal cavity due to tumor cells. Hybridomas grown in the peritoneal cavity of mice produce ascites containing high concentrations of the monoclonal antibody chosen for expression in the selection of the hybridoma cell line.

Ascorbic acid (vitamin c, $C_6H_8O_6$, f.w. 176.12): A water-soluble vitamin present naturally in some plants, and also synthetically produced. Aside from its role as a vitamin, it is used as an antioxidant in plant tissue culture and included in disinfection solutions.

Ascospore: One of the spores contained in the ascus of certain fungi.

Ascus: Reproductive sac in the sexual stage of a type of fungi (Ascomycetes) in which ascospores are produced.

Aseptic: Free from contamination by microorganisms.

Asexual embryogenesis: The sequence of events whereby embryos develop from somatic cells, also known as somatic embryogenesis.

Asexual propagation: Vegetative, somatic, nonsexual reproduction of a plant, without fertilization.

Asexual reproduction: (1) Reproduction that does not involve the formation and union of gametes from the different sexes or mating types. It occurs mainly in lower animals, microorganisms, and plants. In plants, asexual reproduction is by vegetative propagation (e.g., bulbs, tubers, corms) and by formation of spores. (2) Nonsexual means of reproduction which can include grafting and budding.

Asparagine (Asn, N): One of the 20 most common natural amino acids on earth. It has carboxamide as the side chain's functional group. It is not an essential amino acid. Its codons are AAU and AAC.

Aspartic acid (Asp, D): An α-amino acid with the chemical formula $HO_2CCH(NH_2)CH_2CO_2H$. the carboxylate anion of aspartic acid is known as aspartate. The L-isomer of asparatate is one of the 20 proteinogenic amino acids, that is, the building blocks of proteins. Its codons are GAU and GAC.

Assay: A test to detect the presence of some substances in a small amount in solution.

Assay culture: Microbial culture used for the assay of biological activities of microbial products.

Assembly: The process of correctly joining together the DNA sequences from individual sequencing experiments into a contiguous segment.

Assembly factor: A protein that is required for formation of a macromolecular structure but is not itself part of that structure.

Assimilation: Transformation of food into protoplasm.

Assortative mating: The tendency for mates to be chosen nonrandomly.

Assortment: The movement of homologous chromosomes to opposite poles in the first meiotic division and chromatids in the second meiotic division, leading to the segregation of alleles.

Astrocytoma: A type of brain tumor that begins in the astrocyte cells of the brain or spinal chord.

Asymmetrical exon: An exon flanked by introns of different phase classes.

Asymmetric gap LCR (AG-LCR): LCR using reverse transcriptase; specific for detection of RNA.

Asymmetric PCR: The 2 PCR primers are used in a 100:1 ratio so that 1 primer is finished ~10 cycles before the other; yields predominantly single-stranded copies that can be used for DNA sequencing.

Asymmetric somatic hybrid: A somatic hybrid that contains a full chromosome complement of 1 fusion parent plus 1or few chromosomes from the other fusion parent.

Asymmetric unit: In a crystal, the level at which there is no symmetry. For example, the alpha-beta dimer can be considered to be the asymmetric unit of the hemoglobin tetramer in solution.

Asynapsis: The failure or partial failure in the pairing of homologous chromosomes during the meiotic prophase.

ATPase: An enzyme that brings about the hydrolysis of ATP by the cleavage of either 1 phosphate group with the formation of ADP and inorganic phosphate, or 2 phosphate groups, with the formation of AMP and pyrophosphate.

ATP synthase: A group of polypeptides that converts ADP and inorganic phosphate (Pi) to ATP and H_2O.

Attenuated: Weakened with reference to vaccines, made from pathogenetic organisms that have been treated so as to render them avirulent.

Attenuated vaccine: (1) A virulent organism that has been modified to produce a less virulent form, but nevertheless retains the ability to elicit antibodies against the virulent form. (2) A vaccine composed of infectious organisms (typically whole viruses) that exhibit low virulence or that cannot multiply in their host because of their inactivation or attenuation by exposure to radiation, chemical treatment, or genetic manipulation.

Attenuation: Regulation of a bacterial operon by termination of transcription at a site that is located before the first structural gene.

Attenuator: A nucleotide sequence in the fifth region of a prokaryotic gene (or in its RNA) that causes premature termination of transcription, possibly by forming a secondary structure.

Autoantibodies: Antibodies that the immune system erroneously forms against self-antigens.

Autoantigen: A self-protein that is the target of an autoimmune response.

Autocatalysis: Catalysis in which one of the products of the reaction is a catalyst for the reaction. Usually, the catalysis starts slowly and increases as the quantity of the catalyst increases, falling off as the product is used up.

Autoclavable: A substance, which does not deteriorate or undergo alteration due to autoclaving.

Autoclave: An instrument used for sterilization of glassware and culture media.

Autoclaving: Sterilization using moist heat at 15 psi (1.06 kg/cm^2) and 121°C for 5–40 min; the duration of autoclaving depends on the nature of liquid and its volume; efficiency of sterilization depends on temperature and not directly on pressure.

Autograft: Graft obtained from the body of the recipient itself.

Autoimmune disease: (1) A disease in which the body produces antibodies against its own tissues. (2) Disorder in which the immune systems of affected individuals produce antibodies against molecules that are normally produced by those individuals (called self-antigens).

Autoimmune lymphoproliferative syndrome (ALPS): A human disease caused by failure of lymphocytes to die once they have finished doing their job. As a result, the spleen and lymph nodes grow large, and immune cells may attack the body's own tissues, a condition known as autoimmunity.

Autoimmunity: (1) A condition in which the body produces an immune response against its own organs or tissues. (2) An immune response to tissues, cells, or molecular components of the host. The pathological result of this response is an autoimmune disease. (3) A disorder in the body's defense mechanisms in which an immune response is elicited against its own (self) tissues.

Autologous: Term used to denote derivation from self. Tissue from one person returned to the same person is termed an autologous sample.

Autologous cells: Cells that are taken from an individual, cultured (or stored), and, possibly, genetically manipulated before being infused back into the original donor.

Autolysis: The process of self destruction of a cell, cell organelle, or tissue. It occurs by the action of lysosomic enzymes.

Autonomous: A term applied to any biological unit that can function on its own, that is, without the help of another unit, such as a transposable element that encodes an enzyme for its own transposition.

Autonomous agent: An entity with limited perception of its environment that can process information to calculate an action so as to be goal-seeking on a local scale.

Autonomously replicating sequence (ARS): Any cloned DNA sequence that initiates and supports extrachromosomal replication of a DNA molecule in a host cell. It is often used in yeast cells.

Autopolyploid: A polyploidy resulting from the doubling of a single genome.

Autoradiograph: A picture prepared by labeling a substance such as DNA with a radioactive material such as tritiated thymidine and allowing the image produced by decay radiation to develop on a film over a period of time.

Autoradiography: A technique for the detection of radioactively labelled molecules by overlaying the specimen with photographic film. When the film is developed, an image is produced which corresponds to the location of the radioactivity.

Autoregulation: Product of a gene that controls the level of its own synthesis.

Autosomal: A locus on any chromosome other than a sex chromosome. Not sex-linked.

Autosome: A chromosome that is not involved in sex determination.

Autotrophic: Characteristic of plants that are capable of manufacturing their own food, as in photosynthesis.

Autotrophic phase of embryo development: *In ovulo*: embryo capable of synthesizing all the biochemicals it needs; the specific developmental stage varies with the species, for example, in *Capsella*, it is in the late heart stage and later stages.

Autotrophy: The capacity of an organism to use light as the sole energy source in the synthesis of organic material from inorganic elements or compounds. Autotrophic organisms include green photosynthesizing plants and some photosynthetic bacteria.

Auxins: A class of plant growth regulators that stimulates cell division, cell elongation, apical dominance, and root initiation. 2,4-D, IAA, IBA, and NAA are some of the auxins commonly used in plant cell and tissue culture.

Auxin-cytokinin ratio: The relative proportion of auxin to cytokinin present in plant-tissue culture media. Varying the relative amounts of these 2 hormone groups in tissue culture formulas affects the proportional growth of shoots and roots *in vitro*. As the ratio is increased (increased auxin or decreased cytokinin), roots are more likely to be produced, and as it is, decreased root growth declines and shoot initiation and growth are promoted. This relationship was first recognized by C.O. Miller and F. Skoog in the 1950s.

Auxotroph: A mutant microorganism or cell line requiring a substance for growth that can be synthesized by wild-type strains.

Auxotrophic Mutant: A mutant requiring a biochemical for its normal growth and development because it cannot synthesize this biochemical because of a mutation in a gene.

Availability: A reflection of the form and location of nutritional elements and their suitability for absorption. In tissue culture media, this is related to the abundance of each nutritional element, the osmotic concentration and pH of the medium, the stability and solubility of the item in question, the presence of absorbing agents in the media, and other factors.

Avidin: A protein (derived from egg white) that binds very tightly to biotin This interaction is exploited in several biological as says, such as ELISA and Western blots.

Avidity: A measurement of the binding of an antiserum to a complex antigen. The avidity of the antiserum depends on the affinities of the various antibodies present in the antiserum to bind to different epitopes on the antigen.

Avidity model: According to this model, a low avidity interaction between TCR of double positive (CD_4^+ CD_8^+) T cells and antigen peptide-MHC complex/MHC molecules leads to positive selection; in contrast, negative selection is the consequence of this interaction being of a high avidity.

Avirulent: Lacking virulence; a microorganism lacking the properties that normally cause disease.

Axenic culture: Free of external contaminants and internal symbionts, which is generally not possible with surface sterilization alone and is sometimes incorrectly used to indicate aseptic culture.

Axillary: An adjective describing the relative position of a bud, that is, axillary bud, in the axils of leaves.

Axillary bud: A bud found in the axil of a leaf (synonymous with lateral bud).

Axillary bud proliferation: A technique of micropropagation of plants in culture which is achieved primarily through hormonal inhibition of apical dominance and stimulation of lateral branching.

Axiom: A statement that is assumed to be true.

Azoospermia: A lack of motile sperms in the semen.

B: An amino acid that is either aspartic acid or glutamic acid.

B Cells: An important class of white blood cells that mature in bone marrow and produce antibodies. They are largely responsible for the antibody-mediated or humoral immune response. They give rise to the antibody-producing plasma cells and some other cells of the immune system.

B lymphocytes (B cells): Lymphocytes derived from bone marrow cells that produce antibodies.

Back-bulb: In sympodial orchids; an old pseudobulb (often without leaves); alive and bears 1 or more buds behind the actively growing portion.

Back-bulb propagation: Procedure that involves separation of the oldest pseudobulb to force the development of its dormant bud.

Backcross: Hybrid is crossed to one of its 2 parents.

Background selection: Selection for recurrent parent genotype in a backcross program; also used in a pedigree selection scheme to ensure the recovery of a specified level of genome from the desired parent.

Backpropagation: An algorithm for efficiently calculating the error gradient of a neutral network, which can then be used as the basis of learning.

Backtracking algorithm: The process of repeatedly exploring paths until you encounter the solution.

BAC library: A resource that identifies a sequence of bacterial artificial chromosomes (BACs).

Bacillus Calmette-Guerin (BCG): An attenuated strain of *Mycobacterium tuberculosis* that is used as a vaccine for tuberculosis.

Bacillus: A rod-shaped bacterium.

Bacillus thuringiensis (BT): A toxin produced by a bacterium, such as BT toxin by Bacillus thuringiensis.

Bacteria: Structurally simple, single-celled organisms that have no nucleus.

Bacterial artificial chromosome (BAC): A synthetic DNA molecule that contains sequences required for replication and segregation in bacteria; used to amplify 100–200 kb long DNA sequences.

Bacterial lawn: A continuous bacterial growth on an agar plate.

Bacteriocide: A chemical or drug that kills bacterial cells.

Bacteriocin: A protein produced by bacteria of 1 strain that is active against those of a closely related strain.

Bacteriophage: A virus that infects and replicates in bacteria. Also called, simply, phage.

Bacteriostat: A substance that inhibits or slows down the growth and reproduction of bacteria.

Bacteriostatic agent: An agent that has a biostatic effect on bacteria.

Bacterium: (1) A single-celled, microscopic prokaryotic organism; a single-celled organism without a distinct nucleus. (2) A structurally simple single cell with no nucleus. One bacterium, *Bacillus thuringiensis*, produces a protein toxic to certain destructive insects. Having plants produce this protein after genetic engineering is a form of built-in pest resistance in plants.

Bacterization: A technique of seed-dressing with bacteria as a water soluble suspension used for introduction of bacteria in soil.

Bacteroids: The pleomorphic forms of *Rhizobium* present inside root nodules that fix atmospheric nitrogen gas.

Baculovirus: A virus that infects arthropods, including mostly insects.

Bait: In yeast, 2-hybrid (Y2H) system, a protein (say, protein X) expressed as a fusion (a hybrid) protein with the DNA-binding domain of a transcription factor; used to find a protein that interacts with X.

Balanced lethal system: A system for maintaining a recessive lethal allele at each of 2 loci on the same pair of chromosomes. In a closed population with no crossing-over between the loci, only the double heterozygotes for the lethal mutations survive.

Balanced polymorphism: A polymorphism that is stable over time and is maintained by balanced selection.

Balanced selection: A selection regime that results in the maintenance of 2 or more alleles at a locus in a population; for example, over-dominance.

Balanced tree: A common data structure for creating indexes on disk.

Balloon angioplasty: A medical procedure in which a catheter with a balloon at the tip is used to clear or repair damaged blood vessels.

Band: A chromatographic zone, that is, a region where the separated substance is concentrated.

Bar: A unit used for pressure of a liquid. 1 bar = 10^5 Pa. 1 bar is approximately equal to 1 atmosphere.

Bar body: The highly condensed chromatin formed by an inactivated X chromosome.

Barley malt: Barley seeds germinated at 17°C to a specific stage and dried at 65°C and, usually, powdered.

***Barnase*:** A gene isolated from *Bacillus amyloliquefaciens* that encodes an RNase, which is cytotoxic to the cell expressing it; used to produce male sterility that is restored by *barstar*.

Barrier to gene flow: A factor, such as geographic, mechanical, and behavioral isolating mechanisms that restrict gene flow between populations, leading to populations with differing allele frequencies.

***Barstar*:** A gene isolated from *Bacillus amyloliquefaciens* that encodes a specific inhibitor of *barnase* RNase; used for fertility restoration in *barnase* male sterile plants.

Basal: (1) Located at the base of a plant or a plant organ. (2) A fundamental formulation of a tissue culture medium.

Basal body: Small granule to which a cilium or flagellum is attached.

Basal factors: Components of the nucleus that associate with chromosomes and enhance gene transcription.

Basal medium: A growth regulator-free medium; that is, a medium containing inorganic salts, vitamins, sucrose, and agar (if needed).

Basal promoter: The position within a eukaryotic promoter where the initiation complex is assembled.

Basal rate of transcription: The number of productive initiations of transcription occurring per unit time at a particular promoter.

Base: One of 5 molecules that are assembled, along with a ribose and a phosphate, to form nucleotides. Adenine (A), guanine (G), cytosine (C), and thymine (T) are found in DNA; RNA is made from adenine (A), guanine (G), cytosine (C), and uracil (U).

Base analog: A chemical whose molecular structure mimics that of a DNA base; because of the mimicry, the analog may act as a mutagen.

Base collection: A collection of seed stock or vegetative propagating material (ranging from tissue cultures to whole plants) held for long-term security in order to preserve the genetic variation for scientific purposes and as a basis for plant breeding as multiplication and evaluation.

Base excision repair: A DNA repair process that involves excision and replacement of an abnormal base.

Baseless site: A position in a DNA molecule where the base component of the nucleotide is missing.

Base pair (bp): It is a partnership of A with T or C with G in a DNA double helix; other pairs can be formed in RNA under certain circumstances.

Base sequence: The order of nucleotide bases in a DNA or RNA molecule.

Base sequence analysis: A method, sometimes automated, for determining the base sequence of DNA or RNA.

Base stacking: The hydrophobic interactions that occur between adjacent base pairs in a double-stranded DNA molecule.

Base substitution: Replacement of 1 base by another in a DNA molecule.

Basic chromosome number: Gametic chromosome number of a diploid species of the parental diploid species in case of a polyploidy species.

Basic domain: A type of DNA-binding domain.

Basin of attraction: The subset of states of a dynamical systems' state space that converge into an attractor.

Basin portrait: The set of basins of attraction for a dynamical system.

Basophil: (1) A type of white blood cell (leukocyte) produced by stem cells in the red bone marrow. (2) A multinucleated leukocyte (white blood cell) with granules in its cytoplasm containing acid glycoproteins that release histamine and other mediators of the hypersensitivity response.

Batch bioreactor: In such a bioreactor, the medium and inoculum are located into the reactor in the beginning, and the cells are allowed to grow. There is no further addition or replacement of medium thereafter.

Batch culture: A cell suspension culture in which cells are grown in a finite volume of nutrient medium and follow a sigmoid pattern of growth.

Batch digesters: These are suited for digestion of solid wastes, where hydrolysis occurs in a solid bed, and the leachate is circulated through a methanogenic digester and back over the solid waste bed.

Batch fermentation: Fermentation carried out by growing microbes in batch culture.

BB rat: A mutant rat strain that is susceptible to, and therefore acts as a model for, insulin-dependent diabetes.

B chromosome: A chromosome possessed by some individuals in a population, but not all.

B-DNA: The normal form of DNA found in biological systems. It exists as a right-handed helix.

B-form: Right-handed double helix of DNA with 10 base pairs per complete turn (360°) of the helix.

Bead bed reactor: In animal cell culture; column packed with 3–5 mm glass beads, in which cells attach to the bead, surface, and medium is pumped up or down the column.

Beads-on-a-string: An unpacked form of chromatin that consists of nucleosome beads on a string of DNA.

Beer: Undistilled beverage from barley malt fermented by yeast.

Beer wort: Filtered mash in beer making.

Behavioral genetics: The study of genes that may influence behavior.

Behavioral resistance: The ability of an insect population to change its behavior in order to avoid insecticides or other injurious factors.

Behavioral strategy: A type of game-theoretic strategy.

Beige mouse: A mutant strain of mouse that is prone to cancer, due to its lack of Natural Killer (NK) cells.

Bence-Jones proteins: Free monoclonal antibody light chains found in the urine of patients with multiple myeloma. A source of large quantities of antibodies prior to the development of monoclonal antibody (hybridoma) technology.

Bench-scale process: A small- or laboratory-scale process commonly used in connection with fermentation.

Benign tumor: A noncancerous growth that does not spread to other parts of the body.

Bergmann's cell planting technique: Cells are suspended in melted agar medium and spread in a 1-mm thick layer, usually in a petri plate.

BEST map: Map indicating the correlation between bacterial artificial chromosome and expressed sequence tags.

Beta DNA: The normal form of DNA found in biological systems, which exists as a right-handed helix.

Beta function: A type of cell in the pancreas. Within the pancreas, the beta cells are located in areas called the islets of Langerhans where they constitute the predominant type of cell. The beta cells make and release insulin, a hormone that controls the level of glucose (sugar) in the blood. Degeneration of the beta cells is the main cause of type I (insulin-dependent) diabetes mellitus.

Beta lactamase: Ampicillin-resistance gene.

Beta sheet/ Beta strand: A 3-dimensional arrangement taken up by polypeptide chains that consists of alternating strands linked by hydrogen bonds. The alternating strands together form a sheet that is frequently twisted. One of the secondary structural elements, along with the alpha-helix, characteristic of proteins.

Beta-galactosidase: An enzyme that catalyzes the formation of glucose and galactose from lactose.

Bidirectional replication: DNA replication in which 2 replication forks move in opposite directions, away from the same origin of replication.

Biennial: In botany, a plant which completes its life cycle within 2 years and then dies. For most biennial plants, the 2 growing seasons have to be separated by a period of cold temperature sufficient to induce flowering and fruit formation.

Bifunctional Enzyme: An enzyme, which can catalyze 2 reactions, encoded by a recombinant gene.

Binary Vector System: A 2-plasmid system in *Agrobacterium tumefaciens* for gene transfer in plant cells. One plasmid contains the virulence gene for infection, which is responsible for transfer of T-DNA present on other plasmid. The other plasmid contains the T-DNA borders, the selectable marker and the DNA to be transferred.

Binding: The ability of molecules to stick to each other because of the exact shape and chemical nature of parts of their surfaces. Many biological molecules bind extremely tightly and specifically to other molecules: enzymes to their substrates, antibodies to their antigens, DNA strands to their complementary strands so on. Binding can be characterized by a binding constant or association constant (Ka), or its inverse, the dissociation constant (Kd).

Binding affinity: Measure of the strength of interaction between 2 molecules.

Binomial expansion: The probability that an event will occur 0, 1, 2….n times out of n. It is given by the successive terms of the expression $(p + q)n$, where p is the probability of the event occurring, and $q = 1\text{-}p$.

Binomial nomenclature: In biology, each species is generally identified by 2 terms, the first is the genus to which it belongs, and the second is the specific epithet that distinguishes it from others in that genus (e.g., *Quercus suber*, cork oak).

Binucleate pollen: A pollen having 2 nuclei produced by the first pollen mitosis. It is the most suitable stage for another culture.

Bio: A prefix derived from bios and used in scientific words to associate the concept of "living organisms."

Bioaccumulation: Concentration of a chemical agent (e.g., DDT) in the increasing amount in the organism of a food chain.

Bioassay: A procedure for the assessment of a substance by measuring its effect in living cells or on organisms.

Bio-augmentation: Increasing the activity of bacteria that decompose pollutants, a technique used in bioremediation.

Biobutanol: Butanol produced by microorganisms from biomass.

Bio-catalysis: The aim of bio-catalysis is to produce a wide variety of chemical substances efficiently in a single step by means of microorganisms or enzymes. Traditional chemical processes are usually more expensive because they require several synthesis steps.

Biocatalyst: Use of enzymes to catalyze chemical reactions.

Biochemical oxygen demand (BOD): Amount of O_2 used up after 120 ± 1 h incubation of 1 L water at 20°C in complete dark.

Biochemical pesticides: These are naturally occurring substances that control pests by nontoxic means. These pesticides include natural insect growth regulators, for example, neem insecticide and insect pheromones that disrupt mating and reduce reproductive rates.

Biocidal agents: Agents that kill the susceptible microorganisms.

Biocontrol: Any process using the living organism to inhibit the growth and development of the pathogenic organisms.

Biocontrol agents: Microorganisms, for example, viruses, bacteria, fungi, and protozoa, employed for the control of insect pests, pathogens, or weeds.

Bioconversion: Conversion of one chemical into another by living organisms, as opposed to their organisms, as opposed to their conversion (which is biotransformation) or by chemical processes.

Biodegradable: Capable of being biologically decomposed by microorganisms or enzymes.

Biodegradation: The gradual breakdown of a compound to its constituents by a living organism.

Biodiesel: Diesel obtained from conversion of hydrocarbons accumulated by certain plants.

Biodiversity: The variability among the living organism from all sources, soil, water, air, extreme habitat, or associated with organisms.

Bioenergetics: The study of the flow and the transformations of energy that occur in living organisms.

Bioengineering: The use of artificial tissues, organs, and organ components to replace parts of the body that are damaged, lost, or malfunctioning.

Bioenrichment: Adding nutrients or oxygen to increase microbial breakdown of pollutants.

Bioethanol: Ethanol produced from biomass by microorganisms.

Bioethics: (1) The branch of ethics that deals with the life sciences and their potential impact on society. (2) A discipline dealing with the ethical implications of biological research and applications.

Biofertilizers (microbial inoculants): Commercial preparation of microorganisms by which using increases nitrogen and phosphorus levels and, thus, growth of plants increased.

Biofilm: In sewage water treatment; the layer of microorganisms present on filter particles.

Biofilter: A device to remove/breakdown/convert hazardous gases using biological agents maintained either in a solid support or immobilized on a membrane.

Biofortification: It is a method of breeding crops to increase their nutritional value; it can be achieved through conventional selective breeding or through genetic engineering.

Biofortified food: Refers to the food, obtained from those crop plants that have been developed through biofortification.

Biofuel: A gaseous, liquid, or solid fuel that contains energy derived from a biological source.

Biogas: A mixture of methane and carbon dioxide resulting from the anaerobic decomposition of waste such as domestic, industrial, and agricultural sewage, a.k.a. gobar.

Biogenesis: The principle that a living organism can only arise from other living organisms similar to itself and can never originate from nonliving material.

Bioherbicides: Biological agents, mainly fungi and insects, used for control of weeds.

Biohydrogen: H_2 produced by biological agents.

Bioinformatics: The science associated with biomolecular database-like activities involving persistent sets of data that are maintained in a consistent state over essentially indefinite periods of time.

Bioinsecticides: Use of biocontrol agents or their byproducts, including plant extracts, to control insect pests.

Bioleaching: Solubilization of metals in ores by using a solution containing specific microorganisms.

Biolistics: A technique to insert DNA into cells. Also known as microprojectile bombardment or particle bombardment.

Biological containment: Restricting the movement of genetically engineered organisms by arranging barriers to prevent them from growing outside the laboratory.

Biological containment for plants: In the case of plants, biological containment procedures aim at minimizing the risk of dispersal of pollen and seeds of the plants and of dissemination and establishment of plant-associated microorganisms and insects.

Biological control: Use of biological agents to control insect pests, pathogens, or weeds.

Biological diversity: Biological diversity, called biodiversity in short, refers to the diversity present among all living organisms found in the natural world; this concept includes diversity among species, ecosystems, and genetic diversity within a given species.

Biological indicators: Organisms, whose relative abundance/absence in river water serves as a good indicator of the level of pollution.

Biological information: Refers to the information present in the genome of an organism.

Biological invasion: When exotic species introduced in new areas and naïve species are subject to competition for food and space, and the exotic species eliminate the native species.

Biological nutrient removal: A process employed in sewage treatment to remove nitrogen and phosphorous from the effluent coming from secondary treatment.

Biological resistance: Ability of an insect population to develop enzymes that detoxify an insecticide or other injurious substance before it can reach its site of action.

Biological wealth: The life sustaining combination of commercial, scientific, and aesthetic values imparted to a region by its biota.

Biologics: Agents, such as vaccines, that give immunity to diseases or harmful biotic stresses.

Biomagnification: Progressive increase in the concentration of a lipid soluble xenobiotic compound as it passes through the food chain.

Biomass: The cell mass produced by a population of living organisms.

Biomass concentration: The amount of biological material in a specific volume.

Biomass energy: Energy produced by burning plant related materials such as fire wood.

Biome: A major ecological community or complex of communities, extending over a large geographical area and characterized by a dominant type of vegetation.

Biomethane: Methane produced by biological agents.

Biometry: The application of statistical methods to the analysis of biological problems.

Biomining: Solubilization and recovery of metals and ores.

Biopesticides: Commercial preparation of microorganisms by using which insects, termite or nematode pests are killed or their level is minimized and crop yield increased.

Biopharmaceuticals: Biotechnology products having pharmaceutical applications.

Biopolymer: Any polymer, such as proteins, nucleic acids, lipids, polysaccharides, produced by a living organism.

Bioprocess: Any process that uses complete living cells or their components (e.g., enzymes, chloroplasts) to effect desired physical or chemical changes.

Bioreactor: A vessel, that is, fermenter used for fermentation or other processes.

Bioremediation: The process of using organisms to remove contaminants, pollution, or unwanted substances from soil or water.

Biosensor: An electronic device that uses biological molecules or cells to detect specific compounds.

Biosource: A patented technology that permits a rapid and effective elucidation of functions of unknown genes/DNA sequences.

Biosphere: The part of the earth and its atmosphere that is inhabited by living organisms.

Biosphere reserves: Natural areas set aside to protect genetic biodiversity.

Biosurface: The region on a biological protein, enzyme, or receptor that acts as a binding site for ligands or other molecules.

Biosynthesis: Synthesis of compounds by living cells, which is the essential feature of anabolism.

Biostatic agents: Agents that inhibit growth and multiplication of susceptible microorganisms.

Biota: The sum total of al living organisms of a region.

Biotechnology: The scientific manipulation of living organisms, especially at the molecular genetic level, to produce useful products. Gene splicing and use of recombinant DNA (rDNA) are major techniques used.

Biotic elicitor: An elicitor of biological origin; usually, a component of cell wall, for example, pectin, pectic acid, chitin, chitosan.

Biotic factor: Other living organisms that are a factor of an organism's environment, and form the biotic environment, affecting the organism in many ways.

Biotic stress: (1) Living organisms which can harm plants, such as viruses, fungi, bacteria, and harmful insects. (2) Stress resulting from living organisms that can harm plants, such as viruses, fungi, bacteria, parasitic weeds, and harmful insects.

Biotin: A vitamin of the B complex. It is a coenzyme for various enzymes that catalyze the incorporation of carbon dioxide into various compounds.

Biotope: A small habitat in a large community.

Biotoxin: A naturally produced toxic compound that shows pronounced biological activity and presumably has some adaptive significance to the organism that produces it.

Biotransformation or bioconversion: The use of cultured cells to convert substrates into desired organic compounds by virtue of an endogenous enzyme that catalyzes the reaction.

Biphase liquid system: Consists of 2 immiscible liquids in which the enzyme is able to function and remains fairly stable; 2 types: aqueous 2-phase systems and aqueous-organic 2-phase systems.

Bipolar germination: In most angiosperms; development of roots and shoots from different poles of the embryo.

Bispecific monoclonal antibody: A hybrid antibody specific for 2 different epitopes; usually, each epitope belongs to a different antigen.

Bivalent: Having 2 binding sites with 2 free electrons available for binding.

B-lactamase: An ampicillin-resistance gene.

Blastocyst: A mammalian embryo (fertilized ovum) in the early stages of development, approximately up to the time of implantation. It consists of a hollow ball of cells.

Blastomere: Any 1 of the cells formed from the first few cleavages in animal embryology. The embryo usually divides into 2, then 4, then 8 blastomeres, and so on.

Blastula: In animals, an early embryo form that follows the morula stage, typically a single-layered sheet (blastoderm) or ball of cells (blastocyst).

Blot: A spot on membrane filter gel made by adopting a solution.

Blotting: The process of transferring DNA, RNA, or protein from a polyacrylamide or agarose gel to a nitrocellulose filter for further analysis (for examples, see Southern, Northern, and Western blotting).

Blot transfer: Transfer of blot from a gel to a membrane filter.

Blunt end: The end of a DNA duplex molecule in which neither strand extends beyond the other, also known as flush end.

Blunt-end cut: To cleave phosphodiester bonds in the backbone of duplex DNA between the corresponding nucleotide pairs on opposite strands. This cleavage process results in both strands finishing at the same residue, that is, there are no nucleotide extensions on either strand, a.k.a. flush-end cut.

Blunt-end ligation: Joining of nucleotides present at ends of 2 DNA molecules.

Bone marrow: The site of origination of stem cells, which are the precursors of all blood and immune cells.

Bonus trait: When progeny from a somatic hybrid possess some novel feature not present in their fusion parents; such a feature is called bonus trait.

Booster (vaccination): The secondary immunization sometimes required to "boost" or increase the weak immune response obtained from the first immunization.

Boring platform: Sterile bottom half of a petri dish, used for preparing explants with a cork borer.

Botanical insecticide: An insecticide produced from a plant or plant product, for example, pyrethrum.

Bound water: Water held by the cell and not released if fertilizing occurs in the intracellular space.

Boundary: Edge of a zone; as of macromolecule solution next to the solvent.

Bovine somatotrophin (bst): Bovine growth hormone, this protein is found naturally in cattle, and is the bovine counterpart of human growth hormone, one of the earliest biopharmaceutical products. It has been cloned, using recombinant DNA technology, expressed in large amounts, and marketed as an agricultural product to improve the growth rate and protein:fat ratios in farm cattle and to improve milk yield; its use is banned in some countries.

Bovine spongiform encephalopathy (BSE): Mad cow disease.

Bp: Abbreviation for base pair.

Bract: A modified leaf that associated with a reproductive structure such as flowers or inflorescences and may appear to be a petal.

Brca1: A gene located on chromosome 17, that in its defective form predisposes some individuals to breast, ovary, and prostate cancer.

Bradykinin: A kinin involved in inflammatory reactions, generated following tissue injury. Bradykinin is involved in vasodilation, smooth muscle contraction, and increased vascular permeability and is an important target for anti-inflammatory drugs.

BrdU (5-bromodeoxyuridine): A mutagenically active analog of thymidine in which the methyl group at the 5 position is replaced by bromine.

Breed: (1) A sub-specific group of domestic livestock with definable and identifiable external characteristics that enable it to be separated by visual appraisal from other similarly defined groups within the same species or (2) A group of domestic livestock for which geographical and/or cultural separation from phenotypically similar groups has led to acceptance of its separate identity breed at risk, breed not at risk, critical breed, critical-maintained breed, endangered-maintained breed.

Breed at risk: Any breed that may become extinct if the factors causing its decline in numbers are not eliminated or mitigated. Breeds may be in danger to becoming extinct for a variety of reasons. Risk of extinction may result from inter alia, low population size, direct and indirect impacts of policy at the farm, country or international levels, lack of proper breed organization, or lack of adaptation to market demands.

Breed not at risk: A breed where the total number of breeding females and males is greater than 1,000 and 20, respectively, or the population size approaches 1,000 and the percentage of pure-bred females is close to 100%, and the overall population size is increasing.

Breeder's exemption: Under plant breeder rights (PBR) regime; the use of a protected variety (the initial variety) for the development of new varieties is exempted from protection.

Breeding: (1) In classical plant breeding methods, suitable cross-breeding partners are selected in an attempt to provide plants with the desired characteristics. The result, however, is dependent on chance. Modern biotechnology is able to produce organisms with specially selected characteristics. (2) The process of sexual reproduction and production of offspring.

Brewer's yeast: Strain of yeast, often *Saccharomyces cerevisiae*, that is used in the production of beer.

Brewing: The process by which beer is made. In the first stage, the barley grain is soaked in water and allowed to germinate (malting) during which the natural enzymes of the grain convert the seed starch to maltose and then to glucose. Grain is then dried, crushed, and added to water at a specific temperature (steeping) and any remaining starch is converted to sugar. The resulting liquid (wort) is the raw material to which yeast is added to convert sugar to alcohol.

Bridge: A filter paper or other substrate used as a wick and support structure for a plant tissue in culture when a liquid medium is used.

Broad-host-range plasmid: A plasmid that can replicate in a number of different bacterial species.

Broad-sense heritability: In quantitative genetics, the proportion of the total phenotypic variation due to genetic variation.

Broth: In fermentation; the medium containing both products and microorganism biomass.

Browning: Discoloration due to phenolic oxidation of freshly cut surfaces of explant tissue. In later stages of culture, such discoloration may indicate a nutritional or pathogenic problem, generally leading to necrosis.

Brucellosis: Disease caused by infection with organisms of the genus *Brucella*.

BSE: Bovine spongiform encephalopathy.

Buffer: A solution that maintains the specific pH range needed for a physiological reaction.

Bulked segregant analysis (BSA): In this method, 2 bulked DNA samples are drawn from a segregating population originating from a single cross. Each bulk contains DNAs from individuals that are identical for a particular trait but arbitrary at all unlinked regions. The bulks are screened for DNA polymorphisms; a marker that differs between the 2 bulks is expected to be linked to the particular trait for which the bulks were created.

Bulking agents: Substances such as starch that are added to food to increase the quantity of a food without affecting its nutritional value.

Buoyant density: The intrinsic density, which a molecule, virus, or subcellular particle has when suspended in an aqueous solution of a salt, such as CsCl, or a sugar, such as sucrose. DNA from different species has a characteristic buoyant density that reflects the proportion of G = C base pairs. The greater the proportion of G = C, the greater the buoyant density of the DNA.

Bystander effect: The term describing the beneficial therapeutic effects elicited when a primary therapeutic agent triggers events in neighboring cells and tissues via other mediators.

C: Cytosine residue in either DNA or RNA.

C Value: The total amount of DNA in a haploid genome.

CAAT Box: Part of conserved sequence located upstream of the start points of eukaryotic transcription factors. It has the consensus sequence GGCCAATCT; it occurs around 75 bases prior to the transcription initiation site.

CAP binding site: The sequence on DNA to which CAP (catabolite activator protein, for which the presence is necessary for the activation of the lac operon) binds.

CAP (catabolic activator protein): (1) The structure found on the 5′ end of eukaryotic mRNA and consisting of an inverted, methylated guanosine residue. (2) A positive control protein that complexes with cAMP and binds to a specific region of a promoter and which, in turn, stimulates the binding of RNA polymerase at the start of a gene and hence enables transcription of that gene.

Capacitation: The final stage in the maturation process of a spermatozoon taking place inside the female genital tract as the sperm penetrates the ovum.

Capacity factor: The ratio of the elution volume of a substance to the void volume in the column.

Capillary blotting: In this approach, gel is placed on top of a buffer-saturated filter paper, the blotting membrane is placed on top of the membrane. The buffer moves from the bottom filter paper through the gel, carrying macromolecules along with it; the latter get trapped onto the blotting membrane.

Capillary electrophoresis: In this case, separation by electrophoresis is achieved in 20–30 cm long capillaries of fused silica having internal diameter of 50–100 μm.

Capsid: The protein coat surrounding the genetic material (DNA or RNA) of a virus.

Carbohydrate: An organic compound based on the general formula $C_x(H_2O)_y$. performing many vial roles in living organisms. The simplest carbohydrates are the sugars (saccharine), including glucose and sucrose.

Carbon source: Compound used by cells/tissues as the source of energy, as well as carbon, for their metabolic needs; usually, sucrose (2%–5%).

Carboxyl group: The –COOH functional group, acidic in nature, found in all amino acids.

Carboxypeptidases: Two enzymes (A and B) found in pancreatic juice. Their role is to remove the C-terminal amino acid from a peptide. The A form removes any amino acid and the B form removes only lysine or arginine. Used when sequencing peptides.

Carotene: A reddish-orange plastid pigment involved in light reactions in photosynthesis.

Carotenoid: Red to yellow pigments responsible for the characteristic color of many plant organs or fruits such as tomatoes, carrots, etc. Oxidation products of carotene are called xanthophylls. Carotenoids serve as light-harvesting molecules in photosynthetic assemblies and also play a role in protecting prokaryotes from the deleterious effects of light.

Carpel: Female reproductive organ of flowering plants consisting of sigma, style, and ovary. In some plants 1 or more carpel unite to form the pistil.

Carrier: An individual who possesses a mutant allele but does not express it in the phenotype because of a dominant allelic partner.

Carrier DNA: DNA of undefined sequence content which is added to the transforming (plasmid) DNA used in physical DNA-transfer procedures. This additional DNA increases the efficiency of transformation in electroporation and chemically mediated DNA-delivery systems. The mechanism responsible for this effect is not known.

Carrier gas: The gas that carriers the sample in gas chromatography.

Carrier matrix: Materials used for immobilization of cells/enzymes; usually inert polymers or inorganic materials.

Carrier molecule: A molecule that plays a role in transporting electrons through the electron transport chain. Carrier molecules are usually proteins bound to nonprotein groups that are able to undergo oxidation and reduction relatively easily, thus allowing electrons to flow.

Carrier protein: A protein to which a hapten is attached in order to render it more immunogenic.

Cartagena protocol: An international agreement governing trade in genetically modified organisms. This protocol was signed in January 2000.

Cascade: The chain of sequential protein-protein interactions that transmit a signal within a cell or between neighboring cells or tissues.

Casein: A group of proteins found in milk.

Casein hydrolysate: A mixture of amino acids and peptides produced by enzymatic or acid hydrolysis of casein. It is an undefined mixture.

Catabolic pathway: A pathway by which an organic molecule is degraded in order to release energy for growth and other cellular processes.

Catabolism: The metabolic breakdown of large molecules in living organism with accompanying release of energy.

Catabolite repression: The inactivation of an operon caused by the presence of large amounts of the metabolic end product of the operon.

Catalysis: The process of changing the rate of a chemical reaction by the use of a catalyst.

Catalysis by approximation: The substrate bound to an enzyme molecule is held close to another chemical group in the optimal orientation for the reaction to proceed.

Catalyst: A catalyst is a substance that accelerates a (chemical) reaction without being used up itself in the process.

Catalytic antibodies: Antibodies that can act asenzymea to catalyze a chemical reaction. The antibodies are raised against a transition-state that is an analog of the desired substrate.

Catalytic biosensor: Biological component of such a biosensor catalyzes the conversion of an analyte into a new compound; this reaction produces a change in temperature, pH, etc.

Catalytic converter: A device used by US automobile manufacturers to reduce the amount of CO and hydrocarbons in a car's exhaust. Converter contains a catalyst that oxidizes these compounds to CO_2 and water as they pass through the exhaust.

Catalytic RNA: A natural or synthetic RNA molecule that cuts an RNA substrate.

Cation: Positively-charged particle or ion.

cDNA: Complementary DNA, a fragment of DNA that has been produced form an RNA sequence by reverse transcription. Messenger RNA is commonly used to synthesize cDNA.

cDNA clone: A double-stranded DNA complement of mRNA synthesized *in vitro* by using reverse transcriptase and DNA polymerase.

cDNA cloning: A method of cloning the coding sequence of a gene, starting with its mRNA transcript. It is normally used to clone a DNA copy of a eukaryotic mRNA.

cDNA library: A collection of cDNA clones that were generated *in vitro* from the mRNA sequences isolated from an organism or a specific tissue or cell type or population of an organism.

cDNA of satellite RNA-mediated Vvrus resistance: cDNA copies of such a satellite RNA that reduces disease severity of an RNA virus are transferred into plants; expression of the satellite sequences reduces disease symptoms as well as virus accumulation.

Cell: The basic structural unit of all living organisms.

Cell synchronization: Induction of synchronous cell division in cell cultures; achieved through starvation (of a nutrient or a growth regulator; cells arrested in G_1/G_2, inhibition of DNA synthesis by 5-aminouracil FUdR etc.), etc.

Cell wall: A rigid external coat that surrounds plant cells. It is formed outside the plasmalemma and consists primarily of cellulose.

Cell-mediated immune response: The activation of T cells of the immune system in response to the presence of a foreign antigen.

Cell-mediated immunity: The branch of the immune response that is mediated by the T lymphocytes (T cells).

Cellular oncogene (proto-oncogene): A normal mammalian or avian gene that, when mutated or improperly expressed, contributes to the development of cancer.

Cellulase: Enzyme that catalyzes the breakdown of cellulose.

Cellulose: A complex carbohydrate composed of long, unbranched chains of beta-glucose—(1, 4)-linked-b-d-glucose—molecules, which contribute to the structural framework of plant cell walls. It comprises 40%–55% by weight of the plant cell wall.

Cellulose acetate electrophoresis: Cellulose acetate membranes having large pores are used for separation of molecules entirely on the basis of their charge density; resolution and limit of detection is low.

Cellulosome: A multiprotein aggregate that is present in some cellulolytic microorganisms and contains multiple choices of all the enzymes required to completely break down cellulose. This complex is often found on the outer surface of cellulolytic microorganisms.

Central dogma: The basic concept that in nature genetic information generally can flow only from DNA to RNA to protein; it is known that information contained in RNA molecules of certain viruses (called retroviruses) can also flow back to DNA.

Centers of origin: The geographical centers are basic and independent foci where agricultural crops originated but are also geographic areas where cultivated plants are grown.

Centimorgan: One percent recombination between 2 loci (cM).

Changes in chromosomal structure: Environmental factors including radiation, chemicals, and viruses can cause chromosomes to break; if the broken ends do not rejoin in the same pattern, this causes a change in chromosomal structure, for example, deletion, duplication, inversion, translocation.

Chaperone: A protein that binds to other proteins and helps them to fold correctly and also prevents their premature folding.

Character: An observable/estimable, for example, morphological, anatomical, biochemical, physiological, or behavioral, feature of an organism.

Chelate: Complex organic molecule that can combine with cations and does not ionize. Chelates can supply micronutrients to plants at slow, steady rates. Usually used to supply iron to plant cells.

Chemical control of organ differentiation: Hypothesis due to Skoog and Miler (1957); organ differentiation depends on the relative (not absolute) concentrations of auxins and cytokinins; a high auxin to cytokinin ratio favors root regeneration, while the opposite promotes shoot regeneration; in fact, the exogenous need for the 2 GRs depends on their endogenous levels in the species/tissue.

Chemical energy: The potential energy that is contained in chemicals; most importantly, the energy contained in organic compounds, such as food and fuels, that may be released through respiration or burning.

Chemical oxygen demand (COD): Amount of oxygen required for chemical oxidation of the organic matter present in a water sample; provides an estimate of the total oxidizable organic matter.

Chemiluminescence: The emission of light forms a chemical reaction.

Chemoautotrophs: Organisms that derive their energy needs from oxidation of simple inorganic compounds and use CO_2 as carbon source.

Chemostat: An open continuous culture in which cell growth rate and cell density are held constant by a fixed rate of input of growth limiting nutrient.

Chemosynthesis: Process in which microorganisms utilize the chemical energy contained in certain reduced inorganic molecule to produce organic materials.

Chemotherapy: Chemical treatment to eradicate viruses from plant cells/tissues; chemicals used: virazole, (ribavirin), actinomycin D, cyclohexamide, etc.

Chiasma (*pl.* chiasmata): A site at which 2 homologous chromosomes appear to have exchanged material during meiosis.

Chimeric DNA: A recombinant DNA molecule containing unrelated genes.

Chimeric gene: A gene that contains coding sequence from one organism and regulatory sequence from another organism; all prokaryotic genes expressed in eukaryotes have to be chimeric, and *vice-versa.*

Chimera: Recombinant DNA molecules containing sequences from a different organism; a plant or animal that has a population of cells with different genotypes.

Chimeric protein: Fusion protein.

Chlorinated hydrocarbons: These are synthetic organic molecules in which one or more hydrogen atoms have been replaced by chlorine atoms. These hydrocarbons are hazardous compounds because they tend to be nonbiodegradable and therefore bioaccumulate; many have been shown to be carcinogenic.

Chlorination: A disinfection process, in which chlorine is added to drinking water or sewage water to kill pathogenic microorganisms.

Chlorofluorocarbons (CFCs): Synthetic organic molecules that contain one or more of both chlorine and fluorine atoms and that are known to cause ozone destruction.

Chlorophyll: The green pigment in plants responsible for absorbing the light energy required for photosynthesis.

Chloroplast: Plastids containing chlorophyll; enclosed by 2 concentric membranes and contain typical stroma and grana organization; contains DNA which contributes to cytoplasmic inheritance.

Chloroplast DNA (ctDNA): An independent genome (usually circular) found in a plant chloroplast.

Chromatids: The copies of a chromosome produced by replication. The name is usually used to describe the copies in the period before they separate at the subsequent cell division.

Chromatin remodeling: Describes the energy-dependent displacement or reorganization of nucleosomes that occurs in conjunction with activation of genes for transcription.

Chromatin: Describes the state of nuclear DNA and its associated proteins during the interphase (between mitoses) of the eukaryotic cell cycle.

Chromatin: Protein/DNA complex making the chromosome.

Chromatogram: A series of separated bands or zones detected either visually, as in chromatographic or thin layer chromatographic separations or indirectly by a detection system. In the latter case, the detection system usually outputs an electrical signal, which is graphically plotted through time, to display the series of separated zones or bands.

Chromocenter: An aggregate of heterochromatin from different chromosomes.

Chromomeres: Densely, staining granules visible in chromosomes under certain conditions, especially early in meiosis, when a chromosome may appear to consist of a series of chromomeres. A chromosome is a discrete unit of the genome carrying many genes. Each chromosome consists of a very long molecule of duplex DNA and an approximately equal mass of proteins. It is visible as a morphological entity only during cell division.

Chromosomal mutations: include changes in chromosome number and structure.

Chromosomal theory of inheritance: Genes are located on chromosomes; proposed in 1902 by Theodor Boveri and Walter S. Sutton, accounts for the similarity of genes and chromosomal behavior during meiosis and fertilization. Both chromosomes and factors (now called alleles) pair in diploid cells, chromosomes and alleles of each pair separate during meiosis so gametes have one-half chromosomes and alleles that separate independently. Gametes contain all combinations; fertilization restores diploid chromosome number and paired condition for alleles in zygote.

Chromosomal virulence genes: In *Agrobacterium*; chromosomal genes involved in infection; most of these genes are concerned with polysaccharide biosynthesis.

Chromosome: A molecules of DNA, complexed with specific proteins responsible for storage and transmission of genetic information in eukaryotes; a physically distinct unit of the genome.

Chromosome jumping: Use of jumping and linking libraries to determine the correct order in a chromosome of the DNA fragments generated by the restriction enzyme used to prepared these libraries.

Chromosome mapping: Percentage of recombinant phenotype; measures distance between genes to map the chromosomes, 1% of crossing-over equals 1 map unit between genes.

Chromosome pairing: The coupling of the homologous chromosomes at the start of meiosis.

Chromosome walking: A technique that identifies overlapping cloned sequences of about 40 kb long that form 1 continuous segment of a chromosome.

Circularization: A DNA fragment generated by digestion with a single restriction enzyme will have complementary 5'→3' extension (sticky ends). If these ends are annealed and ligated, the DNA fragment will have converted to a covalently closed circle or circularized.

***cis* acting site:** Affects the activity only of sequences on its own molecule of DNA (or RNA); this property usually implies that the site does not code for protein.

***cis* configuration:** Two sites on the same molecule of DNA.

***cis*-dominant site:** Or mutation affects the properties of its own molecule of DNA. *cis* dominance is taken to indicate that a site does not code for diffusible project. (A rare exception is that a protein is *cis*-dominant when it is constrained to act only on the DNA or RNA from it was synthesized.)

***cis* face:** Of the Golgi is juxtaposed to the nucleus.

Cisgenic plants: Refer to plants which have foreign gene(s); belong to the same species/a related species; risk of transfer of genes through pollen will not be a big issue in this case.

Cisternae: Of the Golgi apparatus are the successive stacks, each bounded by a membrane, that makes up individual compartments.

Cistron: Segment of a DNA molecule within which 2 mutations cannot complement each other, coding for a single polypeptide chain, equated with a gene.

Cleave: To break phosphodiester bonds of double stranded DNA, usually with a type II restriction enzyme.

Cleaved amplified polymorphic sequence (CAPs): Partial DNA sequence information for the locus of interest is used to create a set of PCR primers that are used to amplify the genome segment of interest. The amplification product is digested with the selected restriction enzyme, and digest is subjected to gel electrophoresis to detect polymorphism in the lengths of fragments so generated.

Clonal anergy: When extrathymic T-cells encounter antigens presented by such cells that lack the costimulatory molecules, they become nonresponsive; T-cell self-tolerance can also arise from clonal energy.

Clonal deletion: B cells marked by negative selection die by apoptosis.

Clonal propagation: Vegetative propagation of plants considered to be physiologically and/or genetically uniform which originated from a single individual or explant is called clonal propagation.

Clonal selection: Phenomenon of selective proliferation of lymphocytes, for example, B-cells, in response to their interaction with an antigen.

Clone: A collection of genetically identical cells or organisms derived from a common ancestor in which all members have similar genetic composition.

Clone bank: A population of organisms each containing a DNA molecule inserted through cloning vector.

Clone-by-clone sequencing: In this method, the fragments are first aligned into contigs, and each clone of the contig is then sequenced.

Cloning: Incorporation of a DNA molecule into a chromosomal site or a cloning vector.

Cloning site: A location on a cloning vector into which DNA can be inserted.

Cloning vector: A carrier molecule, replicating autonomously, generally a plasmid or phage, into which any foreign DNA can be inserted, introduced into the host cell and many copies of the foreign DNA are produced.

Closed continuous culture: In this approach, a definite volume of culture is withdrawn regularly, cells are separated from the used medium and are added back to the culture so that cell biomass keeps on increasing.

Closed reading frame: It contains termination codons that prevent its translation into protein.

CO_2 fertilization effect: The response of plants to elevated concentrations of CO_2.

CO_2 fixation: The enzymatic reaction in which CO_2 is attached to a receiver compound, such as ribulose-1,5-bisphosphate, thereby adding to the supply of organic carbon; occurs chiefly during photosynthesis.

Cocultivation: Incubation of host cells or tissues with *Agrobacterium* or any other organism acting as a vehicle for a vector.

Coat protein-mediated virus resistance: Transgenic plants expressing the coat protein gene of a virus exhibit resistance to the concerned and related viruses; resistance presumably arises due to the inability of the viral genome to replicate and express.

Coated vesicles: Vesicles whose membrane has on its surface a layer of a protein such as clathrin, COP-I, or COP-II.

Coconut milk: Liquid endosperm from immature nuts (before cellular endosperm formation has begun); boiled (to denature the proteins and enzymes), filteredm and stored at -20°C.

Coconversion: The simultaneous correction of 2 sites during gene conversion.

Coculture: Culture of plant cells/tissues of different species or even of different organisms (e.g., plant cells with bacteria) together in the same culture vessel.

Coculture with *Agrobacterium*: Culture of plant cells/tissues with *Agrobacterium* cells, usually, to achieve genetic transformation.

Coding strand: The strand of DNA in a gene that has the same sequence as the mRNA transcribed from the gene.

Codominance: Both the alleles of a gene express themselves in heterozygotes.

Codominant alleles: A pattern of inheritance in which both alleles of a gene are expressed in the heterozygote, 3alleles A, B, and O determine human blood type. Three possible genotypes are AA, BB, OO that correspond to the phenotypes of blood type A, B, and O respectively. Two other genotypes are AO and BO that correspond to blood A and B, respectively because the O allele is recessive. The remaining genotype is AB, corresponding to blood type AB. Both the A and B alleles contribute to the phenotype of the heterozygote. Thus the alleles A and B are said to be codominant, both the alleles produce effective product.

Codon: A triplet of nucleotides that represents an amino acid or a termination signal.

Codon bias: Preference for certain codons over the others from among the multiple codons specifying the same amino acid; each organism exhibits a different pattern of codon preference.

Coenzyme: In case of enzymes; a specialized molecule that mediates electron transfer during redox catalysis.

Cofactor: In case of enzymes; a transition metal that mediates electron transfer during redox catalysis.

Cognate tRNAs: Cognate tRNAs are those recognized by a particular aminoacyl-tRNA synthetase.

Cohesive ends: Single-stranded complementary nucleotides on the ends of double stranded DNA molecule.

Colony hybridization: A technique for using *in situ* hybridization to identify bacteria carrying chimeric vectors whose inserted DNA is homologous with some particular sequence.

Colorimetric biosensor: It measures the heat released or absorbed during the reaction catalyzed by the biological component of the biosensor.

Color Retention Agent: Agent that are used to preserve a food's existing color.

Combustion: The practice of disposing of wastes by incineration in a special facility designed to handle large amounts of waste; modern combustion facilities also utilize some of the energy by generating electricity on the site.

Cometabolite: The substrate, which induces the enzymes necessary for degradation of a xenobiotic, and serves as a source of energy, carbon, and reducing power.

Comparative genomics: Study of differences and similarities in genome structure and organization of different organisms.

Compatible trait transgenics: Those transgenics which reduce the use of nonsustainable inputs without polluting the environment.

Competent cell: The cells those are capable of taking up DNA.

Complement: Initially, it was defined as "the activity of blood serum that completes the action of antibody"; it consists of over 30 soluble and cell bound proteins and glycoproteins that interact in a highly regulated cascade, following the initial activation and carry out a number of basic functions.

Complementary base pairing: The formation of hydrogen bonds between adenine and thymine and between guanine and cytosine is known as complementary base pairing.

Complementary DNA (cDNA): Synthetic DNA transcribed from a specific RNA through the action of the enzyme reverse transcriptase.

Complementary Strand: In case of nucleic acids; a polynucleotide strand that has complementary bases at every position of another polynucleotide strand.

Complete digestion (of DNA): Complete digestion of DNA signifies that the restriction enzyme has cut the DNA at every recognition site for that enzyme present in the DNA.

Complete linkage: Describes the inheritance patterns for 2 genes on the same chromosome when the observed frequency for crossover between the loci is zero.

Complexity: The total length of different sequences of DNA present in a given preparation.

Complex locus: Complex locus (of *D. melanogaster*) has genetic properties inconsistent with the function of a gene representing a single protein. Complex loci are usually very large (>100 kb) at the molecular level.

Complex medium: In fermentation; it contains undefined constituents like soybean meal, molasses, cornsteep liquors, etc. and gives much higher yields of metabolites.

Complex organic additive: A nutritive mixture of undefined composition, for example, yeast extract, coconut milk, malt extract, tomato juice, etc.

Complex probes: The labelled mRNA used as probe; represents the total mRNA produced by the cell.

Composite transposons: Composite transposons have a central region flanked on each side by insertion sequences, either or both of which may enable the entire element to transpose.

Compost: Fully decomposed organic material.

Compositing: The process of letting organic wastes decompose in the presence of air; a nutrient-rich compost is the resulting product.

Compound: Any substance that is made up of two or more different kinds of atoms bonded together.

Concatemer: Concatemer of DNA consists of a series of unit genomes repeated in tandem.

Conservation: Refers to the management of a resource in such a way as to assure that it will continue to provide maximum benefit to humans in the long run; it includes various degrees of use or protection, depending on what is necessary to maintain the resource over a long period.

Contamination: Presence of unwanted microorganisms in microbial, plant or animal cultures, food, or any products in laboratory or industry.

Contig: A set of overlapping clones that provides a physical map of a protein of chromosome. It refers to contiguous map.

Control of endophytic contamination: In this case, surface sterilization is ineffective; a suitable systemic fungicide and/or an antibiotic is usually added to the culture medium; preferably, explants are taken from greenhouse-grown plants.

Controlling elements: Controlling elements of maize are transposable units originally identified solely by their genetic properties. They may be autonomous (able to transpose independently) or autonomous (able to transpose only in the presence of an autonomous element).

Convection: Mass or bulk movement of one part of a solution relative to the rest, usually because of density differences.

Conventional crop variety: Genetic makeup is not modified by genetic engineering.

Conventional vaccine: Consisting of whole pathogenic organism, which may either be killed or attenuated.

Coordinate regulation: It refers to the common control of a group of genes.

Coppicing: Production of young shoots, especially, after the main shoot has been cut off; the young shoots are juvenile.

Copy number: The average number of molecules of a plasmid of a gene per genome present in a cell.

Cordycepin: 3′ deoxyadenosine, an inhibitor of polyadenylation of RNA.

Core DNA: The 146 bp of DNA contained on a core particle.

Core particle: A digestion product of the nucleosome that retains the histone octamer and has 146 bp of DNA; its structure appears similar to that of the nucleosome itself.

Corepressor: A small molecule that triggers repression of transcription by binding to a regulator protein.

Corn meal medium: It is prepared by grinding degermed (de-embryonated) corn to fine powder, and 8%–10% of this corn meal is heated to gelatinize the starch.

Correct PCR product: At the second cycle of PCR, primers will anneal to the long product, that is, amplification product of first cycle much before 3′ end where sequences complementary to them are located. Primer extension step in this cycle

COS site (ends): The 12-base, single-stranded, complementary extension of phage lambda (λ) DNA.

Cosmid: A hybrid plasmid composed of the *cos* sites of lambda and antibiotic resistance genes of plasmids, enters into the host cell like a phage and replicates like a plasmid.

Cosmids: Plasmids into which phage lambda *cos* sites have been inserted; as a result, the plasmid DNA can be packaged *in vitro* in the phage coat.

Cot: The product of DNA concentration and time of incubation in a reassociation reaction.

$Cot_{1/2}$: The Cot required to proceed to half completion of the reaction; it is directly proportional to the unique length of reassociating DNA.

Cotransfection: The simultaneous transfection of 2 markers.

Counterion: An ion of opposite charge to the one under consideration. Generally a similar ion.

Covalent catalysis: The chemical groups, for example, amino groups, in the side chains of amino acids present in the active site of the enzyme become covalently bound to a group in the substrate molecule; another substrate may then react with the substrate bound to the enzyme to form the product.

Crisscross inheritance: Transmission of a gene from a male parent through female child to male grandchildren, for example, X-linked inheritance.

Cross-contamination: Especially in animal cell cultures; presence of cells from other cell lines of the same species, or from some other species.

Cross-feeding: Cells of 1species (or genotype) provide the biochemicals needed by cells of another species (or genotype); in coculture or feeder layer technique.

Cross reaction: In immunology; interaction of an antibody induced by one antigen with another antigen due to a similarity in their epitopes.

Crossing over: The exchange of corresponding chromosome parts between homologs by breakage and reunion.

Crossover fixation: Refers to a possible consequence of unequal crossing-over that allows a mutation in 1 member of a tandem cluster to spread through the whole cluster (or to be eliminated).

Crown gall: A tumor occurring at the base of certain plants due to infection of the plant by *Agrobacterium tumifaciens.*

Cruciform: The structure produced at inverted repeats of DNA if the repeated sequence pairs with its complement on the same strand (instead of with its regular partner in the other strand of the duplex).

Cryobiology: The branch of science, which deals with preservation of cells, tissues, etc. in frozen state in liquid nitrogen.

Cryogenic injuries: Injuries to cells due to the procedure of cryopreservation; chiefly during freezing and, also during thawing; may be physical (due to ice crystal formation, which may rupture cell organelles or even cells) or chemical (due to toxic solution effect and/or loss of vital solutes from cells through leakage during freezing/thawing).

Cryopreservation: Culture preservation at ultralow temperature of −196°C.

Cryoprotectant: A chemical agent that inhibits freezing and thawing damage to cells.

Cryotherapy: Prolonged exposure (4 months or more) of plants to a low temperature (e.g., 5°C), followed by shoot-tip culture for virus elimination; superior to thermotherapy.

Cryptic satellite: A satellite DNA sequence not identified as such by a separate peak on a density gradient; that is, it remains present in main-band DNA.

Crystal protein: An insecticidal protein present as a crystalline inclusion in *Bacillus thuringiensis* spores; 17 distinct groups of 130 or 70 kDa proteins.

CTAB method: When CTAB (cetyltrimethyl-ammonium bromide) is added to the cell extracts, it forms an insoluble complex with nucleic acids, this precipitate is recovered through centrifugation and resuspended in 1 M NaCl; carbohydrates, proteins, etc., are discarded with supernatant. The nucleic acid solution is treated with RNase, and DNA is concentrated by ethanol precipitation.

ctDNA: Chloroplast DNA.

Cultivar: Refers to the cultivated variety of a plant species. Genetically, all individuals of a cultivar are uniform.

Culture initiation: Surface sterilization of explants and establishing them *in vitro*.

Culture medium: The medium on which cells, tissues and organs are cultured; contains inorganic salts, vitamins, a carbon source, and growth regulators as per need.

Culture room: Room with controlled environment, for example, regulated temperature, light, etc., and used for culture of cells/tissues/organs/organisms.

Cutinization: Impregnation of cell wall with cutin; cutinization of the outer-most cell wall is essential for somatic embryo development.

Cybrid: A cytoplasmically hybrid cell with organelles from both parental sources (obtained through fusion of cytoplast with a whole cell) and a nucleus of only one cell. Nucleus of the other cell denatured.

Cybridization: The process of cybrid production; usually, to transfer cytoplasm from one species into other.

Cyclic AMP (cAMP): A molecule of AMP in which the phosphate group is joined to both the 3′ and 5′ positions of the ribose; its binding activates the CAP, a positive regulator of prokaryotic transcription.

Cyclins: Proteins that accumulate continuously throughout the cell cycle and are then destroyed by proteolysis during mitosis. (*See also* MPF.)

Cytodifferentiation: Development of various cell types from the relatively undifferentiated meristematic cells.

Cytogenetic abnormalities: Changes in chromosome number and/or structure.

Cytogenetic instability: Origin of variation in chromosome number and/or structure; occurs in virtually all cell cultures.

Cytogenetic map: It depicts the locations of various genes in a chromosome relative to specific microscopically visible landmarks in the chromosomes; commonly used in eukaryotes since they have relatively large microscopically observable chromosomes.

Cytokines: Hormone like proteins that act by binding to specific receptors on cell surface; needed to activate B cells and Tc cells.

Cytokinesis: The final process involved in separation and movement apart of daughter cells at the end of mitosis.

Cytokinins: A class of plant growth regulators which causes cell division, cell differentiation, shoot differentiation and breaking of apical dominance for example, BAP, kinetin, zeatin.

Cytoplasm: The material between the plasma membrane and the nucleus.

Cytoplasmic inheritance: Non-Mendelian form of inheritance involving genetic information transmitted by self-replicating cytoplasmic organelles, such as chloroplasts and mitochondria.

Cytoplasmic protein synthesis: The translation of mRNAs representing nuclear genes; it occurs via ribosomes attached to the cytoskeleton.

Cytoplast: A protoplast without an active nucleus.

Cytosine (c): One of the pyrimidine (6-membered single ringed nitrogenous base) in DNA and RNA.

Cytoskeleton: Consists of networks of fibers in the cytoplasm of the eukaryotic cell.

Cytosol: The general volume of cytoplasm in which organelles (such as the mitochondria) are located.

D loop: A region within mitochondrial DNA in which a short stretch of RNA is paired with one strand of DNA, displacing the original partner DNA strand in this region. The same term is used also to describe the displacement of a region of one strand of duplex DNA by a single-stranded invader in the reaction catalyzed by RecA protein.

Dalton: A unit (used in designating molecular weights) of mass equal to that of the hydrogen atom, which is 1.67×10^{-24} gram.

Database: A repository of sequences (DNA or amino acids) which provide a centralized and homogenous view of its contents.

Database mining: Process of database utilization.

Database mining tools: Search engines and analysis tools needed for utilization of various databases.

DDT (dichlorodiphenyl trichloroethane): The first and once most widely used synthetic organic pesticide; belongs to the chlorinated hydrocarbon class.

de Novo: It means anew, fresh.

Decamer (oligonucleotide): An oligonucleotide 10 bases long; generally used in RAPD (randomly amplified polymorphic DNA) analysis.

Decapitation: Removal of the shoot tip/shoot apex, usually of the main shoot; relatives apical dominance.

Dedifferentiation: Reversion of differentiated cell to the meristematic stage; induced by auxin and/or cytokinin.

Deep well injection: Disposal of liquid chemical wastes by putting them into deep dry wells, where they permeate dry strata.

Defective viral genome-mediated resistance: In case of some viruses; a deleted version of their genomes, when integrated into host genome, leads to resistance to the concerned virus.

Defined molecular marker: A DNA fragment whose function or specific features are known.

Degeneracy: Degeneracy in the genetic code refers to the lack of an effect of many changes in the third base of the codon on the amino acid that is represented.

Degenerate primer: Set of primers having the same sequence, except for one or more positions where variation occurs.

Degradative plasmids: These plasmids enable the host bacterium to metabolize unusual molecules like toluene and salicylic acid.

Delaney Clause: Provision of the Federal Food, Drug and Cosmetic Act of 1958 that prohibits the use of any food additives that induces cancer in people or animals.

Deletion (deficiency): It is a type of mutation in which an end of a chromosome breaks off or when two simultaneous breaks lead to the loss of a segment from the chromosome, even if only one member of pair of chromosomes is affected, a deletion can cause abnormalities. Cri du chat syndrome is deletion in which an individual has a small head, is mentally retarded, and has facial abnormalities, and abnormal glottis and larynx resulting in a cry resembling that of a cat.

Deletions: Generated by removal of a sequence of DNA, the regions on either side being joined together.

Demethylation: Removal of methyl ($-CH_3$) groups usually from methylated "C" residues.

Denaturation (of DNA or RNA): Conversion from the double-stranded to the single-stranded state; separation of the strands is most often accomplished by heating.

Denaturation (of protein): Conversion from the physiological conformation to some other (inactive) conformation.

Deoxyribonucleotide: A deoxyribonucleoside having a phosphate residue at 3′- or 5′- position; in vivo, 5′- phosphate is the rule.

Deoxyribose: A 5 C sugar with a hydrogen atom rather than a hydroxyl group attached to 2′ carbon.

Deoxyriboside: Compound obtained by a linkage between deoxyribose sugar and an organic base.

Derepressed: This state describes a gene that is turned on. It is synonymous with *induced* when describing the normal state of a gene; it has the same meaning as *constitutive* in describing the effect of mutation.

Desalinization: Refers to the process, which purifies seawater into high-quality drinking water via distillation or microfiltration.

Descriptor: Character used for evaluation and characterization of different accessions of germplasm.

Desiccated artificial seeds: Somatic embryos hardened by treatment with ABA and then coated with a suitable polymer to enable them to withstand desiccation; encapsulated in a suitable matrix and dried; afford easy handling and storage.

Designer crops: Crop varieties specially designed to meet certain specific consumer, industrial, etc. needs; they possess features that were specified before their development.

Development culture: A research culture producing a compound of sufficient commercial interest and used for developing large-scale fermentation procedures.

Developmental determination: During shoot regeneration; this phase follows morphogenic competence acquisition; during this phase, cells become irreversibly committed to a developmental path in response to the inductive condition.

Dextran: A polysaccharide produced by certain bacteria.

Diazotrophs: The microorganisms that fix atmospheric nitrogen (N_2).

Dicentric chromosome: The product of fusing two chromosome fragments, each of which has a centromere. It is unstable and may be broken when the two centromeres are pulled to opposite poles in mitosis.

Dideoxynucleotide: Nucleotide in which oxygen is absent from the 2′ and 3′ positions of ribose; blocks further extension of a nucleotide chain beyond the point at which it incorporated into the chain.

Differentiation: Formation of various cell types and organs like root, shoot, etc. either *in vivo* or in cell cultures.

Digester: A low technology reactor used for anaerobic bacterial digestion.

Dihaploid: A haploid derived from tetraploid.

Dihybrid Cross: A type of cross involving parents differing in only two traits or in which only 2 traits are considered.

Diplochromosome: Chromosome having 4 chromatids; produced by endoreduplication.

Diploid: Diploid set of chromosomes contains two copies of each autosome and two sex chromosomes.

Direct gene transfer: Vector-independent gene transfer.

Direct repeats: Identical (or related) sequences present in 2 or more copies in the same orientation in the same molecule of DNA; they are not necessarily adjacent.

Direct selection: Cells resistant to the selection pressure survive and divide to form colonies, while the wild type cells are killed by the selection agent.

Direct somatic embryogenesis: Embryos develop directly from explants cells without any intervening callus phase; responding explants cells are themselves either embryogenic or quite close to this state.

Disarmed plasmid: A plasmid from which some portion has been deleted.

Discontinuous replication: The synthesis of DNA in short (Okazaki) fragments that are later joined into a continuous strand.

Discontinuity: The site in a DNA strand at which the phosphodiester bond between 2 adjacent nucleotides is broken.

Discovery: Isolation/identification of things which are already present in nature/ universe; discovery only indicates their presence.

Disinfection: Killing of microorganisms in water or other media where they might otherwise pose a health threat, for example, chlorine is commonly used o disinfect water supplies. Inactivation of microorganisms present on the surface of explants, etc., also see, *Surface sterilization.*

Disjunction: The movement of members of a chromosome pair to opposite poles during cell division. At mitosis and the second meiotic division, disjunction applies to sister chromatids; at first meiotic division it applies to sister chromatid pairs.

Disomic: A cell/an individual having normal somatic chromosome complement of a species; each chromosome of the genome(s) is present in 2 copies.

Dispersed growth digesters: In such digesters, the microbial population is dispersed throughout the sewage being treated.

Dissolved oxygen (D.O.): Oxygen gas dissolved in water; concentration of D.O. is a measure of water quality.

Distant hybridization: Hybridization between distantly related taxa, for example, 2 different species, genera, etc.

Distinctiveness: In case of PBR, the new variety must be distinguishable from other varieties by 1 or more identifiable morphological, physiological or other characteristics.

Distribution Coefficient (K_d): The ratio of the concentration of a sample component in one phase to its concentration in a second phase equilibrium. The two phases may be immiscible liquids or the mobile phase and the stationary phase.

Divergence: The percent difference in nucleotide sequence between two related DNA sequences or in amino acid sequences between 2 proteins.

Divergent transcription: The initiation of transcription at 2 promoters facing in the opposite direction, so that transcription proceeds away in both directions from a central region.

***dna*:** *dna* mutants of bacteria are temperature-sensitive; they cannot synthesize DNA at 42°C, but can do so at 37°C.

DNA: A macromolecule in which deoxyribonucleotides are joined together by phosphodiester linkage; usually, double-stranded.

DNA amplification: Multiplication of a piece of a DNA in a test tube into many thousands or millions of copies by using PCR.

DNA array: A series DNA sequences fixed as distinct spots on a suitable solid support, like a glass chip; basically 2 types: spotted DNA array and printed oligonucleotide chips.

DNA chip: The solid supports of glass or silicon about the size of a microscope slide consisting of DNA attached in highly organized arrays. It is available commercially also, for example, GeneChip® Probe Array.

DNA construct: A suitable DNA which has been prepared for cloning purpose.

DNA fingerprint: Molecular marker pattern of an individual obtained by using highly polymorphic markers; used for unequivocal identification of strains, individuals, criminals, etc.

DNA fingerprinting: Developing the DNA fingerprint of an individual.

DNA footprinting: It enables identification of the site on a DNA molecule to which a protein, such as RNA polymerase, binds.

DNA insert: The DNA segment to be cloned.

DNA ligase: An enzyme that seals nicks in DNA strands by catalizing the formation of phosphodiester bond between the contiguous nucleotides.

DNA methylation: Addition of methyl (-CH_3) residues, chiefly to C residues, usually, located in CG or CNG (N means any one of the four DNA bases) sequences, after they are incorporates in DNA.

DNA microarray: An assembly of a very large number of small dots onto a solid support, each dot containing a short sequence of a different gene of the organism concerned.

DNA polymerase: An enzyme that synthesizes a daughter strand(s) of DNA (under direction from a DNA template). May be involved in repair or replication.

DNA probe: Isolated single radiolabelled ^{32}P-DNA strands (oligonucleotides) used to detect the presence of the complementary (opposite) strands, and as a very sensitive biological detectors.

DNA replicase: A DNA-synthesizing enzyme required specifically for replication.

DNA sequencing: Determination of the base sequence of a DNA fragment.

DNA vaccine: It contains a gene encoding an immunogenic protein from the concerned pathogen.

DNAase: An enzyme that attacks bonds in DNA.

DNAase I hypersensitive site: A short region of chromatin detected by its extreme sensitivity to cleavage by DNAase I and other nucleases; probably comprises an area from which nucleosomes are excluded.

DNA-delivery system: Any procedure that introduce DNA into a recipient cell.

DNA-driven hybridization: The reaction of an excess of DNA with RNA.

Domain: A region of a polypeptide chain with a particular function.

Domain: Domain of a chromosome may refer *either* to a discrete structural entity defined as a region within which supercoiling is independent of other domains; *or* to an extensive region including an expressed gene that has heightened sensitivity to degradation by the enzyme DNAase I.

Domain: Domain of a protein is a discrete continuous part of the amino acid sequence that can be equated with a particular function.

Dominance: It is the over-riding effect of a gene over its allele in such a way that the heterozygote is phenotypically indistinguishable from the dominant homozygote.

Dominant: Dominant allele determines the phenotype displayed in a heterozygote with another (recessive) allele.

Dominant allele: The allele expresses itself in the heterozygous state.

Dominant trait: A trait expressed preferentially over another trait.

Donor plant: Plant from which explants are obtained for *in vitro* culture. Syn. *Mother plant.*

Dosage compensation: Mechanisms employed to compensate for the discrepancy between the presence of two X chromosomes in one sex but only one X chromosome in the other sex.

Dose: It is the mathematical product of the concentration of hazardous material and the length of exposure to it. The effects of any given material or radiation corresponds to the dose received.

Dot blot technique: In this technique, total DNA (or RNA) isolated from each of several test individuals/tissues are placed onto nitrocellulose filter as dots; DNA is denaturated, fixed to the filter, and hybridization is detected by autoradiography or fluorescence.

Double fertilization: In plants; fusion of one sperm with the egg cell and of the other sperm with the secondary nucleus (produces endosperm); both fusion must occur in the same embryo sac to ensure normal seed development.

Double H]haploid (DH): A disomic obtained by doubling the chromosome number (usually, by colchicine treatment) of a haploid; as a rule homozygous.

Doubling time: The time required to double the number of cells.

Down: Down promoter mutations decrease the frequency of initiation of transcription.

Downstream: It identifies sequences proceeding farther in the direction of expression; for example, the coding region is downstream of the initiation codon.

Downstream proccessing: Separation and purification of product(s) from a fermentation process.

Driver mRNA: The biotin-labelled mRNA from nonstressed cell.

Dry heat sterilization: Sterilization by heating in an oven at 160–180°C for 3 h.

Dry weight: Weight of cells/tissues after drying them at 60°C for 12 h.

Duplications: Are a doubling of a chromosomal segment. A broken segment from one chromosome can simply attach to its homologue.

Early development: The period of a phage infection before the start of DNA replication.

Early region replacement vector: SV 40 (mammalian) transducing vector; early region of virus is replaced by DNA insert.

Economic threshold: The level of pest damage below which an application of pesticides to control the pest is more costly than the economic loss caused by the pest.

Ecosystem diversity: The differences among ecosystems including habitat diversity and ecological processes in an ecosystem.

Ectopic expression: The expression of a gene in a tissue in which it is not usually expressed; for example, in a transgenic animal or as the result of injection into an unusual location in an embryo.

Edible vaccine: An edible plant part in which an antigenic protein of a pathogen is produced and accumulated; feeding of this part generates immunity against the concerned pathogen.

Effluent: Mobile phase that has exited from the column.

Electrical polarity in embryogenesis: K^+ ions enter from the end that would later develop into plumule, while H^+ ions pass out of the would be redicular end; established much before the development of morphological bipolarity of embryos; redicular end exhibits a high concentration of calmodulin.

Electroblotting: A couple of sheets of filter paper soaked in anode buffer are placed on the anode plate, and blotting membrane is placed on top of these sheets. The gel is then laid on the top of the membrane and is covered with a couple of sheets of filter paper soaked in cathode buffer; the cathode plate is then placed on the top of these filter papers. The anions migrate from the gel onto the blotting membrane surface at the same speed; as a result, a regular transfer of molecule takes place.

Electrode: Substances in contact with a conductor. The substances are connected to a source of an electric field.

Electrofusion: Fusion of, usually two, protoplasts induced by an exposure to a very short (few milliseconds) duration of a high voltage current.

Electrolysis: Refers to the use of electrical energy to split water molecules into hydrogen and oxygen atoms.

Electro-osmosis: Migration of water toward an electrode as a result of the supporting medium and/or the surface of the separation equipment, for example, of capillaries, also carrying charge.

Electrophoresis: The method of separation of molecules such as DNA, RNA, or protein based on their relative migration applying a strong electric field.

Electrophoretic mobility: The speed of migration of a charged molecule under the influenced of an electric field; depends on the charge to mass ratio of the molecule.

Electroporation: A technique where electric field is applied to facilitate protoplast fusion derived from 2 different plants.

Electroporator: A specially designed equipment used for protoplast fusion.

Element: In genetics; a sequence of DNA that serves as a protein binding site, for example, promoter, operator, etc.; it functions only in *cis* position. In chemistry; a substance made up of one and only 1 distinct kind of atom.

Elicitation: Use of an elicitor to enhance secondary metabolite production.

Elicitor: Compounds, which dramatically enhance the production of secondary metabolites; may be biotic or abiotic; the term was originally used for molecules of pathogen origin, which induced the expression of pathogenesis related proteins in resistant varieties of the host.

ELISA: Enzyme-linked immunosorbent assay; uses an enzyme conjugated to an antibody for detection of a specific antigen/antibody based on antigen-antibody interaction; the enzyme linked to the antibody produces detectable color, which enables an easy and highly sensitive detection.

Elongation factors (EF in prokaryotes, eEF in eukaryotes): Proteins that associate with ribosomes cyclically, during addition of each amino acid to the polypeptide chain.

Eluate: A compound or mixture that has been separated in the exit from the column.

Eluent: Mobile phase.

Elution: Removal of a solute from a stationary phase by passage of suitable mobile phase.

Elution volume: The volume of mobile phase required to eluate a solute from a chromatographic column.

Emasculation: Removal of anthers in the bud condition before the shed of the pollen.

Embryo: A very young individual (animal/plant) developing usually, from zygote.

Embryo culture: Culture of young embryos aseptically excised from developing seeds on suitable media to raise young seedlings.

Embryo rescue: Recovery, by embryo culture, of seedlings by culturing such young embryos that would die due to endosperm abortion; usually, in cases of distant hybridization.

Embryo sac: Female gametophyte; derived from megaspore by, usually, 3 successive mitotic divisions yielding 8 haploid nuclei; 3 nuclei migrate to one pole and form 3 antipodal cells, 3 nuclei move to the opposite pole and form a central egg cell and two synergids, and two nuclei remain in the center and fuse to produce the secondary nucleus.

Embryo splitting: Young animal embryos (4–8 cells) treated with trypsin to separate the individual cells; each such cell can develop into a separate embryo.

Embryo transfer: Implantation of embryos from donor animals or generated by *in vitro* fertilization into the uterus of the recipient animals.

Embryo transplantation: Implantation of young embryos developed *in vitro* or obtained from the uterus of donor females into the womb of surrogate females.

Embryogenesis: Process of development of embryo from a fertilized egg cell.

Embryogenic culture: A culture containing cells capable of producing SEs; usually, has embryogenic clumps and even globular stage SEs.

Embryoids: Embryo-like structure produced as a result of differentiation process such as embryogenesis and androgenesis.

Embryonic clump: A localized group of meristematic and embryogenic cells, which are produced on somatic embryo induction medium; they give rise to somatic embryos when cultured on a suitable medium or proliferate and remain in embryogenic state on the induction medium.

Embryonic stem (ES) cell: Cells of early embryo that give rise to all differentiated cells including germ line cells.

Emulsifiers: Substances which allow water and oils to remain mixed together in form of an emulsion, for example, in ice cream, homogenized milk, and mayonnaise.

Encode: To specify the sequence of amino acids in a protein.

End filling: This method is used with those DNA molecules that have 5′-protruding ends; Klenow fragment is used to extent the recessed 3′-ends, thereby filling up the single-stranded ends.

End labelling: The addition of a radioactively labeled group to one end (5′ or 3′) of a DNA strand.

End-product inhibition: The ability of a product of a metabolic pathway to inhibit the activity of an enzyme that catalyzes an early step in the pathway.

Endangered species: A species whose total population is declining to relatively low levels such that if the trend continues, the species will most likely become extinct.

Endocytosis: A process by which proteins at the surface of the cell are internalized, being transported into the cell within membranous vesicles.

Endolytic vesicles: The membranous particles that transport proteins through endocytosis; also known as clathrin-coated vesicles.

Endonucleases: Cleave bonds within a nucleic acid chain; they may be specific for RNA or for single-stranded or double-stranded DNA.

Endoplasmic reticulum: A highly convoluted sheet of membranes, extending from the outer layer of the nuclear envelope into the cytoplasm.

Endoreduplication: Two or more successive rounds of DNA replication take place in nuclei without attendant cell division.

Endosperm Culture: Culture of endosperm cells, usually, for obtaining triploid plants.

Endotoxin: Cell wall portion (lipopolysaccharide) of Gram-negative bacteria that shows antigenic properties and trigger immune system.

Energy crops: Crops that use solar energy efficiently to convert CO_2 into biomass.

Enhancer: A *cis*-acting sequence that increases the utilization of (some) eukaryotic promoters, and can function in either orientation and in any location (upstream or downstream) relative to the promoter.

Entrapment culture: Cells are entrapped in an open matrix through which medium flows freely.

ENTREZ: It is the integrated information database similarity search.

Enucleate: A cell or protoplast without nucleus.

Envelopes: Surround some organelles (for example nucleus or mitochondrion) and consist of concentric membranes, each membrane consisting of the usual lipid bilayer.

Environment: The combination of all the factors (biotic and abiotic) external to the individual or population of organisms in question.

Environmental impact: The effect on the natural environment caused by human actions; it includes indirect effects, for example, through pollution, as well as direct effects such as cutting down of trees.

Enzyme: Generally, a protein that catalyzes or speeds up a specific chemical reaction; biosynthetic or degradative function; usually, occur in living cells/ living systems.

Enzyme engineering: Improvement in the activity and usefulness of an existing enzyme or creation of a new enzyme activity by making suitable changes in the amino acid sequence of concerned enzyme by recombinant DNA technology.

Enzyme immobilization: Confining the enzyme molecule to a distinct phase from the one in which the substrate and the products are present.

Enzyme-product complex: Enzyme molecule having the product molecule bound to its binding site.

Enzyme-reactor: A vessel employed to carry out the desired conversion using an enzyme.

Enzyme-substrate complex: Enzyme molecule having the substrate molecules bound to its active/binding site; this state precedes catalysis.

Enzyme technology: Use of purified enzymes for generating useful products or services.

EPG fragments: Oligonucleotides produced from cell wall by endopolygalacturonase hydrolysis; affect morphogenic response of cell cultures; included under the general term *oligosaccharins*.

Epidermis: The outermost single (usually) layer of cells covering the entire plant body (primary body); forms a protective layer.

Epigenetic: Epigenetic changes influence the phenotype without alerting the genotype. They consist of changes in the properties of a cell that are inherited but that do not represent a change in genetic information.

Epigenetic variation: When a variant trait shows stability during mitotic division in cell cultures, but disappears either during plantlet regeneration from them or during meiosis in the regenerated plants; may relate to up or down regulation of the promoter of the concerned gene.

Episome: A genetic element in bacteria that can replicate freely in the cytoplasm or can be inserted into the main bacterial chromosome and replicate with the chromosome.

Epistasis: One gene masks the expression of another nonallelic gene, absence of expected phenotype as a result of masking expression of another gene pair. The crossing between two white sweet pea plants (CCpp x ccPP) produces purple flowered (CcPp) plants, F_2 generation results a 9 colored (9C-P-): 7 white (3C-pp + 3ccP- + 1 ccpp) rather than 9:3:3:1 dihybrid ratio.

Epitope: A specific chemical domain present on an antigen recognized by an antibody. Each epitope elicits synthesis of different antibodies.

Equilibrium: A state, in which the forward and the reverse reactions proceed at identical rates and there is no net product formation.

Escape: A susceptible cell/individual being classified as resistant due to errors in evaluation/classification.

Essential fatty acids: Mammals (including humans) cannot synthesize linoleic (18:2) and linolenic (18:3) fatty acids; therefore, they must be supplied in their diet, and are called essential fatty acids.

Essential gene: One for which the deletion is lethal to the organism.

EST (expression sequence tag): A partial gene sequence unique to the gene in question that can be used to identify and position the gene during genomic analysis. It is derived from cDNA molecules.

Established cell lines: Consist of eukaryotic cells that have been adapted to indefinite growth in culture (they are said to be immortalized).

Ethylene ($CH_2 = CH_2$): A plant hormone, which is involved in fruit ripening, leaf abscission, flower senescence, etc.

Eubacteria: Comprises the major line of prokaryotes.

Euchromatin: Comprises all of the genome in the inter-phase nucleus except for the heterochromatin.

Eutrophication: Presence of organic materials and inorganic nutrients in river waters; contributed by waste water.

Evolutionary clock: It defined by the rate at which mutations accumulate in a given gene.

Excise: To remove a tissue or an organ, usually, by cutting with a scalpel, to obtain explants or during subculture, etc.

Excision: Phage or episome or other sequence describes its release from the host chromosome as an autonomous DNA molecule.

Excision-repair: Systems remove a single-stranded sequence of DNA containing damaged or mispaired bases and replace it in the duplex by synthesizing a sequence complementary to the remaining strand.

Exocytosis: The process of secreting proteins from a cell into the medium, by transport in membranous vesicles from the endoplasmic reticulum, through the Golgi, to storage vesicles, and finally (upon a regulatory signal) through the plasma membrane.

Exogenous elicitor: Elicitor produced by an organism, for example, a microorganism, other than the organism in question.

Exon: Any segment of interrupted gene that is represented in the mature RNA product.

Exon shuffling: New combinations of exons produced by recombination within the intervening sequences; produces new genes that encode proteins with altered functions.

Exonuclease: Cleaves nucleotides one at a time from the end of a polynucleotide chain; they may be specific for either the 5′ or 3′ end of DNA or RNA.

Exonuclease III (ExoIII): An enzyme from *E. coli* that removes nucleotides from 3′OH ends of double stranded DNA.

Exotic species: A species introduced into a geographical area to which it is not native.

Explant: The tissue taken from a plant or seed and transferred to a culture medium to establish a tissue culture system or regenerated a plant.

Exponential phase: A phase of cell growth where they undergo maximum rate of cell division.

Expressed sequence tags (ESTs): These are 200–300 bp long cDNA sequences generated by picking thousands of random clones from cDNA libraries and using them for single-pass sequencing; represents a partial gene sequence; used in microarrays.

Expression casette: When an expression vector contains the promoter, ribosome binding site and transcription termination sequences in a cluster, it is termed as an expression cassette.

Expression library: A collection of recombinant clones that together contains all the available cDNAs from a cell type/tissue/organism; cDNAs are integrated in an expression vector so that they are able to generate their protein product.

Expression profiling: Determination of the cell types/tissues in which a gene is expressed as well as when, the given gene is expressed.

Expression vector: A cloning vector in which a cloned gene can be transcribed and also chloroplasts or mitochondria.

***Ex situ* germplasm conservation:** Collection and conservation of germplasm away from the area of their origin as seed banks, field banks, shoot-tip banks, cell and tissue culture banks, and DNA banks.

Extant variety: A notified variety or a farmer variety that is in public domain.

Extinction: Refers to the death of all the individuals of a species; it leads to loss of all the genes of that species.

Extraction: The process of recovering a compound or a group of compounds from a mixture or from cells into a solvent phase.

Extranuclear genes: Reside outside the nucleus in organelles such as mitochondria and chloroplasts.

Extremozymes: Enzymes that function optimally under extreme conditions of temperature, pH, etc.

***Ex vitro* hardening:** Hardening of plantlets outside culture vessels.

F factor: A bacterial sex or fertility plasmid.

F plasmids: These plasmids carry tra genes that promote conjugation; "F' from fertility.

F: Symbolizes filial from the Latin meaning progeny.

F^+ cell: A bacterial cell that contains a fertility (F) factor, acts as a donor in bacterial conjugation.

F_1 generation: First filial generation,oOffspring of a cross involving the F_1 generation.

F_2 generation: Second filial generation, offspring of a cross involving the F_1 generation.

Facultative heterochromatin: The inert state of sequences that also exist in active copies, for example, 1 mammalian X chromosome in females.

Facultative parasites: Refer to those saprophytes which mostly feed on dead organic matter, but if they reach the living plants they also attack the same.

Facultative saprophytes: Refer to those parasites, which mostly feed on their living host plants but when living plants are not available, they feed on the dead residues of their hosts.

Fast component: Of a reassociation reaction is the first to renature and contains highly repetitive DNA.

Fate map: A map of an embryo showing the adult tissues that will develop from the descendants of cells that occupy particular regions of the embryo.

Fed-batch culture: Microbial growth where nutrients are added into the fermenter without removing the product.

Feedback inhibition: Activity of an enzyme is inhibited by the end product of the biosynthetic pathway in which the enzyme participates.

Feminization: The staminodes present in female flowers develop into fleshy carpels around the fruit leading to abnormal fruit development and partial to complete sterility.

Fermentation: The process by which microorganisms turn raw materials such as glucose into products such as alcohol.

Fermentor: A large sized bioreactor/culture vessel used to produce products under controlled conditions.

Fertilization: Fusion of egg cell with a sperm leading to zygote production, which gives rise to embryo (2n).

Fibroblast: Flattened cells of connective tissue.

Field collection: Maintenance of germplasm by growing different accessions in the field; ordinarily, in case of fruit and forest trees, tuber crops, etc.

Field strategies for transgene containment: These strategies are based on spatial isolation of the transgenic crops coupled with planting of continuous borders of a non-transgenic variety to serve as a pollen trap.

Figure eight: Two circles of DNA linked together by a recombination event that has not yet been completed.

Film reactors: These reactors contain a support for attachment of the bacterial cells to avoid their washout due to higher waste flow rates; support forms anaerobic filter through which the waste passes and is degraded; used to treat vegetable processing wastes of medium to high strength (1%–10% solids), and animal wastes.

Filter hybridization: Performed by incubating denatured DNA preparation immobilized on a nitrocellulose filter with a solution of radioactively labeled RNA or DNA.

Filter paper bridge: A platform formed in a culture tube above the surface of liquid medium by inserting the two arms of a filter paper strip into the medium; explants are placed on the platform.

Filter sterilization: Thermolabile compounds are sterilized by passing their solutions through a membrane of 0.45 μ or lower pore size.

Fingerprint (of DNA): A pattern of polymorphic restriction fragments that differ between individual genomes.

Fingerprint (of protein): The pattern of fragments (usually resolved on a two-dimensional electrophoretic gel) generated by cleavage with an enzyme such as trypsin.

Finite cell lines: In animal cell culture; cell lines that die after several subculture.

Fixed film digestors: The biological components or microorganisms are present in the form of a film on filter particles or large discs.

Flame sterilization: Instruments like forceps, scalpels, etc. are ordinarily sterilized by dipping them in 95% ethyl alcohol followed by flaming.

Flanking sequences: The sequences lying upstream and downstream of a given DNA sequence, that is, usually, on either side of the coding region of a gene.

Flash point: It is the temperature at which a substance catches fire.

Floatation: Raising of microbial cells to the surface of medium by fine gas bubbles adsorbed to their surface.

Flocculation: Sticking together of bacterial and other cells induced by inorganic salts, mineral hydrocolloids, etc.

Fluidity: A property of membranes; it indicates the ability of lipids to move literally within their particular monolayer.

Fluorescent *in situ* hybridization (FISH): In situ hybridization using a probe coupled to a fluorescent molecule.

Focus formation: The ability of transformed eukaryotic cells to grow in dense cluster, piled up on one another.

Focus formatting unit (ffu): A quantitative measure of focus formation.

Foldback DNA: DNA consists of inverted repeats that have ranatured by intrastrand reassociation of denatured DNA.

Footprinting: It is a technique to identify the site on DNA which is protected from the attack by nucleases by some protein.

Foreign DNA: A DNA molecule which is incorporated into a cloning vector or chromosome.

Foreground Selection: In a backcross programm; selection for the gene/allele being transferred.

Fortification: Addition of vitamins or minerals to a food at levels higher than those it originally present in the food.

Forward Mutation: Inactivate a wild-type gene.

Fosmid: A useful derivative of BAC vector that contains λ phage cos sites to facilitate its packaging in heads of phage λ.

Fossil fuel: Fuel derived from coal or petroleum.

Founder effect: The presence in a population of many individuals all with the same chromosome (or region of a chromosome) derived from a single ancestor.

Frameshift mutation: The insertion or deletion of a nucleotide pair or pairs, causing a disruption of the translational reading frame.

Free probe: Probe molecules that are not paired with the test nucleic acid sequence.

Freezing: In case of cryopreservation; lowering of the temperature of cells from 0°C to –196°C.

Fresh weight: Weight of cells/tissues, cultures in their native state after removal of medium (by washing) and extra-cellular water (by partial vacuum).

Friable: Crumbling or fragmenting callus. A friable callus easily dissected and readily digested into single cell or clump of cells in solution.

Frictional coefficient: A measure of the resistance a particle offers to movement through a solvent.

Fuel biotechnology: Use of biological agents or their components to generate various energy rich compounds from different substrate/resources, which can be used as fuel.

Functional genomics: The study of the pattern of expression of the genes present in the genome of organisms that leads to the growth and development of the organisms.

Functional proteomics: Use of proteomics techniques to analyze the characteristics of molecular protein networks involved in a living cell.

Fungi: They are numerous species of molds, mushrooms, brackens, and other forms of nonphotosynthetic plants; derive energy and nutrients by consuming organic material.

Fusion protein: A single protein molecule encoded by parts of two or more genes combined together as one unit.

Fusogen: A fusion including agent used for agglutination of protoplast in somatic hybridization.

G Banding: A technique that generates a striated pattern in metaphase chromosomes that distinguishes the members of a haploid set.

G proteins: Guanine nucleotide-binding trimeric proteins that reside in the plasma membrane. When bound by GDP the trimer remains intact and is inert. When the GDP bound to α subunit is replaced by GTP, α subunit is released from the $\beta\gamma$ dimmer. One of the separated units (either α monomer or the $\beta\gamma$ dimer) then activates or represses a target protein.

G1: The period of the eukaryotic cell cycle between the last mitosis and the start of DNA replication.

G2: The period of the eukaryotic cell cycle between the end of DNA replication and the start of the next mitosis.

Gamete: A mature germ cell.

Gameto clonal variation: Phenotypic variation apart from normal segregation occurring during gamete formation arising from the culture of gametophytic cells or tissues such as in anther culture of unfertilized ovule culture.

Gap: The absence of one or more nucleotides in one strand of the duplex DNA.

Gel: A network of interacting fibers, or a polymer that is solid but traps large amounts of solvent in pore channel inside.

Gel electrophoresis: Electrophoresis, in which a variety of chemically inert gel matrices having adjustable and regular pore size and used as matrix.

Gel filtration chromatography: Matrix used is a gel of a neutral cross-linked carrier; molecules smaller than the pore size of the matrix are retained within the matrix and are later recovered.

Gel retardation analysis: A technique used to identify those DNA fragments to which a specific protein, such as DNA polymerase, binds.

Gelatinization: Starch becomes dissolved to form a viscous suspension.

Gene: A particulate hereditary determiner, a unit of inheritance, a unit of DNA, located at a fixed point along a chromosome. Conceptually, a gene is an informational storage unit capable of undergoing replication, expression, and mutation. Biochemically a gene is a length of DNA that specifics a product or action. A region on DNA encoding a particular polypeptide chain or functional RNA such as rRNA or a tRNA.

Gene cloning (molecular cloning): Insertion of a gene into a vector from a new DNA molecule that can be perpetuated into host cell.

Gene cluster: A group of adjacent genes that are identical and related.

Gene conversion: The process of nonreciprocal recombinant by which one allele in a heterozygote is converted into the corresponding allele.

Gene desert: In human genome; gene-poor regions separating gene-rich regions.

Gene disruption: Integration of a DNA sequence within another gene so that the latter is inactivated; used to produce "knockout" mice.

Gene dosage: Gives the number of copies of a particular gene in the genome.

Gene expression: When a gene performs the specific function, which it is capable of, it is called gene expression; it begins with transcription of the gene and ends with the development of the specific phenotype; ordinarily, it refers to transcription of the concerned gene.

Gene family: A set of genes whose exons are related; the members were derived by duplication and variation from some ancestral gene.

Gene farming: Production of proteins encoded by transgene in animals/plants; the protein is recovered from milk, urine, blood (in animals), seeds etc. (in plants).

Gene flow: Introgression of genes from improved varieties of crop plants into their wild relatives.

Gene gun (biolistic): A device to accelerate particles for physically delivering the recombinant DNA, typically precipitated onto microprojectile, into a cell using an explosive propellant.

Gene machine: An automated microprocessor controlled DNA synthesizer.

Gene pyramiding: Bringing two or more different genes conferring resistance to the same pathogen in the same line/variety.

Gene silencing: Suppression of transgene expression in later generations of transgenic plants; usually, due to transgene (including promoter) methylation; more common when the transgene is integrated in higher copy number.

Gene stacking: Many transgenes are transferred into a single variety through a single transformation event.

Gene tagging: Use of *Agrobacterium* T-DNA, transposons, etc. for identification and/or isolation of a gene by first integrating it into the concerned gene and isolating mutants so produced.

Gene therapy: The correction of a gene deficiency in a cell by the addition of new DNA and its insertion into the genome.

Gene transfer: The use of genetic or physical manipulations to introduce foreign genes into host cell to achieve desired characteristics in progeny.

Generalized transduction: The transduction of any gene in the bacterial genome by a phage.

Genetic code: The correspondence between triplets in DNA (or RNA) and amino acids in protein.

Genetic counselling: Analysis of risk for genetic defects in a family and the presentation of options available to avoid or ameliorate possible risks.

Genetic disease: A disease caused by a mutant gene; generally, due to the recessive allele.

Genetic diversity: The variability present within a species, between individuals of a population or different population groups of the same species for one or a group of traits.

Genetic engineering: *In vitro* DNA technologies used to isolate genes from an organism manipulate them in laboratory as per desire and insert them into another cell system for specific genetic trait. It is also called gene cloning.

Genetic fingerprinting: A technique of analyzing DNA of an individual to reveal the pattern of repetition of particular nucleotide sequences through the genome. The unique pattern of DNA is identified by a PCR or Southern hybridization technique.

Genetic immunization: Use of DNA for immunization.

Genetic instability: Origin of genetic variation in a relatively high frequency, usually, much higher than that ascribable to spontaneous mutations.

Genetic map (linkage map): The linear array of gene on a chromosome based on recombination frequencies.

Genetic markers: Alleles used as experimental probes to keep track of an individual a tissue, a cell, a nucleus, a chromosome, or a gene.

Genetic material: The chemical that contains the genetic information, that is, of which genes are composed; in most of the organisms, DNA is the genetic material; in only some viruses, RNA serves this purpose.

Genetic resource: Sum total of all the genes and their alleles present in a species and its wild relatives.

Genetic transformation: Introduction of DNA or recombinant DNA into a suitable cell/organism.

Genetically engineered microorganism: Microorganism, which has one or more genetically engineered traits.

Genetically modified crops: A crop variety expressing a transgene; produced by genetic engineering.

Genetically modified food: Foods are additives obtained from genetically modified organisms.

Genetically modified organism (GMO): An organism that has been modified by the application of recombinant DNA technology.

Genetically modified organism: An organism with a unique gene combination that does not occur naturally, and is produced either by using genetic engineering or protoplast fusion.

Genetics: The branch of biology concerned with the transmission, expression and evolution of genes- the molecules which control the function, development and ultimate appearance of individuals.

Genome: The array of genes carried by an individual.

Genomic (chromosomal): DNA clones are sequences of the genome carried by a cloning vector.

Genomic library: A collection of clones containing sequences of genomic DNA of an organism. These molecules are propagated in bacteria or viruses.

Genomics: The study of genomes, including nucleotide sequence, gene content, organization and gene number. Cloning and molecular characterization of whole genomes. Categorized into 2 basic areas: Structural genomics-characterizing the physical nature of whole genomes; Functional genomics-characterizing the proteome and overall patterns of gene expression.

Genotype: The genetic constitution (genetic make-up) of an organism with respect to a trait. Complete set of genes inherited by an individual. For a single trait on an autosome, an individual can be homozygous for the dominant trait or homozygous for the recessive trait, or heterozygous. Yellow seeds are dominant, but yellow seeded plants could have a genotype of either YY or Yy.

Germ line cells: The cells that produce gamete.

Golgi apparatus: Individual stacks of membrane near the endoplasmic reticulum; involved in glycosylating proteins and sorting them for transport to different cellular locations.

Gradient elution: An elution system where the solvent composition is varied during the run.

Gratuitous inducers: An authentic inducers of transcription but are not substrates for the induced enzymes.

Green manuring: A farming practice where a leguminous plant which has derived benefits from its association with appropriate species of *Rhizobium* is ploughed in the soil and then non-legume is grown and allowed to take the benefits of the already fixed nitrogen.

GT-AG rule: The presence of these constant dinucleotides at the first 2 and last 2 positions of introns of nuclear genes.

Guanine: One of the 2 purines (nine membered double ringed nitrogenous bases) in DNA or RNA.

Gynandromorphs: Part of the body expresses male characters whereas other parts express female characters. A bilateral gynanadromorph is male on one side (right or left) and female on the other. For example in *Drosophila* right and left body halves are determined at the first cleavage of the zygote. Lagging of the X chromosome at first mitosis can result in two daughter cells of the chromosomal composition AAXX giving normal female, AAXO developing into the male (sterile).

Gyrase: A type II topoisomerase of *E. coli* with the ability to introduce negative supercoils into DNA.

3′-hydroxyl end: The hydroxyl group attached to 3′ carbon atom of sugar of the terminal nucleotide of a nucleic acid.

Hairpin loop: A region of double helix found by base pairing within a single strand of DNA or RNA which has folded itself.

Hairpin: A double-helical region formed by base pairing between adjacent (inverted) complementary sequences in a single strand of RNA or DNA.

Haploid: A cell or organism with a single set of unpaired chromosomes.

Haplotype: Haplotype is the particular combination of alleles in a defined region of some chromosome, in effect the genotype in miniature. Originally used to describe combinations of MHC alleles, it now may be used to describe particular combinations of RFLPs.

Hapten: A small molecule that acts as an antigen when conjugated to a protein.

Hardening: Gradual acclimatization of *in vitro* grown plants to *in vivo* conditions just to adapt the outdoor conditions.

HAT (histone acetyltransferase): Enzymes modify his-tones by addition of acetyl groups; some transcriptional coactivators have HAT activity.

HDAC (histone acetyltransferase): Enzymes remove acetyl groups from histones; they may be associated with repressors of transcription.

Heat shock protein (HSP): Protein for which the rate of synthesis increases after an increase in temperature or any other stress.

Helper virus: It provides functions absent from a defective virus, enabling the latter to complete the infective cycle during a mixed infection.

Hemizygote: A diploid individual that has lost its copy of a particular gene (for example, because a chromosome has been lost) and which therefore has only a single copy.

Hemizygous: If there is only one copy of a gene for a particular trait in a diploid organism, the organism is hemizygous for the trait, and can display a recessive phenotype. X-linked genes in fly or human males are hemizygous.

Heterochromatin: It describes regions of the genome that are permanently in a highly condensed condition, are not transcribed, and are late-replicating. May be constitutive or facultative.

Heterocysts: The modified vegetative cells in cyanobacteria associated with N_2 fixation.

Heterodimer: A protein consisting of two polypeptide chains encoded by different genes.

Heteroduplex (Hybrid) DNA: A generated by base pairing between complementary single strands derived from the different parental duplex molecules; it occurs during genetic recombination.

Heterogametic sex: The sex that produces gametes containing unlike sex chromosomes.

Heterogeneous nuclear (hn) RNA: Comprises transcripts of nuclear genes made by RNA polymerase II; it has a wide size distribution and low stability.

Heterogeneous: Large RNA molecules which are unedited mRNA transcripts found in nucleus of an eukaryotic cell nuclear RNA (hnRNA).

Heterokaryon: A somatic cell containing nuclei from 2 different sources.

Heterologous probe: A DNA probe desired for 1 species and used to screen for a similar DNA sequence from another species.

Heteromultimeric proteins: It consists of nonidentical subunits (coded by different genes).

Heterozygote: An individual with different alleles at some particular locus.

Heterozygous: Differing alleles for a trait in an individual, such as Yy. A zygote is said to be heterozygote, if the two members of an allelic pair are unlike.

HETP (height equivalent to a theoretical plate): The number obtained by dividing the column length by the theoretical plate number.

Hfr strain (high frequency recombination): A bacterial stain with chromosomally integrated plasmid that helps in transfer of part of chromosome to a recipient F cell during conjugation.

Highly repetitive DNA: A first component to reassociate and is equated with satellite DNA.

Histones: Five kinds of proteins forming complexes with eukaryotic DNA.

Holandric: In humans the genes present on Y chromosomes, transmitted from males to males.

Homeobox: A sequence of about 180 nucleotides that encodes a 60-amino acid sequence called a homeodomain, which is part of a DNA-binding protein that acts as a transcription factor.

Homeotic genes: It defined by mutations that convert one body part into another; for example, an insect leg may replace an antenna.

Homodimer: A protein consisting of two identical polypeptide chains encoded by the same gene.

Homogametic sex: It diploid chromosome constitution 2A + XX.

Homokaryon: A cell with two or more identical nuclei as a result of fusion.

Homologous: Having the same evolutionary function or structure.

Homologous chromosomes: The pair of chromosomes in a diploid individual that have the same overall genetic content, centromere placement and pair during meiosis. One member of each homologous pair of chromosomes is inherited from each parent.

Homomultimeric protein: It consists of identical subunits.

Homopolymer: A nucleic acid strand composed of one kind of nucleotide.

Homozygote: An individual with the same allele at corresponding loci on the homologous chromosomes.

Homozygous: Both alleles for a trait are the same in an individual. They can be homozygous dominant (YY), or homozygous recessive (yy). A zygote is said to be a homozygote, if the two members of an allelic pair are alike.

Hot spot: A region in DNA that is more prone to mutagenesis by a particular mutagen.

Housekeeping (constitutive) genes: Those (theoretically) expressed in all cells because they provide basic functions needed for sustenance of all cell types.

Hox: Genes are clusters of mammalian genes containing homeoboxes; the individual members are related to the genes of the complex loci ANT-C and BX-C in *D. Melanogaster.*

Human immunodeficiency virus (HIV): It is a retrovirus, RNA containing virus that replicates by making a DNA intermediate. It consists of an outer lipid envelope derived from the host cell.

Hybrid: Heterozygous, usually referring to the offspring of two true-breeding (homozygous) individuals differing in the traits of interest.

Hybrid dysgenesis: It describes the inability of certain strains of *D. Melanogaster* to interbreed, because the hybrids are sterile (although otherwise they may be phenotypically normal).

Hybrid-arrested translation: A technique that identifies the cDNA corresponding so an mRNA by relying on the ability to base pair with RNA *in vitro* to inhibit translation.

Hybridization: The pairing of complementary RNA and DNA strands to give and RNA-DNA hybrid.

Hybridoma: A cell line produced by fusing a myeloma with a lymphocyte; it continues indefinitely to express the immunoglobulins of both parents.

Hydrogen uptake (Hup) gene: A gene found in N_2 fixing microbes that converts H_2 to $2H^+$ and harvests energy of the organisms.

Hydrolytic reaction: One in which a covalent bond is broken with the incorporation of a water molecule.

Hydropathy plot: A measure of the hydrophobicity of a protein region and therefore of the likelihood that it will reside in a membrane.

Hydrophilic: Hydrophilic groups interact with water, so that hydrophilic regions of protein or the faces of a lipid bilayer reside in an aqueous environment.

Hydrophobic: Hydrophobic groups repel water, so that they interact with one another to generate a nonaqueous environment.

Hyperchromicity: The increase in optical density that occurs when DNA is denatured.

Hypervariable regions: An immunoglobulin are the parts of the variable region that show maximum alteration when different antibodies are compared.

Ideogram: A diagrammatic representation of the G-banding pattern of a chromosome.

Idling reaction: The production of pppGpp and ppGpp by ribosomes when an uncharged tRNA is present in the A site; triggers the stringent response.

Immobilization: Attachment of cells or protoplasts into a matrix.

Immobilized enzymes: An enzyme physically localized in a defined region enabling it to be reused in a continuous process.

Immortalization: It describes the acquisition by a eukaryotic cell line of the ability to grow through an indefinite number of divisions in culture.

Immunity: Plasmids describes the ability of a plasmid to prevent another of the same type from becoming established in a cell. It results usually from interference with the ability to replicate.

Imprinting: It describes a change in a gene that occurs during passage through the sperm or egg with the result that the paternal and material alleles have different properties in the very early embryo. May be caused by methylation of DNA.

***In Situ*:** It means "in original or natural place."

***In situ* hybridization:** A performed by denaturing the DNA of cells squashed on a microscope slide so that reaction is possible with an added single-stranded RNA or DNA; the added preparation is radioactively labeled and its hybridization is followed by autoradiography.

***In Vitro*:** Any process carried out in sterile cultures or measurement of biological processes outside the intact organism such as enzyme reaction.

***In vitro* complementation assay:** It consists of identifying a component of a wild-type cell that can confer activity on an extract prepared from a mutant cell. The assay identifies the component rendered inactive by the mutation.

***In vitro* mutagenesis:** The production of either random or specific mutations in a piece of cloned DNA. Typically, the DNA will then be repackaged and introduced into a cell or an organism to assess the results of the mutagenesis.

***In Vitro* translation:** Protein synthesis directed by purified DNA with bacterial extracts or mRNA that provide ribosomes, tRNA and protein synthesis factors. The reaction mixture is supplemented with ATP, GTP, and amino acids.

***In Vivo*:** The natural conditions in which organisms live.

Incompatibility: The inability of certain bacterial plasmids to coexist in the same cell. It is a cause of plasmid immunity.

Incomplete dominance: Intermediate phenotype in F_1, one allele of a heterozygous pair only partially dominates expression of its partner, red (RR) and white (rr) flowered snapdragon plants produce pink-flowered (Rr) offspring, this does not support a blending theory, parental phenotypes reappear in F_2 generation, color is determined at a single locus, level of gene-directed protein production may be between that of the 2 homozygotes.

Incomplete linkage: Occasional separation of two genes on the same chromosome by a recombinational event.

Indirect end-labelling: A technique for examining the organization of DNA by making a cut at a specific site and isolating all fragments containing the sequence adjacent to one side of the cut; it reveals the distance from the cut to the next break(s) in DNA.

Induced mutations: Induced mutations result from the addition of a mutagen.

Inducer: A small molecule (for example lactose) whose introduction into the medium increases the transcription of an operon (lac operon).

Induction: It refers to the ability of bacteria (or yeast) to synthesize certain enzymes only when their substrates are present; applied to gene expression, refers to switching on transcription as a result of interaction of the inducer with the regulator protein.

Induction of prophage: It describes its excision from the host genome and entry into the lytic (infective) cycle as a result of destruction of the lysogenic repressor.

Initiation codon: AUG; 3-base sequence in mRNA which specifies methionine, the first amino acid.

Initiation factors (IF in prokaryotes, eIF in eukaryotes): Proteins that associate with the small subunit of the ribosome specifically at the stage of initiation of protein synthesis.

Inoculum: A culture containing viable cells of microorganism.

Insert DNA: A DNA molecule incorporated into a cloning vector.

Insertions: It identified by the presence of an additional stretch of base pairs in DNA.

Insertion sequence: A small DNA sequence carrying gene for transposase enzyme for its transposition. A mobile piece of bacterial DNA (several hundred nucleotide pairs in length) that is capable of inactivating a gene into which it inserts.

Integral membrane protein: A protein (noncovalently) inserted into a membrane; it retains its membranous association by means of a stretch of ~25 amino acids that are uncharged and/or hydrophobic.

Integration: Viral or another DNA sequences is its insertion into a host genome as a region covalently linked on either side to the host sequences.

Inteins: Selfish DNA elements (genetic elements) found within coding regions (protein coding sequences). They are translated with their host protein, but then catalyze their own excision and the formation of a peptide bond between their flanking protein regions. The intein protein domain is approximately 140 amino acid long and is sufficient for carrying out the protein splicing reaction.

Interallelic complementation: It describes the change in the properties of a heteromultimeric protein brought about by the interaction of subunits coded by two different mutant alleles; the mixed protein may be more or less active than the protein consisting of subunits only of one or the other type.

Interbands: The relatively dispersed regions of polytene chromosomes that lie between the bands.

Intercistronic region: The distance between the termination codon of one gene and the initiation codon of the next gene.

Interference: A measure of the degree to which one crossover interferes with the chances of another crossover in an adjacent region of the same chromatid. Positive interference reduces the probability of another crossover; negative interference increases the chances of a second crossover event.

Interferon: A glycoprotein produced by a virus-infected cells which protect another cell from attack of the virus.

Intermediate component(s): A reassociation reaction is those reacting between the fast (satellite DNA) and slow (nonrepetitive DNA) components; contain moderately repetitive DNA.

Interphase: The period mitotic cell divisions; divided into G1, S, and G2.

Intervening: Intervening sequence is an intron.

Intron: A portion of DNA between coding regions in a gene that is transcribed, but does not appear in the final mRNA product.

Inversion: A segment that gets separated from the chromosome and is reinserted at the same place but in reverse order; the position and sequence of the genes are altered.

Inverted repeat: Two nearby DNA sequences in opposite strands are same when read in 5′ to 3′ direction.

Inverted terminal repeats: The short related or identical sequences present in reverse orientation at the ends of some transposons.

IS: An abbreviation for insertion sequence, a small bacterial transposon carrying only the genetic functions involved in transposition.

Isoaccepting tRNAs: Isoaccepting tRNAs represent the same amino acid.

Isocratic elution: Elution with a solvent mixture of constant composition.

Isoelectric focusing: The ordering and concentration of substances according to their isoelectric points.

Isoelectric point: The pH at which a substance has a 0 net charge.

Isotype: A group of closely related immunoglobulin chains.

Isozyme: Multiple molecular forms of an enzyme that exhibit similar catalytic properties.

Karyotype: The entire chromosome complement of an individual or cell, as seen during mitotic metaphase. A Karyotype is a photomicrograph of an individual's chromosomes arranged in a standard format showing the number, size and shape of each chromosome type; low resolution physical mapping to correlate gross chromosomal abnormalities with the characteristics of specific diseases.

Kilobase (kb): A unit of length of DNA consisting of 1,000 nucleotides.

Kinase: An enzyme that phosphorylates (add a phosphate group) to a substrate; the substrates for protein.

Kinases: Amino acids in other proteins, and they are divided into those specific for tyrosine and those specific for threonine/serine (and histidine in prokaryotes).

Kinetic complexity: The complexity of a DNA component measured by the Kinetics of DNA reassociation.

Kinetochore: Structure formed at centromere during mitosis for binding microtubules.

Klenow fragment: A part of bacterial DNA polymerase with polymerase activity but lacks exonuclease acitivity.

Label: A radioactive compound attached with DNA/RNA to indicate the presence of a complementary DNA strand in a sample.

Lag phase: Time required by inoculated cells in fresh medium to adapt the new environment before the start of cell division.

Lagging strand: DNA must grow overall in the 3′ to 5′ direction and is synthesized discontinuously in the form of short fragment (5′-3′) that are later connected covalently.

Lampbrush chromosomes: The large meiotic chromosomes found in amphibian oocytes.

Lariat: An intermediate in RNA splicing in which a circular structure with a tail is created by a 5′-2′ bond.

Late period: Phage development is the part of infection following the start of DNA replication.

Late-replicating: Material does not replicate until the end of S phase. Typically this consists of the heterochromatin.

Law of independent assortment of alleles: Alleles of different genes are assorted independently of one another during the formation of gametes. (Genes for different characters are inherited independently of one another). The different pairs of homologous chromosomes arrange themselves on the metaphase I equator in an independent manner and remain independent throughout meiosis. As consequence, genes that are located on nonhomologous chromosomes (in other words, genes that are not linked) undergo independent assortment in meiosis.

Law of segregation: Alleles segregate from one another during the formation of gametes. When two different alleles are brought in a hybrid, both of them coexist without contaminating each other and are separated in meiosis as pure as brought from the parents. During the formation of gametes, the paired unit factors separate or segregate randomly so that each gamete receives one or the other. The principle of segregation applies to homologous chromosomes during meiosis.

Leader: The nontranslated sequence at the 5′ end of mRNA that precedes the initiation codon.

Leader sequence: A protein is a short N-terminal sequence responsible for passage into or through a membrane.

Leading strand: DNA is synthesized continuously in the 5′-3′ direction.

Leaky mutations: Allow some residual level of gene expression.

Left splicing junction: The boundary between the right end of an exon and the left end of an intron.

Leghemoglobin: A globular protein containing heme expressed in legume root nodules.

Lethal alleles: Mutated genes that are capable of causing death.

Lethal Locus: Any gene in which a lethal mutation can be obtained (usually by deletion of the gene).

Library: A set of cloned fragments together representing the entire genome.

Ligase: An enzyme used by a genetic engineer to join the cut ends of the double stranded DNA.

Ligation: The formation of a phosphodiester bond to link two adjacent bases separated by a nick in one strand of a double helix of DNA. (The term can also be applied to blunt-end ligation and to joining of RNA.)

Lignocellulose: The composition of woody biomass consisting of lignin and cellulose.

Lines: A long-period interspersed sequences in mammalian genomes that are retroposons generated from RNA polymerase II transcripts.

Linkage: Tendency of genes to be inherited together because of their location on the same chromosome; measured by percent recombination between loci. Linked genes do not obey Mendel's law of independent assortment because they tend to go into the gametes together. Crosses involving linked genes do not give same results as unlinked genes. Linkage can be incomplete or complete.

Linkage disequilibrium: It describes a situation in which some combinations of genetic markers occur more or less frequently in the population than would be expected from their distance apart. It implies that a group of markers has been inherited coordinately. It can result from reduced recombination in the region or from a founder effect, in which there has been insufficient time to reach equilibrium since one of the markers was introduced into the population.

Linkage groups: All genes on a single chromosome form one linkage group. Number of linkage groups is equal to the haploid chromosome number except in sexes with dissimilar sex chromosomes. Fruit flies have four pairs of chromosomes to hold thousands of genes.

Linker DNA: All DNA contained on a nucleosome in excess of the 146 bp core DNA.

Linker fragment: A short synthetic duplex oligonucleotide containing the target site for some restriction enzyme; may be added to ends of a DNA fragment prepared by cleavage with some other enzyme during reconstruction of recombinant DNA.

Linkers: A synthetic double stranded oligonucleotide carrying the sequence for one or more restriction sites.

Linker scanner mutations: Introduced by recombining two DNA molecules *in vitro* at a restriction fragment added to the end of each; the result is to insert the linker sequence at the site of recombination.

Linking number: The number of times that 2 strands of a closed, circular DNA duplex cross over each other.

Linking number paradox: It describes the discrepancy between the existence of 2supercoils in the path of DNA on the nucleosome compared with the measurement of 1 supercoil released when histones are removed.

Lipid bilayer: The form taken by concentration of lipids in which the hydrophobic fatty acids occupy the interior and the polar heads face the exterior.

Lipids: Lipids have polar heads, containing phosphate (phospholipid), sterol (such as cholesterol), or saccharide (glycolipid) connected to a hydrophobic tail consisting of fatty acid(s).

Liposomes: Artificial vesicles of phospholipids.

Liquid (solution) hybridization: A reaction between complementary nucleic acid strands performed in solution.

Locus: Position on a chromosome at which the gene for a particular trait resides; a locus may be occupied by any one of the alleles for the gene.

LOD Score: A measure of genetic linkage, defined as the $\log_{10}$ ratio of the probability that the data would have arisen if the loci are linked to the probability that the data could have arisen from unlinked loci. The conventional threshold for declaring linkage is a LOD score of 3.0, that is, a 1,000:1 ratio (which must be compared with the 50:1 probability that any random pair of loci will be unlinked).

Long-period interspersion: A pattern in which long stretches of moderately repetitive and nonrepetitive DNA alternate.

Loop: A single-stranded region at the end of a hairpin in RNA (or single-stranded DNA); corresponds to the sequence between inverted repeats in duplex DNA.

LTR: An abbreviation for long-terminal repeat, a sequence directly repeated at both ends of a retroviral DNA.

Lumen: The interior of a compartment bounded by membranes, usually the endoplasmic reticulum or the mitochondrion.

Luxury genes: Those coding for specialized functions synthesized (usually) in large amounts in particular cell types.

Lysis: Death of bacteria at the end of a phage infective cycle when they burst open to release the progeny of an infecting phage.

Lysogen: A strain of bacterium that harbors a prophage.

Lysogenic immunity: The ability of a prophage to prevent another phage genome of the same type from becoming established in the bacterium.

Lysogenic life cycle: The life cycle of a temperate bacteriophage in which the virus can either replicate via the lytic life cycle or become a latent prophage within the infected host bacterium and not replicate.

Lysogenic repressor: The protein responsible for preventing a prophage from re-entering the lytic cycle.

Lysogeny: The ability of a phage to survive in a bacterium as a stable prophage component of the bacterial genome.

Lysosomes: Small bodies, enclosed by membranes that contain hydrolytic enzymes in eukaryotic cells.

Lytic infection: Lytic infection of bacteria by a phage cells in destruction of bacteria and release of progeny phage.

M13: A single stranded DNA bacteriophage used as vector for DNA sequencing.

Main band (of genomic DNA): A broad peak on a density gradient, excluding any visible satellite DNAs that form separate bands.

Major histocompatibility locus: A large chromosomal region containing a giant cluster of genes that code for transplantation antigens and other proteins found on the surface of lymphocytes.

Map distance: Measured as cM (centiMorgans) = percent recombination (sometime subject to adjustments).

Map unit: A distance between genetic markers corresponding to 1% recombination frequency between the markers.

MAR: A region (matrix attachment site; also known as SAR for scaffold attachment site) of DNA that attaches to the nuclear matrix.

Marker (DNA): A fragment of known size used to calibrate an electrophoretic gel.

Marker (genetic): Any allele of interest in an experiment.

Maternal inheritance: The preferential survival in the progeny of genetic markers provided by one parent.

Megabase (Mb): A length of DNA consisting of 10^6 bp. 1 Mb = 10^3 kb = 10^6 bp.

Meiosis: Occurs by 2 successive divisions (meiosis I and II) that reduce the starting number of $4n$ chromosomes to $1n$ in each of four product cells. Products may mature to germ cells (sperm or eggs).

Melting: DNA denaturation.

Melting temperature (T_m): The midpoint of the temperature range over which DNA is denatured.

Membrane proteins: Proteins have hydrophobic regions that allow part or all of the protein structure to reside within the membrane; the bonds involved in this association are usually noncovalent.

Membranes: An asymmetrical lipid bilayer that has lateral fluidity and contains proteins.

Meristem: Undifferentiated but determined tissue, the cells of which are capable of active cell division and differentiation into specialized and permanent tissue such as shoot and root.

Messenger RNA: RNA transcript with coding sequence for at least 1 polypeptide.

Metabolites: The products produced through biochemical activities.

Metastasis: The ability of tumor cells to leave their sites of origin and migrate to other locations in the body, where a new colony is established.

Micrococal nuclease: An endonuclease that cleaves DNA; in chromatin, DNA is cleaved preferentially between nucleosomes.

Microinjection: A technique of injecting DNE or RNA into the nucleus of protoplast of cell with the help of a micropipette.

Micropropagation: Use of small piece of tissue such as meristem grown in culture to produce large number of plants.

Microsomes: Fragmented pieces of endoplasmic reticulum associated with ribosomes.

Microtubule associated proteins (MAPs): Proteins associated with microtubules and responsible for influencing their stability and organization.

Microtubule organizing center (MTOC): A structure from which microtubules may be extended. The major MTOCs in a mitotic cell are the chromosomes.

Microtubules: Filaments consisting of dimmers of tubulin; interphase microtubules are recoginized into spindle fibers at mitosis, when they are responsible for chromosome movement.

Minicell: An annucleate bacterial (*E. coli*) cell produced by a division that generates a cytoplasm without a nucleus.

Minichromosome (of SV 40 or polyoma): The nucleosomal form of the viral circular DNA.

Mismatch: When normal base pairing is not observed in DNA, say for example pairing between T and G.

Missense mutation: A mutation that alters a codon so that it encodes a different amino acid.

Mitosis: The division of a eukaryotic somatic cell.

Mobile phase: The mixture of solvents that is percolated through the column.

Mobility: The velocity a particle or ion attains for a given voltage. A relative measure of how quickly an ion moves in an electric field.

Modification (of DNA or RNA): Includes all changes made to the nucleotides after initial incorporation into the polynucleotide chain.

Modified bases: All those except the usual four from which DNA (T, C, A, G) and RNA (U, C, A, G) are synthesized; they result from postsynthetic changes in the nucleic acid.

Molecular chaperone: A protein that is needed for the assembly of proper folding of some other protein, but which is not itself a component of the target complex.

Molecular farming: Growing and harvesting genetically modified crops, with the object of producing not foodstuffs but pharmaceuticals. It refers to the use of genetic engineering-putting specific genes from other organisms into a plant-to create plants which will produce products such as drugs, chemicals, vaccines and even biodegradable plastics. More than 150 proteins and peptides (linked amino acids) have been produced using tobacco mosaic virus as a vector. In the past, molecular farming of drugs from transgenic livestock such as pigs, sheep and cows required the death of the animal so the drug could be harvested form their blood. Now the focus has shifted to producing useful proteins secreted into the mills of transgenic livestock.

Molecular markers: The markers (DNA sequences) revealing variation at DNA level which can be detected and monitored in the subsequent generations, for example, RFLP, RAPD, SSR, etc.

Molecular sieving: Separation of molecules on the basis of their effective sizes.

Monocistronic mRNA: An mRNA that codes for 1 protein.

Monoclonal antibodies: Antibodies derived from a single clone of cells which recognize only 1 kind of antigen.

Monohybrid cross: A type of cross involving parents differing in only one trait or in which only one trait is considered.

Monolayer: It describes the growth of eukaryotic cells in culture as a layer only one cell deep.

Monosomy (2n-1): Occurs when an individual has only 1 type of a particular chromosome.

Morphogen: A factor that induces development of particular cell types in a manner that depends on its concentration.

Morphogenesis: The development of a structure from an unorganized state to a differentiated and organized state.

MPF (maturation- or M phase-promoting factor): A dimeric kinase, containing the p34 catalytic subunit and a cyclin regulatory subunit, whose activation triggers the onset of mitosis.

mtDNA: Mitochondrial DNA.

Multiforked chromosome (in bacteria): More than 1 replication fork, because a second initiation has occurred before the first cycle of replication has been completed.

Multimeric proteins: It consists of more than 1 subunit.

Multiple Alleles: More than two alleles for 1 locus, but each individual inherits only 2 alleles, ABO system of human blood type.

Mushrooms: Edible fruiting bodies of ascomycetous and basidiomycetous fungi.

Mutagen: Any agent that causes mutations by damaging DNA.

Mutant: An organism differing from the normal or wild type due to a change in its DNA sequence.

Mutation: Change in the DNA sequence of a gene to some new, heritable form generally, but not always a recessive allele. A forward mutation converts a wild-type allele to mutant allele.

Mutation frequency: The frequency at which a particular mutant is found in the population.

Mutation rate: The rate at which a particular mutation occurs, usually given as the number of events per gene per generation.

Mycoherbicide: Fungal preparations that kill the abnoxious weeds/herbs.

mycoRhiz®: A commercial preparation of *Pisolithus tinctorius* mycelia produced by Abbott Lab (U.S.A.) used for raising tree seedling through tailoring of roots.

Myeloma: A tumor cell line derived from a lymphocyte; usually produces a single type of immunoglobulin.

Nanometer (nm): A unit of length equal to 1×10^{-9} meter.

NCBI (National Centre of Biology Information): It was established in 1988 for development of information system in molecular biology. It is the foremost repository of publically available genomic and proteomic data.

Negative complementation: Occurs when interallelic complementation allows a mutant subunit to suppress the activity of a wild-type subunit in a multimeric protein.

Negative regulation: When an active protein inhibits or turns off transcription of an operon.

Negative regulators: Function by switching off transcription or translation.

Negative supercoiling: Comprises the twisting of a duplex of DNA in space in the opposite sense to the turns of the strands in the double helix.

Neutral substitutions: Proteins are those changes of amino acids that do not affect activity.

Nick (in duplex DNA): The absence of a phosphodiester bond between two adjacent nucleotides on 1 strand.

Nick Translation: It describes the ability of *E.coli* DNA polymerase I to use a nick as a starting point from which one strand of a duplex DNA can be degraded and replaced by resynthesis of new material; is used to introduce radioactively labeled nucleotides into DNA *in vitro*.

***Nif* (nitrogen fixing) genes:** A 24 kb long nucleotide sequence organized in 7 oberons (transcriptional units) which take part in nitrogen fixation.

Nitrogenase: An enzyme expressed by *nif* genes that converts N_2 to NH_4^+ under O_2-tense conditions in prokaryotes.

Nod genes: Genes that encodes nodulins.

Nodulation: Formation of root nodules in legumes by *Rhizobium.*

Nonpermissive conditions: Do not allow conditional lethal mutants to survive.

Nonrepetitive transposition: It describes the movement of a transposon that leaves a donor site (usually generating a double-strand break) and moves to a new site.

Nonautonomous controlling elements: Defective transposons that can transpose only when assisted by an autonomous controlling element of the same type.

Nondisjunction: is the failure of chromosomes to separate, it is more common during meiosis I than meiosis II; it can also occur in mitosis.

Nonsense codons: UAA, UAG, UGA codons do not specify any amino acid. Also known as termination or stop codons.

Nonsense mutation: Any change in DNA that causes a (termination) codon to replace a codon representing an amino acid.

Nonsense suppressor: A gene coding for a mutant tRNA able to respond to one or more of the termination codons.

Nontranscribed spacer: The region between transcription units in a tandem gene cluster.

Normal phase: A chromatographic mode in which the mobile phase is less polar than the stationary phase.

Northern blotting: A technique for transferring RNA from an agarose gel to a nitrocellulose filter on which it can be hybridized to a complementary DNA. Hybridization of a labeled DNA probe to RNA fragments that have been transferred from an agarose gel to a nitrocellulose filter. It is also called Northern Hybridization.

Nuclear envelope: A layer of 2 membranes surrounding the nucleus. It is penetrated by nuclear pores. The inner membrane is bound on the interior by the nuclear laminin. The outer membrane extends in the cytosol into the network of the endoplasmic reticulum.

Nuclear lamina: It consists of a proteinaceous layer on the inside of the nuclear envelope. It consist of (up to) 3 lamin proteins.

Nuclear matrix: A network of fibers surrounding and penetrating the nucleus.

Nuclear pores: Large structures that extend across the nuclear envelope and are to transport macromolecules both into and out of the nucleus.

Nucleoid: The compact body that contains the genome in a bacterium.

Nucleolar organizer: The region of a chromosome carrying genes coding for rRNA.

Nucleolus: A discrete region of nucleus created by the transcription of rRNA genes.

Nucleolytic: Nucleolytic reactions involve the hydrolysis of a phosphodiester bond in a nucleic acid.

Nucleosome (Nu body): Is the basic structural subunit of chromatin, consisting of ~200 bp of DNA and an octamer of histone proteins.

Null mutation: Completely eliminates the function of a gene, usually because it has been physically deleted.

Ochre codon: The triplet UAA, 1 of 3codons that cause termination of protein synthesis.

Ochre mutation: Any change in DNA that creates a UAA codon at a site previously occupied by another codon.

Ochre suppressor: A gene coding for a mutant tRNA able to respond to the UAA codon to allow continuation of protein synthesis; ochre suppressors also suppress amber codons.

Okazaki fragments: The short stretches of 1,000–2,000 bases produced during discontinuous replication; they are later joined into a covalently intact strand.

Oligomer: A molecule formed from a small number of monomers.

Oligonucleotides: A short synthetic single stranded DNA.

Oncogene: Genes whose products have the ability to transform eukaryotic cells so that they grow in a manner analogous to tumor calls. Oncogenes carried by retroviruses have names of the form *v-onc.* (*see also* proto-oncogenes.) A gene that contributes to the production of cancer. Oncogenes are generally mutated forms of normal cellular gene.

Opal codon: UGA codon, also known as umber codon, stops proteins synthesis.

Open reading frame (ORF): A series of triplets coding for amino acids without any termination codons; sequence is (potentially) translatable into protein.

Operator: The site on DNA at which a repressor protein binds to prevent transcription from initiating at the adjacent promoter.

Operon: A unit of bacterial gene expression and regulation, including structural genes and control elements in DNA recognized by regulator gene product(s).

Opine: The condensation product of an amino acid with *keto* acid or sugar.

Organ culture: Culture of an organ *in vitro.*

Organelles: Compartments located in the cytoplasm and surrounded by a membrane.

Organogenesis: The process of initiation and development of a structure which show natural organ form and/or function, the initiation of which is temporally separated from the initiation of other organs.

Origin (*ori*)**:** A sequence of DNA at which replication is initiated.

Origin for replication (ori): A site of nucleotide sequence from which replication begins.

Orphans: Isolated individual genes found in isolated locations, but related to membrane of a gene cluster.

Osmaticum: Agents such as polyethylene glycol, mannitol, glucose, sucrose which maintains the osmotic potential of a nutrient medium equivalent to that of cultured cells.

Overwinding: Overwinding of DNA is caused by positive supercoiling (which applies further tension in the direction of winding of the two strands about each other in the duplex).

P1 clones: A cloning system for isolating genomic DNA that uses elements from phage P1, that is, *loxP* and *pax* sites.

Packing ratio: The ratio of the length of DNA to the unit length of the fiber containing it.

Pairing: Pairing of chromosomes – *see* synapsis.

Palindrome: A sequence of DNA that is the same when one strand is read left to right or the other is read right to left; consists of adjacent repeats.

Palindromic sequences: Complementary DNA sequences that are the same when each strand is read in the same direction (e.g., 5′→3′). These sequences act as recognition sites for type II restriction endonuclease.

Papovaviruses: A class of animal viruses with small genomes, including SV 40 and polyoma.

Paranemic joint: A region in which two complementary sequences of DNA are associated side by side instead of being intertwined in a double helical structure.

Parental types: Progeny of a genetic cross that is genetically identical to 1 or the other of the parents.

Partial diploid: A bacteria with two copies of part of its genome. It is due to the presence of plasmid or prophage carrying some part of bacterial DNA.

Partition: Process by which a solute is distributed between two immiscible phases.

Patent: A legal document issued by a Government that permits the holder the exclusive right to manufacture, use of sell an invention for a defined period, which varies country to country.

pBR322: One of the standard plasmid cloning vectors.

PCR (polymerase chain reaction): A technique in which cycles of denaturation, annealing with primer, and extension with DNA polymerase, are used to amplify the number of copies of a target DNA sequence by $>10^6$ times.

Peak: A band or zone in a chromatogram showing a maximum concentration between 2 minima.

Peptide bond: A covalent bond between the amino (NH_2) group of 1 amino acid and the carboxyl (COOH) group of another amino acid.

Perinuclear space: Lies between the inner and outer membranes of the nuclear envelope.

Periodicity: Periodicity of DNA is the number of base pairs per turn of the double helix.

Permissive conditions: It allow conditional lethal mutants to survive.

Petite: Petite strains of yeast lack mitochondrial function.

pH: The negative logarithm of hydrogen ion concentration of any solution for measuring acidity or alkalinity.

Phage (bacteriophage): A bacterial virus.

Phage variation: An alteration in the type of flagella produced by a bacterium.

Phagemid: A cloning vector that contains components derived from both phage DNA and plasmid.

Pharmacogenomics: The study of association between genetics and drug response.

Phenocopies: The strength of environmental changes may modify several genes. In some instances specific environmental changes may modify the development of an organism so that its phenotype simulates the effect of a particular gene, although this effect is not inherited. Such individuals are known as phenocopies. The normal body color of the vinegar fly *(Drosophia melanogaster)* is light brown color. If larvae from normal flies are raised on food containing silver salts, they develop into yellow flies. The yellow-bodied but genotypically normal fly is a phenocopy of the yellow mutant. The offspring of a phenocopy have the normal brown body color unless raised again on food containing silver salts; however, the offspring of the yellow mutant have yellow bodies.

Phenotype: The physical appearance of an organism with respect to a trait, for example, yellow (Y) or green (y) seeds in garden peas. The dominant trait is normally represented with a capital letter, and the recessive trait with the same lower case letter.

Phosphatase: An enzyme that removes phosphate groups from substrates.

Plant growth promoting rhizobacteria (PGPR): The bacteria that live in rhizosphere and promote plant growth through complex mechanisms.

Plaque: Clear zone in a bacterial culture resulting due to killing of bacterial cells by phage.

Plasma membrane: The continuous membrane defining the boundary of every cell.

Plasmid: A small circular DNA molecule, which can replicate independently of the bacterial chromosome.

Plating efficiency: The percentage of cells plated on medium which develops colonies.

Playback: Playback experiment describes the retrieval of DNA that has hybridized with RNA to check that it is nonrepetitive by a further reassociation reaction.

Plectonemic winding: The inter winning of the two strands in the classical double helix of DNA.

Pleiotropic: Gene affects more than one (apparently unrelated) characteristic of the phenotype.

Pleiotropy: The phenomenon where a single gene is responsible for number unrelated phenotypic effects (multiplies or manifold effects of a gene). For example in garden peas, the gene which controls the flower color also controls the color of the seed coat and the presence of red spots in the leaf axils. It is not essential that the traits are equally influenced.

Ploidy: The number of copies of the chromosome set present in a cell; a haploid has 1 copy, a diploid has 2 copies, etc.

Point mutations: Changes involving single base pairs.

Polarity: The effect of a mutation in one gene in influencing the expression (at transcription or translation) of subsequent genes in the same transcription unit.

Polyacrylamide gel electrophoresis (PAGE): A method for separation of nucleic acids or protein molecules according to their size. The molecules migrate through the inert gel matrix under the influence of an electric field.

Polyadenylation: The addition of a sequence of polyadenylic acid to the 3′ end of a eukaryotic RNA after its transcription.

Polycistronic mRNA: mRNA with more than one translational region so that same RNA can be translated into more than one polypeptide chains.

Polyclonal antibodies: A serum containing antibodies that binds to different antigenic determinants of an antigen.

Polygenic inheritance: When a trait is controlled by several allelic pairs at different loci on a chromosome or on different chromosomes, contributing alleles have additive intermediate phenotypes; so they produce continuous variations whose frequency distribution forms a normal (bell-shaped) curve, for example, kernel color in wheat and skin color and height in humans.

Polylinker: A segment of DNA that contains several restriction sites. It is also called multiple cloning sites.

Polymerase chain reaction (PCR): The action of the enzyme polymerase to produce many copies of a polynucleotide sequence of DNA at high temperature.

Polymorphism: The simultaneous occurrence in the population of genomes showing allelic variations (as see either in alleles producing different phenotypes or (for example, in changes in DNA affecting the restriction pattern).

Polypeptide: The product of a single gene with a long chain of amino acids linked together by peptide bonds.

Polyploid cell: It has more than 2 sets of the haploid genome.

Polyploidy: Offspring end up with more than 2 complete sets of chromosomes. Polyploidy is a major evolutionary mechanism in plants; probably involved in 47% of flowering plants including major crops.

Polyprotein: A gene product that is cleaved into several independent proteins.

Polysome (polyribosome): An mRNA associated with a series of ribosomes engaged in translation.

Polytene chromosomes: Generated by successive replications of a chromosome set without separation of the replicas.

Position effect: A change in the expression of a gene brought about by its translocation to a new site in the genome; for example, a previously active gene may become inactive if placed near heterochromatin.

Position effect variegation: The consequences when a gene is inactivated in some cells but not in others as the result of variable spread of inactivity from heterochromatin.

Positive regulation: Expression of a gene in the presence of an active regulatory protein.

Positive regulator proteins: Required for the activation of a transcription unit.

Positive supercoiling: The coiling of the double helix in space in the same direction as the winding of the 2 strands of the double helix itself.

Postmeiotic segregation: The segregation of 2 strands of a duplex DNA that bear different information (created by heteroduplex formation during meiosis) when a subsequent replication allows the strands to separate.

Posttranscriptional regulation: Regulation of gene expression after synthesis of mRNA from the gene.

Primary cells: Eukaryotic cells taken into culture directly from the animal.

Primary culture: A culture started from cell, tissue, organ taken directly from an organism. It is regarded as such until subcultured for the first time. Then it becomes cell line.

Primary structure: The sequence of amino acids in a polypeptide chain.

Primary transcript: The original unmodified RNA product corresponding to a transcription unit.

Primer: A short sequence (often of RNA) that is paired with one strand of DNA and provides a free 3′ – OH end at which a DNA polymerase starts synthesis of a deoxyribonucleotide chain.

-10 Sequence: The consensus sequence TATAATG centered about 10 bp before the start point of a bacterial gene. It is involved in the initial melting of DNA by RNA polymerase.

-35 Sequence: The consensus sequence centered about 35 bp before the start point of a bacterial gene. It is involved in the initial recognition by RNA polymerase.

Primosome: The complex of proteins involved in the priming action that initiates synthesis of each Okazaki fragment during discontinuous DNA replication; the primosome may move along DNA to engage in successive priming events.

Prion: A proteinaceous infectious agent, which behaves as an inheritable trait, although it contains no nucleic acid. Examples are PrP^{Sc}, the agent of scrapie in sheep and bovine spongiform encephalopathy, and Psi, which confers an inherited state in yeast.

Pro-Embryo: Early embryonic stage.

Probe: A known sequence complementary to the sequence to be identified from among many sequences.

Probiotics: Health-giving properties derived from microorganisms.

Procentriole: An immature centriole, formed in the vicinity of a mature centriole.

Processed pseudogene: An inactive gene copy that lacks introns, contrasted with the interrupted structure of the active gene. Such genes presumably originate by reverse transcription of mRNA and insertion of a duplex copy into the genome.

Processive enzymes: Enzymes continue to act on a particular substrate, that is, do not dissociate between repetitions of the catalytic event.

Prokaryotic: Organisms (bacteria) lack nuclei.

Promoter: A DNA segment lying upstream (5′ to) a gene onto which RNA polymerase gets attached.

Promoter sequence: A regulatory sequence of DNA that initiates the expression of a gene.

Proofreading: Any mechanism for correcting errors in protein or nucleic acid synthesis that involves scrutiny of individual units *after* they have been added to the chain.

Propagule: A part of plant that serve for propagation.

Prophage: A bacteriophage that has integrated its DNA into the nucleoid of a host bacterium.

Protein engineering (enzyme engineering): Generation of protein with subtly modified structure that confer special properties (as present before) such as high pH stability, thermal stability, high catalytic specificity, etc.

Proteolytic: The hydrolysis of peptide bonds in protein.

Proteome: Is the complete set of proteins that is expressed by the entire genome. Because some genes code for multiple proteins, the size of the proteome is greater than the number of genes. (Study of the complete protein complement of the genome).

Proteomics: The study of proteome or the array of proteins that an organism produce.

Proto-oncogenes: The normal counterparts in the eukaryotic genome to the oncogenes carried by some retroviruses. They are given names of the form *c-onc*.

Protoplast: Microbial or plant cells whose cell wall has been removed so that the cells assume a spherical shape.

Provirus: A duplex DNA in the eukaryotic chromosome corresponding to the genome of an RNA retrovirus.

Pseudodominance: Expression of a recessive trait due to absence of a dominant allele.

Pseudogenes: Inactivate but stable components of the genome derived by mutation of an ancestral active gene.

Puff: An expansion of a band of a polytene chromosome associated with the synthesis of RNA at some locus in the band.

Pulse field gel electrophoresis (PFGE): A procedure used to separate large DNA molecules by alternating the direction of electric current in a pulsed manner across a gel.

Pulse-chase: Pulse-chase experiments are performed by incubating cells very briefly with a radioactively labeled precursor (of some pathway or macromolecule); then the fate of the label is followed during a subsequent incubation with a nonlabelled precursor.

Pure line: A pure line is a population that breeds true for the particular character being studied; that is; all offspring produced by selfing or crossing within the population show the same form for the character.

Purine: A 9-membered double ringed nitrogenous base in DNA or RNA.

Pyrimidine: A 6-membered single ringed nitrogenous base in DNA or RNA.

Pyrimidine dimer: Two adjacent pyrimidines are covalently joined by chemical bonds.

Quantitative trait locus: A locus that affects a quantitative trait.

Quaternary structure: A protein refers to its multimeric constitution.

Quick-stop: *dna* mutants of *E. coli* cease replication immediately when the temperature is increased to 42°C.

R loop: The structure formed when an RNA strand hybridizes with its complementary strand in a DNA duplex, thereby displacing the original strand of DNA in the form of a loop extending over the region of hybridization.

Random amplified polymorphic DNA (RAPD): A PCR-based technique for detecting polymorphism at DNA level. A single short (usually 10-mer) synthetic nucleotide primers are used in PCR.

Rapid lysis (r) mutants: Display a change in the pattern of lysis of *E. coli* at the end of an infection by a T-even phage.

Reading frame: One of 3 possible ways of reading a nucleotide sequence as a series of triplets.

Reassociation of DNA: The pairing of complementary single strands to form a double helix.

RecA: The product of the *recA* locus of *E. coli*; a protein with dual activities, activating proteases and also able to exchange single strands of DNA molecules. The protease-activating activity controls the SOS response; the nucleic acid handling facility is involved in recombination-repair pathways.

Receptor: A transmembrane protein, located in the plasma membrane, that binds a ligand in a domain on the extracellular side, and as a result has a change in activity of the cytoplasmic domain. (The same term is sometimes used also for the steroid receptors, which are transcription factors that are activated by binding ligands that are steroids or other small molecules.)

Recessive: Allele is obscured in the phenotype of a heterozygote by the dominant allele, often due to inactivity or absence of the product of the recessive allele.

Recessive lethal: An allele that is lethal when the call is homozygous for it.

Recessive trait: The opposite of dominant. A trait that is preferentially masked. A mutation or other genetic marker that does not exert its phenotype due to the presence of corresponding wild type marker in a diploid organism. A recessive allele is obscured in the phenotype of a heterozygote by the dominant allele, often due to inactivity or absence of the product of the recessive allele.

Reciprocal cross: Using male and female gametes for 2 different traits, alternating the source of gametes (♀A×♂B) (♀B×♂A).

Reciprocal recombination: The production of new genotypes with the reverse arrangements of alleles according to maternal and paternal origin.

Reciprocal translocation: Exchanges part of new genotypes with part of another chromosome.

Recombinant: Progeny have a different genotype from that of either parent.

Recombinant DNA: Joining of 2 different DNA sequences artificially.

Recombinant joint: The point at which two recombining molecules of duplex DNA are connected (the edge of the heteroduplex region).

Recombination nodules (nodes): Dense objects present on the synaptonemal complex; could be involved in crossing-over.

Recombination-repair: A mode of filling a gap in one strand of duplex DNA by retrieving a homologous single strand from another duplex.

Recon: Smallest genetic unit between which recombination can occur.

Regeneration: Development of new organs or plantlets from a tissue, callus culture or from a bud.

Regulator gene: That codes for a product, typically protein that controls the expression of other genes, usually at the level of transcription.

Relaxed mutants: Relaxed mutants of *E. coli* do not display the stringent response to starvation for amino acids (or other nutritional deprivation).

Relaxed replication control: The ability of some plasmids to continue replicating after bacteria cease dividing.

Release (termination) factors: It respond to termination codons to cause release of the completed polypeptide chain and the ribosome from mRNA.

Renaturation: The reassociation of denaturated complementary single strands of a DNA double helix.

Repeated DNA sequence: A sequence of nucleotides which occurs more than once in a genome. It may be present in a few to many millions of copies. The length of individual repeated sequence may be a few nucleotides to several kb.

Repeating: Unit in a tandem cluster is the length of the sequence that is repeated; appears circular on a restriction map.

Repetition frequency: The (integral) number of copies of a given sequence present in the haploid genome; equals 1 for nonrepetitive DNA, >2 for repetitive DNA.

Repetitive DNA: It behaves in a reassociation reaction as though many (related or identical) sequences are present in a component, allowing any pair of complementary sequences to reassociate.

Replacement sites: Replacement sites in a gene are those at which mutations alter the amino acid that is coded.

Replacement therapy: The administration of metabolites, co-factors or hormones that are deficient due to genetic disease.

Replica planting: Transfer of cells from bacterial colonies growing on the peti plate to the other plate corresponding to the location of the first plate (master plate).

Replication-defective Virus: Lost one or more genes essential for completing the infective cycle.

Replication Eye: A region in which DNA has been replicated within a longer, unreplicated region.

Replication fork: The point at which strands of parental duplex DNA are separated so that replication can proceed.

Replicative transposition: The movement of a transposon by a mechanism in which first it is replicated, and then 1 copy is transferred to a new site.

Replicon: A unit of the genome in which DNA is replicated; contains an origin for initiation of replication.

Replisome: The multiprotein structure that assembles at the bacterial replicating fork to undertake synthesis of DNA. Contains DNA polymerase and other enzymes.

Reporter gene: A coding unit whose product is easily assayed (such as chloramphenicol transacetylase); it may be connected to any promoter of interest so that expression of the gene can be used to assay promoter function.

Reporter gene: A gene encoding a product that can readily be assayed during genetic transformation, for example, GUS, chloramphenicol transacetylase, etc. It may be connected to any promoter of interest so that it may express.

Repression: The ability of bacteria to prevent synthesis of certain enzymes when their products are present; more generally, refers to inhibition of transcription (or translation) by binding or repressor protein to a specific site on DNA (or mRNA).

Repressor: An active protein binds to operator sequences nearby to promoter of an operon and prevents transcription of the operon. It exerts a negative control.

Resolution: The degree of separation between 2 components by chromatography.

Resolvase: Enzyme activity involved in site-specific recombination between two transposons present as direct repeats in a cointegrate structure.

Restriction endonuclease (type II): An enzyme that recognizes a specific duplex DNA and phosphodiester bonds on both the strands between specific nucleotides.

Restriction enzymes: Recognize specific short sequences of DNA and cleave the duplex (sometimes at target site, sometimes elsewhere, depending on type).

Restriction fragment length polymorphism (RFLP): A difference in the size of restriction fragment obtained by cutting DNA from 2 different strains with the same restriction endonuclease.

Restriction map: A linear array of sites on DNA cleaved by various restriction enzymes.

Restriction site: The sequence of nucleotide pairs of double stranded DNA that is recognized by a type II restriction endonuclease.

Retention time: The time that has elapsed from the injection of the sample into the chromatographic system to the recording of the peak maximum of the component in the chromatogram.

Retrograde transport: Movement of proteins in the reverse direction in the reticuloendothelial system, typically from Golgi to endoplasmic reticulum.

Retroposon: A transposon that mobilizes via an RNA form; the DNA element is transcribed into RNA, and the reverse-transcribed into DNA, which is inserted at a new site in the genome.

Retrovirus: Viruses with RNA as genetic material that use enzyme reverse transcriptase during their life cycle.

Reverse transcriptase: A polymerase that uses RNA as temple to transcribe a single-stranded DNA molecule as a product.

Reverse transcription: Synthesis of DNA on a template of RNA; accomplished by reverse transcriptase enzyme.

Reverse translation: A technique for isolating genes (or mRNAs) by their ability to hybridize with a short oligonucleotide sequence prepared by predicting the nucleic acid sequence from the known protein sequence.

Reversed phase: A chromatographic mode in which the mobile phase is more polar than the stationary phase.

Reversion: Reversion of a mutation is a change in DNA that either reverse the original alteration (true reversion) or compensates for it (second site reversion in the same gene).

Revertants: Derived by reversion of a mutant cell or organism.

Rf: A ratio used in paper chromatography and thin-layer chromatography that is the distance from the origin to the center of the origin to the solvent front.

Rhizosphere: The zone of immediate vicinity of growing plant roots that harbors increased microbial population.

Rho factor: A protein involved in assisting *E. coli* RNA polymerase to terminate transcription at certain (rho-dependent) sites.

Rho-independent terminators: Sequences of DNA that cause *E. coli* RNA polymerase to terminate *in vitro* in the absence of rho factor.

Ribonuclease: An enzyme that hydrolyses RNA.

Ribonucleoside triphosphate: A nitrogenous base (A, G, C, or U) attached to a ribose sugar with 3 phosphate groups attached in a tandem to the 5′ carbon of the sugar.

Ribosomal Binding Site: A nucleotide sequence near 5′ end of mRNA that facilitates tha binding of the mRNA to the small ribosomal sub-unit. It is also called Shine-Dalgarno sequence.

Ribosomal RNA (rRNA): RNA makes up the structure of ribosome. There are 3 rRNAs (16S, 23S, 5S) in prokaryotes and 4 rRNAs (28S, 5.8S, 18S, 5S) in eukaryotes.

Ribosome: It is 1 of the cellular organelle, which is the site of protein synthesis.

Rifamycins: (including rifampicin) inhibit transcription in bacteria.

Right splicing junction: The boundary between the right end of an intron and the left end of the adjacent exon.

RNA editing: Information that is contained in the DNA is not always found in the RNA products used to make proteins. Mitochondria and chloroplasts contain the biochemical machinery to alter the sequence of the final transcription product. For example, majority of the events involve a C to U transition. So editing can convert a tryptophan codon to a arginine codon (CGG to UGG). Stop codons can be created by editing CAG. CAA and CGA codons.

RNA interference: It is a phenomenon in which the introduction of double-stranded RNA (dsRNA) into a diverse range of organisms and cell types causes degradation of the complementary mRNA. In the cell, long ds RNA (typically>200 nt) are cleaved into short 21–25 nucleotide small interfering RNAs, or siRNAs, by a ribonuclease known as Dicer.

RNA polymerase: An enzyme that synthesizes RNA using a DNA template (formally described as DNA-dependent RNA polymerase).

RNA replicase: An enzyme that synthesizes RNA using an RNA template (used for replication by RNA viruses).

RNAase: An enzyme whose substrate is RNA.

RNA-driven hybridization reactions: Use an excess of RNA to react with all complementary sequences in a single-stranded preparation of DNA.

Rolling circle: A mode of replication in which a replication fork proceeds around a circular template for an indefinite number of revolutions; the DNA strand newly synthesized in each revolution displaces the strand synthesized in the previous revolution, giving a tail containing a linear series of sequences complementary to the circular template strand.

Rot: The product of RNA concentration and time of incubation in an RNA-driven hybridization reaction.

Rough ER: Endoplasmic reticulum associated with ribosomes.

S Phase: The restricted part of the eukaryotic cell cycle during which synthesis of DNA occurs.

S1 Nuclease: An enzyme that specifically degrades unpaired (single-stranded) sequences of DNA.

Saccharification: Enzymatic hydrolysis (by glucoamylase) of polysaccharides resulting in maltose and glucose.

Saltatory replication: A sudden lateral amplification to produce a large number of copies of some sequence.

Satellite DNA: Consists of many tandem repeats (identical or related) of a short basic repeating unit.

Saturation density: The density to which cultured eukaryotic cells grow *in vitro* before division is inhibited by cell-cell contacts.

Saturation hybridization: Experiment has a large excess of one component, causing all complementary sequences in the other component to enter a duplex form.

Scaffold: A proteinaceous structure in the shape of a sister chromatid pair, generated when chromosomes are depleted of histones.

Scale-Up: (1) Expansion of laboratory experiments to full size industrial processes, (2) transition from small scale to large scale (industrial) production.

Scarec (complex) mRNA: Consists of a large number of individual mRNA species, each present in very few copies per cell.

scRNA: Any one of several small cytoplasmic RNA molecules present in the cytoplasm and (sometimes) nucleus.

scRNPs: Small cytoplasmic ribonucleoproteins (scRNAs associated with proteins).

Secondary metabolite: The metabolite that is not required by a cell for its growth or metabolism but produce as a by-product for protection against microorganisms during stationary phase of growth.

Segmentation genes: Concerned with controlling the number or polarity of body segments in insects.

Selectable: Having a gene product that enables to identify and propagate a particular cell type.

Selection: A method to isolate and identify a specific organisms in a mixed culture. The use of particular conditions to allow survival only of cells with a particular phenotype.

Selectivity: The ratios of the capacity factors for two substances measured under identical chromatographic conditions; sometimes termed separation factor.

Semiconservative replication: A type of DNA replication in which the daughter DNA strands is composed of one old and one newly synthesized strand.

Semidiscontinuous replication: the mode in which one new strand is synthesized continuously while the other is synthesized discontinuously.

Septum: The material that forms in the center of a bacterium to divide it into 2 daughter cells at the end of a division cycle.

Sequence characterized amplified regions (SCARs): A fragment of genomic DNA present at a single locus identified by PCR amplification using a pair of specific oligonucleotide primers.

Serum dependence: The need of eukaryotic cells for factors contained in serum in order to grow in culture.

Sex chromosomes: Those whose contents are different in the two sexes; usually labelled X and Y (or W and Z), one sex has XX (or WW), the other sex has XY (or WZ).

Sex linkage: The pattern of inheritance shown by genes carried on a sex chromosome (usually the X).

Sex plasmid: An episome; it is able to initiate the process of conjugation, by which chromosomal material is transferred from one bacterium to another.

Sex-linked: A gene coded on a sex chromosome, such as the X-chromosome linked genes of flies and man.

Shine-Dalgarno sequence: A sequence (AGGAGG) about 10 bp before the AUG initiation codon in 5′ end of bacterial mRNA is complementary to 3′ end of 16S rRNA and helps in the positioning of ribosome for initiation of translation.

Short-period interpersion: A pattern in a genome in which moderately repetitive DNA sequences of $\sim$300 bp alternate with nonrepetitive sequences of $\sim$1000 bp.

Shotgun experiment: Cloning of an entire genome in the form of randomly generated fragments.

Shuttle vector: A plasmid constructed to have origins for replication for 2 hosts (for example, *E. coli* and *S. cerevisiae*) so that it can be used to carry a foreign sequence in either prokaryotes or eukaryotes.

SI nuclease: An enzyme that degrades single stranded DNA.

Siderophore: A low molecular weight Fe-chelating protein synthesized by several soil microorganisms.

Sigma factor: The subunit of bacterial RNA polymerase needed for initiation; is the major influence on selection of binding sets (promoters).

Signal hypothesis: The role of the N-terminal sequence of a secreted protein in attaching nascent polypeptide to membrane; that is, mRNA and ribosome are attached to membrane via the N-terminal end of the protein under synthesis.

Signal sequence: A segment of about 15–30 amino acids at the n terminus of a protein that enables the protein to be secreted. The signal sequence is removed as the protein secreted through the membrane. It is also called signal peptide and leader peptide.

Signal transduction: The process by which a receptor interacts with a ligand at the surface of the cell and then transmits a signal to trigger a pathway within the cell.

Silent mutation: A mutation which does not change the product of a gene.

Simple-sequence DNA: Equals satellite DNA.

SINES: A class of retroposons found as short interspersed repeats in mammalian genomes; derived from transcripts of RNA polymerase III.

Single cell protein (SCP): Cells or protein extracts of microorganisms produced in large quantities for use as human or animal protein supplements.

Single X hypothesis: The inactivation of one X chromosome in female mammals.

Single-copy plasmids: Maintained in bacteria at a ratio of 1 plasmid for every host chromosome.

Single-strand exchange: A reaction in which one of the strands of a duplex of DNA leaves its former partner and instead pairs with the complementary strand in another molecule, displacing its homolog in the second duplex.

Sister chromatids: The copies of a chromosome produced by its replication.

Site directed mutagenesis: The process of nucleotide changes in cloned genes by site specific mutagenesis.

Site-specific recombination: It occurs between 2 specific (not necessarily homologous) sequences, as in phage integration/excision or resolution of cointegrate structure during transpositions.

Slow component: Of a reassociation reaction is the last to reassociate; usually consists of nonrepetitive DNA.

Slow-stop: *dna* mutants of *E. coli* complete the current round of bacterial replication but cannot initiate another at 42°C.

Smooth ER: A region of endoplasmic reticulum devoid of ribosomes.

Somaclonal variation: Variation that occurs in cell cultures and tissues which may be genetic or epigenetic.

Somatic cell gene therapy: Gene delivery to a tissue other than reproductive cells of an individual to correct gene defect.

Somatic cells: All the cells of an organism except those of the germline.

Somatic mutation: A mutation occurring in a somatic cell, and therefore affecting only its daughter cells; it is not inherited by descendants of the organism.

SOS box: The DNA sequence (operator) of ~20 bp recognized by LexA repressor protein.

SOS response: The coordinate induction of many enzymes, including repair activities, in response to irradiation or other damage to DNA; results from activation of protease activity by RecA to cleave LexA repressor (in *E. coli*).

Southern blot: A kind of electroblot, in which the substances separated and transferred are nucleic acids.

Southern blotting: The procedure for transferring denatured DNA from an agarose gel to a nitrocellulose filter where it can be hybridized with a complementary nucleic acid.

Specialized transduction: Transfer of only specific host gene with the help of transducing phages.

Spheroplast: A bacterial or yeast cell whose wall has been largely or entirely removed.

Spindle: The reorganized structure of a eukaryotic cell passing through division; the nucleus has been dissolved and chromosomes are attached to the spindle by microtubules.

Splice sites: The sequences immediately surrounding the exon-intron boundaries.

Splicing: The removal of introns and joining of exons in RNA; thus introns are spliced out, while exons are spliced together.

Split genes: In eukaryotes structural, genes are split up by a number of non-coding regions called introns.

Spontaneous mutation: Mutations that occur in the absence of any added mutagen to increase the mutation rate, as the result of errors in replication.

Sporulation: The generation of a spore by a bacterium (by morphological conversion) or by a yeast (as the product of meiosis).

SSB: The single-strand protein of *E. coli*, a protein that binds to single-stranded DNA.

Stacking: Ordering or arranging and concentrating macromolecules according to their effective mobilities.

Staggered cuts: Cuts in duplex DNA are made when two strand two strands are cleaved at different points near each other.

Startpoint (startsite): The position on DNA corresponding to the first base incorporated into RNA.

Stationary phase: The portion of separation system that is immobilized in the column.

Stem: The base-paired segment of a hairpin.

Stem cell: A precursor cell that undergoes division and gives rise to different lineages of differentiated cells.

Sterilization: The process for elimination of microorganisms.

Sticky ends: Complementary single strands of DNA that protrude from opposite ends of a duplex or from ends of different duplex molecules; can be generated by staggered cuts in duplex DNA.

Stirred tank fermentor: A bioreactor in which cells or microorganisms are mixed by mechanically driven impellers.

Stop codons: The 3 triplets (UAA, UAG, UGA) which terminate protein synthesis.

Strain: A subdivision of a species. A group of organisms that are identical to each other but differ genetically from other organisms of the same species.

Strand displacement: A mode of replication of some viruses in which a new DNAstrand grows by displacing the previous (homologous) strand of the duplex.

Streptolydigins: Inhibit the elongation of transcription by bacterial RNA polymerase.

Stringent replication: The limitation of single-copy plasmids to replication *pari passu* with the bacterial chromosome.

Stringent response: The ability of a bacterium to shut down synthesis of tRNA and ribosomes in a poor-growth medium.

Structural gene: A DNA sequence that encodes a polypeptide molecule.

Structural genomics: The study of structure of all gene sequences in a fully sequenced genome.

Subculture: Subdivision of a culture and its transfer to a fresh medium.

Substrate: Food source for growing cells of an organism.

Superbug: A recombinant *Pseudomonas putida* prepared by Dr. A.M. Chakrebarty.

Supercoiling: The coiling of a closed duplex DNA in space so that it crosses over its own axis.

Super-repressed: The occurrence of changes that eliminate the effects of a mutation without reversing the original change in DNA.

Support: The particles on which the stationary phase is held.

Suppressor (extragenic): Usually a gene coding a mutant tRNA that reads the mutated codon either in the sense of the original codon or to give an acceptable substitute for the original meaning.

Suppressor (intragenic): A compensating mutation that restores the original reading frame after a frameshift.

Surveillance: System that checks nucleic acids for errors, system to degrade aberrant mRNAs. mRNA biosynthesis can produce RNAs with premature stop codons, unspliced introns and other errors. To minimize the potentially harmful effects of these mRNAs a process called mRNA surveillance has evolved to recognize and destroy defective mRNAs in organisms from yeast to humans.

Suspension culture: A culture consisting of cell aggregates initiated by placing callus tissue or sometimes a seedling in an agitated liquid medium.

SWI/SNF: A chromatin remodeling complex; it uses hydrolysis of ATP to change the organization of nucleosomes.

Symbiosis: A symbiotic relationship between two organisms for mutual benefits.

Synapsis: The association of the two pairs sister chromatids representing homologous chromosomes that occurs at the start of meiosis; the resulting structure is called a bivalent.

Synaptonemal complex: The morphological structure of synapsed chromosomes.

Synchronous culture: A symbiotic in which a specified proportion of the cells are at the same instead phase of the cell cycle, enter the same indicated phase of the cell cycle simultaneously, and/or exhibit the same approximate duration of cell cycle.

Syntenic: Genetic loci lie on the same chromosome.

T-DNA: A fragment of Ti-plasmid of *Agrobacterium tumifaciens* that is transferred and integrated into chromosome of plant cells.

T cells: Lymphocytes of the T (thymic) lineage; may be subdivided into several functional types. They carry a TCR (T-cell receptor) and are involved in the cell-mediated immune response.

T4 bacteriophage: Consists of DNA molecule tightly packed in a protein head. The virus attaches to the surface of a bacterial cell and injects its DNA into the cell through its tail.

T4 DNA ligase: An enzyme produced by *E. coli* after infection by bacteriophage T4. It catalyzes the joining of DNA duplex and repairs nick in DNA molecules. It uses ATP as source of energy.

Tailing: Addition of some nucleotides *in vitro* by the enzyme terminal transferase to 3′-OH ends of a duplex DNA molecule. It is also called homopolymer tailing.

Tandem array: A DNA molecule containing two or more identical sequences in series.

Tandem repeats: Multiple copies of the same sequence lying in series.

Taq polymerase: A heat stable DNA polymerase isolated from a thermophilic bacterium *Thermus aquaticus* and used in PCR.

TATA box: A-T rich septamer consensus sequence found about 10 bp before the star point of each prokaryotic gene.

Telomerase: The ribonucleoprotein enzyme that creates repeating units of one strand at the telomere, by adding individual bases.

Telomere: The natural end of a chromosome; the DNA sequence consists of a simple repeating unit with a protruding single-stranded end that may fold into a hairpin.

Temperate bacteriophage: A bacteriophage having a lysogenic life cycle and capable of inserting its DNA into that of the host bacterium.

Temperature-sensitive mutation: Creates a gene product that is functional at low temperature but inactive at higher temperature (the reverse relationship is usually called cold-sensitive).

Template strand: The polynucleotide strand being used by DNA polymerase for determining nucleotide sequence during synthesis of a new strand.

Template: An RNA or single stranded DNA molecules on which a complementary strand is synthesized. It is also called antisense strand.

Terminal redundancy: The repetition of the same sequence at both ends of (for example) a phage genome.

Terminal transferase: An enzyme that adds nucleotides to the 43′ terminus of DNA molecules.

Termination codon: One of 3 triplet sequences, UAG (amber), UAA (ochre), or UGA, that cause termination of protein synthesis; they are also called "non-sense" codons.

Terminator: A sequence of DNA, represented at the end of the transcript, that causes RNA polymerase to terminate transcription.

Tertiary structure: A protein describes the organization in space of its polypeptide chain.

Test cross: A cross between a heterozygous individual or one of the unknown genotype and an individual homozygous recessive for the gene in question. When a single trait is being studied, a test cross is a cross between an individual with the dominant phenotype but of unknown genotype (homozygous or heterozygous) with a homozygous recessive individual. If the unknown is heterozygous (Yy), then approximately 50% of the offspring should display the recessive phenotype (Yy x yy gives 1:1 ratio).

Thalassemia: Disease of red blood cells resulting from lack of either α or β globin.

Thawing: A slight melting of pre-cooled specimen before long-term storage at −196°C.

Theoretical plate number: A number defining the efficiency of the chromatographic column.

Therapeutic agent: A chemical compound used for treatment of disease. It is also called as drug.

Thymine dimer: A chemically cross-linked pair of adjacent thymine residues in DNA, a result of damage induced by ultraviolet irradiation.

Ti plasmid: An extrachromosomal, double-stranded, and self-replicating DNA molecules found in *Agrobacterium tumifaciens* that causes crown gall disease in plants.

Tissue culture: The rearing of cells or tissue in an artificial medium under controlled condition.

T_m: The abbreviation for melting temperature.

Tobacco mosaic virus (TMV): consists of a single long RNA molecule (6,000 nucleotides) enclosed in a cylindrical protein coat (2,130 copies of the same coat protein). It is approximately 300 nm long and 15 nm wide.

Topoisomerase: An enzyme that can alter the topology of a DNA molecule by cutting one or both strands of DNA.

Topological isomers: Molecules of DNA that are identical except for a difference in linking number.

Totipotency: A property of normal cells that they have the genetic potential to give rise to a complete individual, except terminally specialized cells.

Totipotent: The ability of a cell to respond according to environmental stimuli and divide to form differentiated cell types.

Tracer: A radioactively labeled nucleic acid component included in a reassociation reaction in amounts too small to influence the progress of reaction.

Trailer: A nontranslated sequence at the 3′ end of an mRNA following the termination codon.

***Trans*:** Configuration of two sites refers to their presence on two different molecules of DNA (chromosomes).

Transcribed spacer: The part of an rRNA transcription unit that is transcribed but discarded during maturation; that is, it does not give rise to part of rRNA.

Transcript: RNA made from a region of DNA.

Transcription: Synthesis of RNA molecule catalyzed by RNA polymerase using a DNA strand as a template.

Transcription unit: The distance between sites of initiation and termination by RNA polymerase; may include more than one gene.

Transduction: Virus mediated gene transfer from one bacterium to another.

Transducing phage: Bacterial DNA instead of viral genetic material is incorporated in viral protein coat.

Transfection: (of eukaryotic cells) The acquisition of new genetic markers by incorporation of added DNA.

Transformation (of bacteria): The acquisition of new genetic markers by incorporation of added DNA.

Transformation (of eukaryotic cells): Their conversion to a state of unrestrained growth in culture, resembling or identical with the tumorigenic condition.

Transformylase: The enzyme that transfers a formyl (CHO) group to the amino group of methionine to make formylmethionine.

Transgene: Gene that is introduced into a cell or animal from an external source.

Transgenesis: Introduction of gene into animal or plant cells that is successively transmitted to onward generations.

Transgenic: Animals are created by introducing new DNA sequences into the germline via addition to the egg.

Transgenic plant: A fertile plant consisting of a foreign gene in its germ line.

Transgenic organisms: Plants, microorganisms or animals in which novel DNA has been incorporated into their germ line cells.

Transient: Of short duration.

Transit peptide: The short leader sequence cleaved from proteins that are imported into cellular organelles by post-translational passage of the membrane.

Transition: Mutation in which one purine is replaced by another purine, or a pyrimidine is substituted for other pyrimidine.

Translation: Synthesis of protein on the mRNA template.

Translocation: Chromosomal segment is removed from one chromosome and inserted into another, nonhomologous chromosome. Translocation heterozygotes usually have reduced fertility due to production of abnormal gametes.

Translocation (of gene): The appearance of a new copy at location in the genome elsewhere from the original copy.

Translocation (of protein): Its movements across a membrane.

Translocation (of ribosome): Its movement one codon along mRNA after the addition of each amino acid to the polypeptide chain.

Transmembrane protein: A component of a membrane; a hydrophobic region or regions of the protein resides in the membrane, and hydrophilic regions are exposed on one or both sides of the membrane.

Transplantation antigen: Protein coded by a major histocompatibility locus, present on all mammalian cells, involved in interactions between lymphocytes.

Transposase: The enzyme activity involved in insertion of transposon at a new site.

Transposition: The movement of a transposon to a new site in the genome. (*See also* nonreplicative transposition, replicative transposition and conservative transposition.)

Transposition immunity: The ability of certain transposons to prevent others of the same type from transposing to the same DNA molecule.

Transposon: A DNA sequence able to insert itself at a new location in the genome (without any sequence relationship with the target locus).

Transposon tagging: Insertion of transposable element into nearby a gene, thereby making that gene with a known DNA sequence.

Transvection: The ability of a locus to influence activity of an allele on the other homolog only when two chromosomes are synapsed.

Transversion: A mutation in which a purine is replaced by a pyrimidine or vice versa.

Triploidy: The condition in which a cell or organism possesses 3 haploid sets of chromosomes.

Trisomy: When an individual has three of a particular type of chromosome. Monosomy and trisomy occur in plants and animals; in autosomes of animals, it is generally lethal. Nonlethal human monosomics and trisomics include Turner syndrome: monosomy where the individual has single X chromosome; Down syndrome is most common trisomy among humans; it involves chromosome 21.

True-breeding: Organisms are homozygous for the trait under consideration.

Twisting number (of DNA): The number of base pairs divided by the number of base pairs per turn of the double helix.

Ultrasonication: Disruption of cells or DNA molecules by high frequency sound waves.

Umber codon (opal codon): The UGA codon, one of the three termination codons that end protein synthesis.

Unctional genomics: The study of the function of all specific gene sequences and their expression in time and space in an organism.

Underwinding (of DNA): Produced by negative supercoiling (because the double helix is itself coiled in the opposite sense from the interwinding of the strands).

Unequal crossing-over: A recombination event in which the two recombining sites lie at non-identical locations in the two parental DNA molecules.

Unidirectional replication: The movement of a single replication fork from a given origin.

Uninducible mutants: Cannot be induced.

Unscheduled DNA synthesis: Any DNA synthesis occurring outside the S phase of the eukaryotic cell.

Up promoter mutations: Increase the frequency of initiation of transcription.

Upstream: Sequences proceeding in the opposite direction from expression.

Uracil: One of the pyrimidine (single-ringed nitrogenous) base.

URF: An open (unidentified) reading frame, presumed to code for protein, but for which no product has been found.

V Gene: Sequence coding for the major part of the variable (N-terminal) region of an immunoglobulin chain.

Variable region: An immunoglobulin chain is coded by the V gene and varies extensively when different chains are compared, as the result of multiple (different) genomic copies and changes introduced during construction of an active immunoglobulin.

Variegation: Variegation of phenotype is produced by a change in genotype during somatic development.

Vectors: Vehicles for transferring DNA from one cell to another or a DNA molecule that can carry inserted DNA and be perpetuated in a host cell.

Vegetative propagation: Reproduction of plants using a non-sexual process involving the culture of plant parts such as stem and leaf cuttings.

Vesicles: Small bodies bounded by membrane, derived by budding from one membrane, often able to fuse with another membrane.

***vir* gene:** A set of genes present on Ti-plasmid which facilitate the transfer of T-DNA segment into a plant cell.

Virion: The physical virus particle (irrespective of its ability to infect cells and reproduce).

Virulent: Virulent phage mutants are unable to establish lysogeny.

VNTRs (minisatellites): Polygenic markers.

Void volume: The interstitial volume of the chromatographic column; that is the volume of mobile phase imbibed in the pores and around the stationary phase in a column.

Western blot: A technique in which protein is transferred from an electrophoretic gel to a cellulose or nylon support membrane following electrophoresis. A particular protein is then identified by probing the blot with a radiolabelled antibody which binds on the specific protein to which the antibody was prepared.

Wild type: The organism as it was first isolated from nature. The normal type in a genetic experiment the strain from which the mutants were derived.

Wild-type allele: The nonmutant form of a gene, encoding the normal genetic function. Generally, but not always a dominant allele.

Wobble: Difference of codons of the same amino acid at the third base of mRNA. Wobble hypothesis accounts for the ability of a tRNA to recognize more than one codon by unusual (non-G.C, non-A.T) pairing with the third base of a codon.

Writhing number: The number of times a duplex axis crosses over itself in space.

Xenobiotics: A chemically synthesized compound not produced by living organisms.

Yeast artificial chromosome (YAC): It is a synthetic DNA molecule (for propagation of genes in yeast cells), with an origin of replication, a centromere for segregation, and telomeres to seal the ends, also marker genes from yeast. YACs are used to clone long stretches of eukaryotic DNA.

YEP (yeast episomal plasmid): A cloning vector for yeast that uses 2 μm plasmid origin of replication (*ori*) and maintained as an extachromosomal nuclear DNA molecule.

Zero time-binding DNA: It enters the duplex form at the start of a reassociation reaction; results from intramolecular reassociation of inverted repeats.

Zeta potential: The potential produced by the effective charge of a macromolecule, usually taken at the boundary between what is moving with the macromolecule and the rest of the solution.

Zinc finger protein: It has a repeated motif of amino acids with characteristic spacing of cysteines that may be involved in binding zinc; is characteristic of some proteins that bind DNA and/or RNA.

Zone: A particular region or space within a large one, generally distinguished by some property, such as its occupancy by a protein.

Zoo blot: The use of Southern blotting to test the ability of a DNA probe from one species to hybridize with the DNA from the genomes of a variety of other species.

Zygote: Produced by fusion of 2 gametes, that is, it is a fertilized egg.

Bibliography

Adams, J. M. and Corry, S. 1998. The Bcl-2 protein family: Arbiters of cell survival. *Science* 281, 1322–5.

Adams, M. D. et al. 1991. Complementary DNA sequencing: Expressed sequence tags and human genome project. *Science* 252(5013), 1651–6.

Aebersold, R. 1991. High sensitivity sequence analysis of proteins separated by polyacrylamide gel electrophoresis. In *Advances in Electrophoresis* Vol 4 (eds, Chrambach, A., Dunn, M. J., and Radola, B. J.). VCH-Verlag, Weinheim, Germany, pp. 81–168.

Akiyama, S. K., Olden, K., and Yamada, K. M. 1996. *Cancer Metastasis Rev.* 14, 173.

Amo, M. and Tsugita, A. 1999. N-terminal amino acid sequencing of 2-DE spots. In *2-D Proteome Analysis Protocols* (ed. Link, A. J.). Humana Press, New Jersey, p. 461.

An, G., Ebert, P. R., Mitra, A., and Ha, S. B. 1988. Binary vectors. In *Plant Molecular Biology Manual* (eds. Selvin, S. B. and Schilperoort, R. A.). Kluwer Academic Publishers, Dordrecht, The Netherlands, A3, pp. 1–19.

Andrews, A. T. 1986. *Electrophoresis: Theory, Techniques, and Biochemical and Clinical Applications* (2nd edn). Oxford Science Publications, Oxford, p. 452.

Ausubel, F. M., Bent, R., Kingston, R. E., Moore, D. D., M Seidman, J. G., Smith, J. A., and Struhl, K. 1987. *Current Protocols in Molecular Biology.* Greene Publishing Associates and Wiley-Interscience. John Wiley and Sons, New York.

Baldo, B. A. 1994. *Adv. Electrophoresis* 7, 407.

Bansal, K. C., Viret, J., and Haley, J. 1992. Transient expression from *cab-ml* and *rbcs-m3* promoter sequences in maize leaves. *Proc. Natl. Acad. Sci. (USA)* 89, 3654–8

Beimann, K. 1990. Applications of tandem mass spectrometry to peptide and protein structure. In *Biological Mass Spectrometry* (eds. Burlingame, A. L. and McCloskey, J. A.). Elsevier, Amsterdam, pp. 176–96.

Bell, S. P. 1995. Eukaryotic replicators and associated protein complexes. *Curr. Opin. Genet. Dev.* 5(2), 162–7.

Benson, D. A., Boguski, M. S., Lipman, D. J., and Ostell, J. 1997. *Nucleic Acids Res.* 25, 1–6.

Bentley, D. L. and Groudine, M. 1986. A block to elongation is largely responsible for decreased transcription of c-myc in differentiated HL60 cells. *Nature* 321, 702–6.

Berkelman, T. and Stenstedt, T. 1998. *2-D Electrophoresis Using Immobilized pH Gradients. Principles and Methods.* Amersham Pharmacia Biotech Inc., New Jersey, p. 50.

Betts, J. C., Blackstock, W. P., Ward, M. A., and Anderton, B. H. 1997. *J. Biol. Chem.* 272, 12922.

Beynon, R. J. and Bond, J. S. 1989. *Proteolytic Enzymes: A Practical Approach.* IRL Press, Oxford, p. 259.

Biffo, S., Marti, E., and Fasolo, A. 1992. Carnosine, nerve growth factor receptor and tyrosine hydrolase expression during the ontogeny of the rat olfactory system. *J. Chem. Neuroanat.* 5, 51–62.

Birnboim, H. C. and Doly, J. 1979. *Nucleic Acids Res.* 7, 1513–23.

Black, R. L. and Olson, E. N. 1998. Transcriptional control of muscle development by myocyte enhancer factor-2 (MEF2) proteins. *Annu. Rev. Cell Dev. Biol.* 14, 167–96.

Blackstock, W. P. and Weir, W. P. 1999. Proteomics: Quantitative and physical mapping of cellular proteins. *Tibtech* 17, 121.

Bogdanova, E., Matz, M., Tarabykin, V., Usman, N., Shagin, D., Zaraisky, A., and Lukyanov, S. 1998. Inductive interactions regulating body patterning in planarian, revealed by analysis of expression of novel gene scarf. *Dev. Biol.* 194, 172.

Bolton E., Higgisson, B., Harrington, A., and Gara, F. O. 1986. Dicarboxylic acid transport in *Rhizombium meliloti*: Isolation of mutant and cloning of dicarboxylic acid transport gene. *Arch. Microbiol.* 144, 142–6.

Borges, A. E. and Lineberger, A. S. 1984. Malignant melanoma without metastasis in a giant nevus. *Ann. Plast. Surg.* 12(5), 454–60.

Botstein, D., White, R. L., Skolnick, M. H., and Davies R. W. 1980. Construction of a genetic linkage map in man using restriction fragment length polymorphisms. *Am. J. Hum. Genet.* 32, 314.

Bowtell, D. D. 1999. Options available—from start to finish—for obtaining expression data by microarray. *Nat. Genet.* 21(Suppl. 1), 25–32.

Campisi, J. 1994. Molecular and cell biology of replicative senescence. *Cold Spring Harb. Symp. Quant. Biol.* 59, 67–73.

Castellanos-Serra, L. R., Fernandez-Patron, C., Hardy, E., Santana, H., and Huerta, V. 1997. High yield elution of proteins from sodium dodecyl sulfate-polyacrylamide gels at the low-picomole level. Application to N-terminal sequencing of a scarce protein and to in-solution biological activity analysis of on-gel renatured proteins. *J. Prot. Chem.* 16(5), 415-9.

Chang, D. and Denny, C. 1998. Gene cloning and analysis by RT-PCR. In *Molecular Laboratory Methods Number '1* (eds. Siebert, P. D. and Larrick, J. W.). Biotechniques Books (Division of Eaton Publishing), Massachusetts, p. 193.

Chen, C. S., Ostuni, E., Whitesides, G. M., and Ingber, D. E.2000 Using self-assembled monolayers to generate patterned substrates of ECM for cell culture. *Methods Mol. Biol.* 139, 209–19.

Chen, J. J. 1998. Profiling expression patterns and isolating differentially expressed genes by cDNA microarray system with colorimetry detection. *Genomics* 51(3), 313–24.

Chen, Y., Dougherty, E., and Bittner, M. 1997. Ratio based decisions and the quantitative analysis of cDNA microarray images. *J. Biomed. Opt.* 2, 364.

Chenchik, A., Mogadam, L., and Siebert, P. D. 1996. A new method for full-length cDNA cloning by PCR. In *A Laboratory Guide to RNA: Isolation, Analysis, and Synthesis* (ed. Krie, P.). Wiley-Liss, New York, pp. 273–321.

Chiba, A. and Keshnishian, H. 1996. Neuronal pathfinding and recognition: Roles of cell adhesion molecules. *Dev. Biol.* 180, 424–32.

Chiquet, M. 1999. Regulation of extracellular matrix gene expression by mechanical stress. *Matrix Biol.* 18, 417–26.

Christou P. 1996. *Particle Bombardment for Genetic Engineering of Plants*. Academic Press, UK.

Collins, F. S., Patrinos, A., Jordan, E., Chakravarti, A., Gesteland, R., and Walters, L. 1998. New goals for the U.S. Human Genome Project: 1998-2003. *Science* 282 (5389), 682–9.

Conlon, I. and Raff, M. 1999. Size control in animal development. *Cell* 96, 235–44.

Cooper, G. M. and Hausman, R. E. 2003. *The Cell: A Molecular Approach*. Sinauer Associates Inc./Palgrave Macmillan, Sunderland, Massachusetts.

Cordwell, S. J., Basseal, D. J., Bjellqvist, B. J., Shaw, D. C., and Humphery-Smith, I. 1997. *Electrophoresis* 18, 1393.

Cordwell, S. J., Wasinger, V. C., Cerpa-Poijak, A., Duncan, M., and Humphery-Smith, I. 1997. Conserved motifs as the basis for recognition of homologous proteins across species boundaries using peptide mass fingerprinting. *J. Mass Spectrometry* 32, 373–8.

Cordwell, S. J., Wilkins, M. R., Cerpa-Poljak, A., Cooley, A. A., Duncan, M., Williams, K. L., and Humphery-Smith, I. 1995. Cross-species identification of proteins separated by two-dimensional gel electrophoresis using matrix-assisted laser desorption ionisation/time-of-flight mass spectrometry and amino acid composition. *Electrophoresis* 16, 438–43.

Courchesne, P. L. and Patterson, S. D. 1999. In *2D Froteorne Artaisited* (ed. Unk, A. J.). Humana Press, New Jersey, p. 487.

Creighton, T. E. (ed.) 1997. *Protein Structure. A Practical Approach*. IRL Press, Oxford, p. 383.

Daher, R. and Beaini, M. 1998. *Clin. Chem. Lab. Med.* 36(9), 671.

Dainese, P., Staudenmann, W., Quadrom, M., Korostensky, C., Gannet, G., Kertesz, M., and James, P. 1997. *Electrophoresis* 18, 432.

Davis, L. G., Dibner, M. D., and Battey, J. F. 1986. *Basic Methods in Molecular Biology.* Elsevier, New York.

Deckner, M. L., Frisén, J., Verge, V. M., Hökfelt, T., and Risling, M. 1993. Localization of neurotrophin receptors in olfactory epithelium and bulb. *Neuro Report* 5, 301–4.

DeRisi, J., Penland, L., Brown, P. O., Bittner, M. L., Meltzer, P. S., Ray, M., Chen, Y., Su, Y. A., and Trent, J. M. 1996. Use of a cDNA microarray to analyse gene expression patterns in human cancer. *Nat. Genet.* 14(4), 457–60.

Diatchenko, L., Chenchik, A., and Siebert, P. D. 1998. Gene cloning and analysis by RT-PCR. In *Molecular Laboratory Methods Number 1* (eds. Siebert, P. D. and Larrick, J. W.). Biotechniques Books (Division of Eaton Publishing), p. 213.

Diatchenko, L., Lau, Y. F. C., Campbell, A., Chenchick, A., Moqadam, F., Huang, B., Lukyanov, S., and Lukyanov, K. 1996. *Proc. Natl. Acad. Sci. USA*, 93, 6025.

Diatchenko, L., Lukyanov, S., Lau, Y.-F. C., and Siebert, P. D. 1999. cDNA preparation and characterization. In *Methods in Enzymology* (ed. Weissman, S. M.). Academic Press, San Diego, CA, pp. 303, 349–380.

Dixit, R. and Cyr, R. 2002. Golgi secretion is not required for marking the preprophase hand site in cultured tobacco cells. *Plant J.* 29, 99–108.

Dominov, J. A., Dunn, J. J., and Miller, J. B.1998 Bcl-2 expression identifies an early stage of myogenesis and promotes clonal expansion of muscle cells. *J. Cell Biol.* 142, 537–44.

Dove, A. 1999. *Nat. Biotechnol.* 17, 233.

Dunbar, B. S. 1987. *Two-Dimensional Electrophoresis and Immunological Techniques.* Plenum Press, New York, p. 372.

Dunn, M. J. 1987. *Adv. Electrophoresis* 1, 1.

Dunn, M. J. 1993. *Gel Electrophoresis of Proteins.* Bios Scientific Publishers, Oxford, p. 176.

Dunn, M. J. 1997. *Methods Mol. Biol.* 64, 25.

Edlund, T. and Jessel, T. M. 1999. Progression from extrinsic to intrinsic signaling in cell fate specification. A view from the nervous system. *Cell* 96, 211–24.

Edman, P. 1950. *Acta Chem. Scand.* 4, 283.

Edman, P. and Begg, G. 1967. *Eur. J. Biochem.* 1, 80.

Eisen, M. B., Spellman, P. T., Brown, P. O., and Botsteln, D. 1998. Cluster analysis and display of genome-wide expression patterns. *Proc. Natl. Acad. Sci. USA* 95, 14863–8.

Eng, J. K., McCormack, A. L., and Yates, J. R., III. 1994. *J. Amer. Soc. Mass Spectrom.* 5, 976.

Ermoiaeva, O., Rastogi, M., Pruitt, K. D., Schuler, G. D., Bittner, M. L., Chen, Y., Simon, R., Meltzer, P., and Trent, J. M. 1998. Data management and analysis for gene expression arrays. *Nat. Genet.* 20, 19–23.

Etchegaray, J. P., Martinez, M. A., Krauskopf, M., and Leon, G. 1991. Molecular cloning of *Renibacterium salmoninarum* DNA fragments. *FEMS Microbial. Lett.* 63(1), 61–4.

Ezer, S., Schlesthiger, D., Srivastava, A., and Kere, J. 1997. Anhidrotic ectodermal dysplasia (EDA) protein expressed in MCF-7 cells associates with cell membrane and induces rounding. *Hum. Mol. Genet.* 6, 1581–8.

Fenn, J. B., Mann, M., Meng, C. K., Wong, S. F., and Whitehouse, C. M. 1989. Electrospray ionization for mass spectrometry of large biomolecules. *Science* 246, 64–91.

Figeys, D., Ducret, A., Yates, J. R., III, and Aebersold, R. 1996. *Nat. Biotechnol.* 14, 579.

Finan, T. M., Wood, J. M., and Jordon, D. C. 1981 Succinate transport in *Rhizobium leguminossarum. J. Bacsteriol.* 148, 193–202.

Findlay, J. B. C. and Geisow, M. J. (eds.) 1989. *Protein Sequencing. A Practical Approach.* IRL Press, Oxford, p. 199.

Fink, R. D. and Cooper, M. S. 1996. Apical membrane turnover is accelerated near cell-cell: Contacts in an embryonic epithelium. *Devel. Biol.* 174, 180–9.

Forsburg, S. L. and Nurse, P. 1991. Cell cycle regulation in the Yeasts *Saccharomyces cerevise* and, *Sthizosaccharomvces pombe*. *Ann. Rev. Cell Biol.* 7, 227–56.

Franceschini, I. A. and Barnett, S. C. 1995. Low affinity NGF-receptor and E-N-CAM expression define two types of olfactory nerve unsheathing cells that share a common lineage. *Dev. Biol.* 173(1), 327–43.

Fulda, S. and Debatin, K. M. 2004. Targeting apoptosis pathways in cancer therapy. *Curr. Drug Targets* 4(7), 569–76.

Gasser, S. M. 1995. Chromosome structure. Coiling up chromosomes. *Curr. Biol.* 5(4), 60.

Gianctti, F. G. and Ruooslahti, E. 1990. *Cell* 60, 849.

Glenn, A. R. and Brewin, N. J. 1981. Succinate—resistant mutant of Rhizobium *legumin psarum*. *J. Gen. Microbiol.* 126, 237–41.

Gong, Q. Z., Bailey, M. S., Pixley, S. K., Ennis, M., Liu, W., and Shipley, M. T. 1994. Localization and regulation of low affinity nerve growth factor receptor expression in the rat olfactory system during development and regeneration. *J. Comp. Neurol.* 344, 336–48.

Gooley, A. A. and Packer, N. H. 1997. The importance of protein co- and post-translational modifications in Proteome projects. In *Proteome Research: New Frontiers in Functional Genomics* (eds. Wilkins, M. R., Williams, K. L., Appel, R. D., and Hochstrasser, A. F.). Springer-Verlag, Berlin, p. 65.

Gooley, A. A. and Williams, K. L. 1997. How to find, identify and quantitate the sugars on proteins. *Nature* 385, 557.

Gorman, C. M., Moffat, L. F., and Howard, B. H. 1982. Recombinant genomes which express chloramphenicol acetyl transferase in mammalian cells. *Mol Cell Biol.* 2, 1044–51.

Grant, P. A., Schieltz, D., Pray-Grant, M. G., Yates, J. R. III, and Workman, J. L. 1998. *Mol. Cell* 2, 863.

Gray, N. K. and Wickens, M. 1998. Contol of translation initiation in animals. *Ann. Rev. Dev. Biol.* 14, 399–458.

Grodzicker, T., Williams, J., Sharp, P., and Samrrook, J. 1974. Physical mapping of temperature-sensitive mutants of adenovirus. *Cold Spring Harbor Symposia on Quantitative Biology* 39, 439–46.

Gurskaya, N. G. et al. 1996. *Anal. Biochem.* 240, 90.

Guzman, N. A., Park, S. A. S., Schaufelberger, D., Hernandez, L., Paez, X., Rada, P., Tomlinson, A. J., and Naylor, S. 1997. *J. Chromatography B* 697, 37.

Haier, J. and Nicolson, G. L. 2000. Role of the cytoskeleton in adhesion stabilization of human colorectal carcinoma cells to extracellular matrix components under dynamic conditions of laminar flow. *Glin. Expl. Metastasis* 17, 713–21.

Hames, B. D. (ed.) 1998. *Gel Electrophoresis of Proteins. A Practical Approach* (3rd edn). IRL Press, Oxford, p. 352.

Hara, E., Kato, T., Nakada, S., Sekiya, S., and Oda, K. 1991. *Nucleic Acids Res.* 19, 7097.

Harnngton, M. G. 1991. *Methods: A Companion to Methods in Enzymology*. Academic Press, London, p. 98.

Harris, E. L. V. and Angal, S. (ed.) 1990. *Protein Purification Methods*. IRL Press, Oxford, p. 336.

Hatt, P. D., Quadroni, M., Staudenmann, W., and James, P. 1997. *Eur. J. Biochem.* 246, 336.

Haynes, P. A., Gygi, S. P., Figeys, D., and Aebersold, R. 1998. *Electrophoresis* 19, 1862–71.

Heegaard, N., Nilsson, S., and Guzman, N. 1998. *J. Chromatogr. B* 715, 29.

Hellman, U., Wernstecit, C., Gonez, J., and Heldin, C.-H. 1995. *Anal. Biochem.* 224, 451.

Hewick, R. M., Hunkapiller, M. W., Hood, L. E., and Dreyer, W. J. 1981. *J. Biol. Chem.* 256, 7990.

Hirschi, K. K., Xu, C., Tsukamaoto, T., and Sager, R. 1996. *Cell Growth Differ.* 7, 861.

Hong, Z., Delauney, A., and Verma, D. 2001. A cell plate-specific callow synthase and its interaction with phragmoplastin. *Plant Cell* 13, 755–68.

Hooykaas, P. J. J. and Schilperoort, R. A. 1992. Agrobacterium and plant genetic engineering. *Plant Mol. Biol.* 19, 15–38.

Howard, G. C. and Brown, W. E. (ed.) 1999. *Basic Methods in Protein Chemistry.* CRC, Florida, In Press.

Huang, S. and Axelrod, D. E. 1991. Altered post-translational modification of ras p21 in a transformation-suppressed cell line. *Oncogene* 6, 1211–8.

Hung, M. C. and Wensink, P. C. 1984. Different restriction enzyme-generated sticky DNA ends can be joined in vitro. *Nucleic Acids Res.* 12(4), 1863–74.

Huang, S., Chen, C. S., and Ingber, D. E. 1998. Control of cyclin D1, p27Kip 1 and cell cycle progression in human capillary endothelial cells by cell shape and cytoskeletal tension. *Mol. Biol. Cell* 9, 3179–93.

Hubank, M. and Schatz, D. G. 1994. *Nucleic Acids Res.* 22(25), 5640–8.

Humphery-Smith, I. 1999. *Electrophoresis* 20, 653.

Humphery-Smith, I. and Blackstock, W. P. 1997. *J. Protein Chem.* 16, 337.

Humphery-Smith, I., Cordwell, S. J., and Blackstock, W. P. 1997. *Electrophoresis* 18, 1217.

Hunt, T. and Scherr, C. J. 1994. *Curr. Opin. Cell Biol.* 6, 833.

Itoh, H., Takahashi, A., Adachi, K., Noji, H., Yasuda, R., Yoshida, M., Kinosita, K. 2004. Mechanically driven ATP synthesis by F1-ATPase. *Nature* 427(6973), 465–8.

Iyer, V. R. et al. 1999. *Science* 283(5398), 83–7.

James, P. 1997. *Biochem. Biophys. Res. Comm.* 231, 1.

Jensen, O. N., Podtelejnikov, A. V., and Mann, M. 1997. *Anal. Chem.* 69, 4741.

Jensen, O. N., Wilm, M., Shevchenko, A., and Mann, M. 1999. Peptide sequencing of 2-DE gel-isolated proteins by nano-electrospray tandem mass spectrometry. In *In 2-D Proteome Analysis Protocols* (ed. Link, A. J.). Humana Press, New Jersey, p. 571.

Jensen, R. A., Page, D. L., and Holt, J. T. 1994. *Proc. Natl Acad. Sci. USA* 91, 9257.

Jin, H., Cheng, X., Diatchenko, L., Siebert, P. D., and Huang, C. C. 1997. *BioTechniques* 23, 1084.

Johnston, M. and Davis, R. W. 1984. Sequences that regulate the divergent GAL1-GAL10 promoter in Saccharomyces cerevisiae. *Mol. Cell. Biol.* 8, 1440–8.

Jungblut, P., Thiede, B., Zimny-Arndt, U., Muller, E. C., Scheler, C., Wittmann-Lieboid, B., and Otto, A. 1996. *Electrophoresis* 17, 839.

Kambadur, R., Sharma, M., Smith, T. P. L., and Bass, J. J. 1997. Mutations in myostatin (GDF8) in double-muscled Belgian blue and Piedmontese cattle. *Genome Res.* 7, 910–5.

Karaoglu, H. and Humphery-Smith, I. 2000. Signature peptides: From analytical chemistry to functional genomics. In *Methods in Molecular Biology, Vol. 146: Protein and Peptide Analysis: New Mass Spectrometric Applications.* (ed. Chapman, J. R.). Humana Press, Totowa, New Jersey, pp. 63–87.

Karas, M. and Hillenkarnp, F. 1989. *Anal. Chem.* 60, 2299.

Karp, G. 2003. *Cell & Molecular Biology: Concepts and Experiments.* J. Wiley & Sons.

Kellner, R., Lottspeich, F., and Meyer, H. E. (ed.) 1994. *Microcharacterisation of Proteins.* VCH, Weinheim, p. 267.

Kellner, R., Lottspeich, F., and Meyer, H. E. (ed.) 1994. In *Microcharacterisation of Proteins.* VCH, Weinheim, pp. 93–110.

Kellogg, D. E., Rybalkin, I., Chen, S., Mukhamedova, N., Vlasik, T., Siebert, P. D., and Chenchik, A. 1994. *BioTechniques* 16, 1134.

Kere, J., Grzeschik, K. H., Limon, J., Gremaud, M., Schlessinger, D., and de la Chapelle, A. 1993. Anhidrotic ectodermal dysplasia gene region cloned in yeast artificial chromosomes. *Genomics* 16, 305–10.

Khan, J. 1998. *Cancer Res.* 58, 3009–13.

Kiss, T., Antal M., and Solymosy F. 1987. Plant small nuclear RNAs. II. U6 RNA and a 4.5 SI-like RNA are present in plant nuclei. *Nucleic Acids Res.* 15(2), 543–60.

Klose, J. 1999. *Electrophoresis* 20, 643.

Klose, J. and Zeindi, E. 1984. *Clin. Chem.* 30, 2014.

Kott, J. N., Westrum, L. E., Raines, E. W., Sasahara, M., and Ross, R. 1994. Olfactory ensheathing glia and platelet-derived growth factor b-chain reactivity in the transplanted rat olfactory bulb. *Int. J. Dev. Neurosci.* 12, 315–23.

Kozian, D. H. and Kirschbaum, B. J. 1999. *Tibtech* 17, 73.

Krens, F. A., Molendijk, L., Wullema, G. J., and Shilperoot, R. A. 1982. *In vitro* transformation of plant protoplasts with Ti-plasmid DNA. *Nature* 296, 72–4.

Krieg, P. A. and Melton, D. A. 1987. In vitro RNA synthesis with SP6 RNA polymerase. *Methods Enzymol.* 155, 397–415.

Kriger, M. S., Cook, K. D., and Ramsey, R. S. 1995. *Anal. Chem.* 67, 385.

Kristensen, D. B., Inamatsu, M., and Yoshizato, K. 1997. *Electrophoresis* 18, 2078.

Kuang, W. W., Thompson, D. A., Hoch, R. V., and Weigel, R. J. 1998. Differential screening and suppression subtractive hybridization identified genes differentially expressed in an estrogen receptor-positive breast carcinoma cell line. *Nucleic Acids Res.* 26(4), 1116–23.

Kuge, H., Brownless, G. G., Gershon P. D., and Richter J. D. 1998. Cap ribose methylation of c-mos mRNA stimulates translation and oocyte maturation in *Xenopus laevis*. *Nucleic Acids Res.* 26, 3208–14.

Kuster, B. and Mann, M. 1998. *Curr. Opinion Struct. Biol.* 8, 393.

Kutsuna, N. and Hasezawa, S. 2002. Dynamic organization of vacuolar and microtubule structures during cell cycle progression in synchronized tobacco BY-2 cells. *Plant Cell Physiol.* 43, 965–73.

Ladish, H. F. et al. 2003. *Molecular Cell Biology* (5th edn). W.H. Freeman.

Lakka, K., Mohananl, S., Nicolson, G. L., and Rao, J. S. 2000. In vitro modulation of human lung cancer cell line invasiveness by antisense cDNA of tissue pathway inhibitor-2. *Clin. Expl. Metastasis.* 18, 239–44.

Lamond, A. I. and Mann, M. 1997. *Trends Cell Biol.* 7, 139.

Lareef, M. H. et al. 2004. Chromosome 17p13.2 transfer reverts transformation phenotypes and Fas-mediated apoptosis in breast epithelial cells. *Mol. Carcinog.* 39(4), 234–46.

Laursen, R. A. 1971. *Eur. J. Biochem.* 20, 89.

Lebowitz, R. J. and Sanderfoot, A. A. 1991. An instructional unit on the use of *Agrobacterium* for plant genetic transformation. *Biotechnol. Educ.* 2, 115–21.

Leckband, D. and Sivasankar, S. 2000. Mechanisms of homophilic cadherin adhesion. *Curr. Opin. Cell Biol.* 12, 587–92.

Lee, M. C., Miller, E. A., Goldberg, J., Orci, L., and Schckman, R. 2004. Bi-directional protein trans-port between the ER and Golgi. *Annu. Rev. Cell Dev. Biol.* 20, 87–123.

Levenson, R. M., Anderson, G. M., Cohn J. A., and Blackshear, P. J. 1990. *Electrophoresis* 11, 269.

Li, H., Linke, W. A., Oberhauser, A. F., Carrion-Vazquez, M., Kerkvlict, J. G., Lu, H., Marszalck, P. E., and Fernandez, J. M. 2002. Reverse engineering of the giant muscle protein titin. *Nature* 418(6901), 998–1002.

Lin, X., Lee, C.-G., Casale, E. S., and Shih, J. C. H. 1992. Purification characterization of a keratinase from a feather degrading Bacillus licheniformis strain. *Appl. Environ. Microbiol.* 58, 3271–5.

Liang, P. and Pardee, A. B. 1992. *Science* 257(5072), 967–71.

Link, A. J. (ed.) 1999. *2-D Proteome Analysis Protocols.* Humana Press, New Jersey, p. 601.

Liphardt, J., Onoa, B., Smith, S. B., Tinoco, I. J. R., and Bustamante, C. 2001. Reversible unfolding of single RNA molecules by mechanical force. *Science* 292(5517), 733–7.

Lisitsyn, N., Lisitsyn, N., and Wigler, M. 1993. *Science* 259, 946.

Lockhart, D. J. et al. 1996. Expression monitoring by hybridization to high-density oligonucleotide arrays. *Nat. Biotechnol.* 14, 1675–80.

Loewenstein, W. R. and Rose, B. 1992. *Semin. Cell Biol.* 3, 59.

Lottspeich, F., Houthaeve, T., and Kellner, R. 1994. In *Microcharactensarion of Proteins* (eds. Kellner, R., Lottspeich, F., and Meyer, H. E.). VCH, Weinheim, p. 117.

Lukyanov, K. A., Launer, G. A., Tarabykin, V. S., Zaraisky, A. G., and Lukyanov, S. A. 1995. *Anal. Biochem.* 229, 198.

Maniotis, A. J., Chen, C. S., and Ingber, D. E. 1997. Demonstration of mechanical connections between integrins, cytoskeletal filaments, and nucleoplasm that stabilize nuclear structure. *Proc. Nat. Acad. Sci. USA* 94, 849–54.

Mann, M. and Wilm, M. 1994. *Anal. Chem.* 66, 4390.

Mann, M. and Wilm, M. 1995. *TIBS* 20, 219.

Marsh, R. E. 1997. *J. Mass Spectrometry* 32, 351.

Marshall, B. T., Long, M., Piper, J. W., Yago, T., McEver, R. P., and Zhu, C. 2003. Direct observation of catch bonds involving cell-adhesion molecules. *Nature* 423(6936), 190–3.

Martin, G. R. 1998. The roles of FGFs in the early development of vertebrate limbs. *Genes Dev.* 12, 1571–86.

Marzluff, W. F. and Huang, R. C. C. 1984. Transcription of RNA in isolated nuclei. *In Transcription and Translation: A Practical Approach*, (Eds.Hames, B. D., Higgins, S. J.). IRL Press, Washington. DC, pp. 89–129.

Matsuoka, K. and Neuhaus, J. M. 1999. Cis-element of protein transport to the plant vacuoles. *J. Exp. Bot.* 50, 165–73.

Matsudaira, P. 1987. *J. Biol. Chem.* 262, 10035.

Maxam, A. M. and Gilbert, W. 1977. A new method for sequencing DNA. *Proc. Nctl. Acad. Sci. USA* 74(2), 560–4.

McLafferty, F. W., Fridriksson, E. K., Horn, D. M., Lewis, M. A., and Zubarev, R. A. 1999. *Science* 284, 1289.

Melton, D. A., Krieg, P. A., Rebagliati, M. R., Maniatis, T., Zinn, K., and Green, M. R. 1984. Efficient in vitro synthesis of biologically active RNA and RNA hybridization probes from plasmids containing a bacteriophage SP6 promoter. *Nucleic Acids Res. 25.* 12(18), 7035–56.

Meng, C. K., Mann, M., and Fenn, J. B. 1988. *Zeit. Physik D* 10, 361.

Meyer, H. E. 1994. Analys post-translational protein modifications. In *Microcharacterisation of Proteins* (eds. Kellner, R. F. and Meyer, H. E.). VCH, Weinheim, p. 131.

Michallet, M. C., Saltel, F., Flacher, M., Revillard, J. P., and Genestier, L. 2004. Cathepsin. dependent apoptosis triggered by supraoptimal activation of T lymphocytes: A possible mechanism of high dose tolerance. *J. Immunol.* 1.72(9), 5405–14.

Misteli, T. and Warren, G. 1994. COP-coated vesicles are involved in the mitotic fragmentation of Golgi stacks in a cell-free system. *J. Cell Biol.* 125(2), 269–82.

Mitchell, E. P. 1998. *Swim Oncol.* 25(5 Suppl. 11), 12.

Monteiro, P., Oliveira, P. J., Goncalves, L., and Providencia, L. A. 2003. Mitochondria: Role in ischemia, reperfusion and cell death. *Rev. Port. Cardiol.* 22(2), 233–54.

Mora, E., Norm, O., Mann, M., and Roepstortf, P. 1994. *Biol. Moss Spectrom.* 23, 249.

Moscow, J. A., Schmidt, L., Ingram, D. T., Gnarra, J., Johnson, B., and Cowan, K. H. 1994. *Carcinogenesis* 15(12), 2769.

Mueller, C. G. F., Rissoan, M.-C., Salinas, B., Ait-Yahia, S., Raval, O., Bridon, J.-M., Briere, F., Lebecque, S., and Liu, Y. J. 1997. *J. Exp. Med.* 186(5), 655.

Mullen, R., Lisenbee, C., Miernyk, J., and Trelease, R. 1999. Peroxisomal membrane ascorbate peroxidase is sorted to a membranous network that resembles a subdomain of the endoplasrnic reticulum. *Plant Cell* 11, 2167–85.

Muller, B. M., Yu, Y. B., and Lang, W. E. 1995. *Proc. Natl Acad. Sci. USA* 92, 205.

Nagahara S. and Matsuda, T. 1996. Cell-substrate and cell – cell interactions differently regulate cytoskeletal and extracellular matrix protein gene expression. *J. Biomed Mater Res.* 32, 677–86.

Nakamura, Y. et al. 1987. *Science* 235, 2616.

Nebenfuhr A., Frohlick J., and Staehelin L. 2000. Redistribution of Golgi stacks and other organdies during mitosis and cytokinesis in plant cells. *Plant Physiol.* 124, 135–51.

Nemeth, E., Bole-Feysot, C., and Tashima, L. S. 1998. *J. Mol. Eruirocrinol.* 20, 151.

Neubauer, G., King, A., Rappsilber, J., Calvio, C., Watson, W., Ajuh, P., Sleeman, J., Angus Lamond, A., and Mann, M. 1998. Mass spectrometry and EST-database searching allows characterisation of the multi-protein spliceosome complex. *Nat. Genet.* 520, 46.

Neubauer, G. and Mann, M. 1999. *Anal. Chem.* 71, 235.

Nickoloff, J. A. 1995. *Plant Cell Electroporation and Electrofusion Protocols.* Human Press, Totowa, NJ, USA.

Niessen, G. M. and Gumbiner B. M. 2002. Cadherin-mediated cell sorting not determined by binding or adhesion specificity. *J. Cell Biology* 156, 389–99.

Nishiyama, M., Higuchi, H., and Yanagida, T. 2002. Chemomechanical coupling of the forward and backward steps of single kinesin molecules. *Nat. Cell Biol.* 4(10), 790–7.

O'Hare, M. J. 1991. *Differentiation* 43, 209.

Opiteck, G. J., Ramirez, S. M., Jorgenson, J. W., and Moseley, M. A. 1998. *Anal. Biochem.* 258, 349.

Otto, A., Thiede, B., Muller, E.-C., Scheler, C., Wittman-Liebold, B., and Jungblut, P. 1996. *Electrophoresis* 17, 1643.

Packer, N. H., Ball, M. S., and Devine, P. L. 1999. Glycoprotein detector of 20 separated proteins. In *2-D proteome analysis protocols.* (ed. Laic, A. I.). Humana. Press, New Jersey, p. 341.

Pappin, D. J. C., Rahman, D., Hansen, H. F., and Barlet-Jones. A. I. 1995. In *Mass Spectrometry in the Biological, Sciences* (eds. Carr, S. A.). Humana Press, New Jersey, p. 135.

Paszowski, J., Shillito, R. D., Saul, M., Mandak, V., Hohn, T., Hohn, B., and Potrykus, I. 1984. Direct gene transfer in plants. *EMBO J.* 3, 2717–22.

Patterson, S. D. and Aebersold, R. 1995. *Electrophoresis* 16, 1791.

Patton, W. F., Pluskal, M. G., Skea, W. M., Buecker, J. L., Lopez, M. F., Zimmermann, R., Belanger, L. M., and Hatch, P. D. 1990. *BioTechniques* 8, 318.

Pedraza, C., Gcbcrhiwot, T., Ingerpuu, S., Assefa, D., Wondimu, Z., Kortesmaa, J., Tryggvason, K., Virtanen, I., and Patarroyo, M. 2000. Monocytic cells synthesize, adhere to, and migrate on laminin-8 (alpha 4 beta 1 gamma 1). *J. Immunol.* 165(10), 5831–8.

Persidis, A. 1999. *Nat. Biotechnol.* 16, 393.

Pollard, T. D. and Earnshaw, W. C. 2002. *Cell Biology.* Saunders (Elsevier Group).

Potrycus, I. 1990. Gene transfer to plants: Assessment and perspectives. *Physiol. Plant* 79, 129–34.

Quadroni, M. and James, P. 1999. *Electrophoresis* 20, 664.

Rabilloud, T. (ed.) 2000. *2D Gel Electrophoresis and Detection Methods*, Springer-Verlag, Berlin, p. 248.

Rasmussen, H. H., Van Damme, J., Puype, M., Gesser, G., Celis, J. E., and Vandekerckhove, J. 1991. *Electrophoresis* 12, 873.

Reddy, P. and McKenney, K. 1996. Improved method for the production of M13 phage and single-stranded DNA for DNA sequencing. *BioTechniques* 20, 854–9.

Richter, J. D. and Theurkauf, W. E. 2001. The message is in the translation. *Science* 293, 60–2.

Rider, M. H., Puype, M., Van Damne, J., Gevaert, K., De Boeck, S., D'Alayer, J., Rasmussen, H. H., Cells, J. E., and Vandekerckhove, J. 1995. *Eur. J. Biochem.* 230, 258.

Righetti, P. G., Gianazza, E., Gelfi, C., and Chiari, M. 1990. Isoelectric focusing, In *Gel Electrophoresis of Proteins, a Practical Approach* (ed. B. D. Hames and D. Rickwood), IRL Press, Oxford, pp. 149–216.

Roepstorff, P. 1997. *Curr. Opin. Biotechnol.* 8, 6.

Roepstorff, P. and Fohlman, J. 1984. *Biomed. Mass Spectrometry* 11, 601.

Rommens, J. M. et al. 1989. Identification of the cystic fibrosis gene: Chromosome walking and jumping. *Science* 245, 1059–65.

Rosenfeld, J., Capdeveille, J., Guillemot, J. C., and Ferrara, P. 1992. *Anal. Biochem.* 203, 173.

Rudnicki, 1993. MyoD and Myf-5 is required for the formation of skeletal muscle. *Cell* 75, 1351–9.

Sambrook, J., Fritsch, E. F., and Maniatis, T. 1989. *Molecular Cloning: A Laboratory Manual.* Cold Spring Harbar Laboratory Press, New York.

Sanchez, J.-C., Ravier, F., Pasquali, C., Frutiger, S., Paquet, N., Bjellqvist, B., Hochstrasser, D. F., and Hughes, G. J. 1992. *Electrophorests* 13, 715.

Sanger, F., Nicklen, S., and Coulson, A. R. 1977. DNA sequencing with chain-terminating inhibitors. *Proc. Natl. Acad. Sci. U.S.A.*, 74(12), 5463–7.

Schena, M., Shalon, D., Davis, R. W., and Brown, P. O. 1995. *Science* 270(5235), 467–70.

Seale, P. and Rudnicki, M. A. 2000. A new look at the origin, function, and stem-cell status of muscle satellite cells. *Dcv. Biol.* 218, 115–24.

Sechi, S. and Chait, B. T. 1998. *Anal. Chem.* 70, 5150.

Settler-Stevenson, W. G., Aznavoorian, S., and Liotta, L. A. 1993. *Annu. Rev. Cell. Biol.* 9, 541.

Shevchenko, A., Wilrn, M., Vorm, O., and Mann, M. 1996. *Anal. Chem.* 68, 850.

Shillito, R. D., Saul, M. W., Paszowski, J., Muller, M., and Potrykus, I. 1985. High efficiency direct gene transfer to plants. *Bio./Technol.* 3, 1099–103.

Shively, J. E. 1998. Structural analysis of proteins, p. 100. In *Bioorganic Chemistry. Peptides and Proteins* (ed. Hecht, S. M.). Oxford University Press, Oxford, p. 532.

Siebert, P. D., Chenchik, A., Kellogg, D. E., Lukyanov, K. A., and Lukyanov, S. A. 1995. *Nucleic Acids Res.* 23, 1087.

Simpson, R. J. and Nice, E. C. 1984. *Biochem. Int.* 8, 787.

Smith, T. W., Writer, D. G., Nicolson, G. L., and McIntire, L. V. 1996. Regulation of melanoma cell adhesion stabilization to fibronectin. *Melanoma Res.* 6, 351–62.

Smith, T. W., Yun, Z., Menter, D. G., McIntire, L. V., and Nicolson, G. L. 1996. Computerized analysis of tumor cell interactions with extracellular matrix proteins, peptides and endothelial cells under laminar flow. *Biotech. Bioengineer.* 50, 598–607.

Strick, T. R., Croquette, V., and Bensimon, D. 2000. Single-molecule analysis of DNA uncoiling by a type II topoisomerase. *Nature* 404(6780), 901–4.

Strupat, K., Karas, M., and Hillenkamp, F. 1994. *Anal. Chem.* 66, 464.

Svab, Z. and Maliga, P. 1993. High frequency plastid transformation in tobacco by a chimeric *aad* A gene. *Proc. Natl. Acad. Sci. (USA)* 90, 913–7.

Swissheim, K., Ryan, K., and Sager, R. 1995. *Proc. Natl Acad. Sci. USA* 92, 4472.

Takahashi, N., Nishihira, J., Sato, Y., Kondo, M., Ogawa, H., Ohshima, T., Une, Y., and Todo, S. 1998. Involvement of macrophage migration inhibitory factor (MIF) in the mechanism of tumor cell growth. *Mol. Med.* 4(11), 707–14.

Tanksley, S. D., Young, N. D., Paterson, A. H., and Bonierbale, M. W. 1989. *Biotechnol.* 7, 257.

Tomlinson, A. J., Guzman, N. A., and Naylor, S. 1995. *J. Cap. Electrophresis* 2, 247.

Trent, J. M., Bittner, M., Zhang, J., Wiltshire, R., Ray, M., Su, Y., Gracia, E., Meltzer p., De Risi, J. Penland, L., and Brown, P. 1997. Use of microgenomic technology for analysis of alterations in DNA copy number and gene expression in malignant melanoma. *Clin. Exp. Immunol.* 107(Suppl. 1), 33–40.

Van der Flier, A. and Sonnenberg, A. 2001. Function and interactions of integrins. *Cell Tissue Res.* 305, 285–98.

Velculescu, V. E., Zhang, L., Vogelstein, B., and Kinzler, K. W. 1995. *Science* 270(5235), 484–7.

Velleman, S. G. 2000. The role of the extracellular matrix in skeletal development. *Poultry Sci.* 79, 985–9.

Vickland, H., Westrum, L., Kott, J., Patterson, S., and Bothwell, M. 1991. Nerve growth factor receptor expression in the young and adult rat olfactory system. *Brain Res.* 565, 269–79.

Vieira, J. and Messing, J. 1987. Production of single-stranded plasmid DNA. *Methods Enzymol.* 153, 3–11.

Von Stein, O. D., Thies, W.-G., and Hoffmann, M. 1997. *Nucleic Acids Res.* 25(13), 2598.

Voris, B. P. and Young, D. A. 1980. *Anal. Biochem.* 104, 478.

Vos, P. et al. 1995. *Nucleic Acids Res.* 23, 4407.

Walter, P. and Blobel, G. 1981. Translocation of proteins across the Endoplasmic Reticulum III. *J. Cell Biol.* 91, 557–61.

Wang, J. and Hewick, R. 1999. *DDT* 4, 129.

Wang, X. F. and Herman, B. (ed.) 1996. *Fluorescence Imaging Spectroscopy and Microscopy.* John Wiley & Sons., New York, pp. 179-251.

Wang, Z. and Brown, D. D. 1991. *Proc. Natl Acad. Sci. USA* 88, 11505.

Waseem, N. H. and Lane, D. P. 1990. Monoclonal antibody analysis of the proliferating cell nuclear antigen (PCNA). Structural conservation and the detection of a nucleolar form. *J. Cell Sci.* 96(Pt 1), 121–9.

Wasinger, V. C. et al. 1995. *Electrophoresis* 16, 1090.

Wasinger, V. C., Bjellqvist, B. J., and Humphery-Smith, I. 1997. *Electrophoresis* 18, 1373.

Wasinger, V. C. and Humphery-Smith, I. 1998. *FEMS Microbiol. Lett.* 169, 375.

Wasserman, P. M. 1999. Mammalian fertilization: Molecular aspects of gamete adhesion, exocytosis. *Cell* 96, 175–83.

Weaver JC. 1993. Electroporation: A general phenomenon for manipulating cells and tissues. *J. Cell Biochem.* 51, 426–35.

Westermeier, R. 1993. Suspended farming of Gracilaria chilensis (Rhodophyta, Gigartinales) at Cariquilda River, Maullín, Chile. *Aquaculture* 113(3), 215–29.

Whitehouse, C. M., Dreyer, R. N., Yamashita, M., and Fenn, J. B. 1985. *Anal. Chem.* 57, 675.

Wilkins, M. R., Williams, K. L., Appel, R. D., and Hochstrasser, D. F. (eds.) 1997. *Proteome Research: New Frontiers in Functional Genomics.* Springer-Verlag, Berlin, p. 243.

Wilkins, M. R., Yan, J. X., and Gooiey, A. A. 1999. 2-DE spot amino acid analysts with 9- fluorenylmethyl chloroformate. In *2-D Proteome Analysis Protocols* (ed. Link, A. J.). Humana Press, New Jersey, p. 445.

Wilm, M., Shevchenko, A., Houthaeve, T., Breit, S., Schweigerer, L., Fotsis, T., and Mann, M. 1996. *Nature* 379, 466.

Wirth, P. J. and Romano, A. 1995. *J. Chromatography* 698, 123.

Wise, M. J., Littlejohn, T., and Humphery-Smith, I. 1997. *Electrophoresis* 18, 1399.

Wong, B. R. et al. 1997. *J. Biol. Chem.* 272(40), 25190.

Yan, Q. and Johnson, E. M. 1988. An immunohistochemical study of the nerve growth factor receptor in developing rats. *J. Neurosci.* 8, 3481–98.

Yates, J. R. 1998. *J. Mass Spectrometry* 33, 1.

Young, D. A. 1984. *Clin. Chem.* 30, 2104.

Yuan, J. 1993. The *C. elegans* cell death gene ced-3. *Cell* 75, 641–52.

Zabarovsky, E. R. and Allikmets, R. L. 1986. An improved technique for the efficient construction of gene library by partial filling-in of cohesive ends. *Gene 42*, 119–23.

Zabeau, M. and Vo, P. 1993. European patent application, publication number EP0534858.

Zhcng, H., von Mollard, G., Kovaleva, V. Stevens, T., and Raikhel, N. 1999. The plant vesicle-associated SNARE AtVTI1A likely mediates vesicle transport from the trans-Golgi network to the prevacuolar compartment. *Mol. Biol. Cell.* 10, 2251–64.

Index

H

I

K

L

M

N

O

P

T

U

V

W

X

Y